Methods in Molecular Biology™

AF768720

Series Editor
John M. Walker
School of Life Sciences
University of Hertfordshire
Hatfield, Hertfordshire, AL10 9AB, UK

For further volumes:
http://www.springer.com/series/7651

DNA Cloning and Assembly Methods

Edited by

Svein Valla and Rahmi Lale

Norwegian University of Science and Technology, Trondheim, Norway

Humana Press

Editors
Svein Valla
Norwegian University of Science and Technology
Trondheim, Norway

Rahmi Lale
Norwegian University of Science and Technology
Trondheim, Norway

ISSN 1064-3745 ISSN 1940-6029 (electronic)
ISBN 978-1-4939-6057-6 ISBN 978-1-62703-764-8 (eBook)
DOI 10.1007/978-1-62703-764-8
Springer New York Heidelberg Dordrecht London

Printed on acid-free paper

Humana Press is a brand of Springer
Springer is part of Springer Science+Business Media (www.springer.com)

Preface

DNA cloning and assembly methods are essential tools in molecular biology research. Current advancements in the fields, such as structural genomics and proteomics, and especially the growing discipline of synthetic biology, require the assembly of large DNA constructs involving numerous number of genes with relative ease. While conventional DNA cloning strategies involving the use of type II restriction enzymes to generate appropriate DNA fragments are still applicable, they are no longer suited for parallel cloning and/or assembly of multiple DNA fragments. High-throughput methods for cloning, protein expression, and purification necessitate protocols that are fast and reliable. This book aims to serve the molecular biology community with a collection of DNA cloning and assembly protocols that make cloning procedures faster, more reliable, and also suitable for high-throughput handling.

The methods described in this book are based on several mechanisms including type II and IIS restriction enzymes, single-stranded annealing, sequence overlap, and recombination. Furthermore, software programs suitable for primer design, a feature crucial for the functionality of the described methods, are also explained in this book.

We are convinced that the methods and protocols listed in this book will provide a valuable and useful resource for wet lab researchers within life sciences.

Trondheim, Norway

Rahmi Lale
Svein Valla

Contents

Contributors

RINJI AKADA • *Department of Applied Molecular Bioscience, Yamaguchi University Graduate School of Medicine, Ube, Japan*
YAAKOV BENENSON • *Department of Biosystems Science and Engineering, Eidgenössische Technische Hochschule (ETH) Zürich, Basel, Switzerland*
LOUISE E. BIRD • *Oxford Protein Production Facility-UK, Oxfordshire, UK; Division of Structural Biology, Henry Wellcome Building for Genomic Medicine, University of Oxford, Oxford, UK*
JONATHAN M. DIPROSE • *Oxford Protein Production Facility-UK, Oxfordshire, UK; Division of Structural Biology, Henry Wellcome Building for Genomic Medicine, University of Oxford, Oxford, UK*
JULIAN J. EATON-RYE • *Department of Biochemistry, University of Otago, Dunedin, New Zealand*
WINFRIED EDELMANN • *Department of Cell Biology, Albert Einstein College of Medicine, Bronx, NY, USA*
CAROLA ENGLER • *Icon Genetic GmbH, Halle, Germany*
ARIEL ERIJMAN • *Department of Biological Chemistry, The Alexander Silberman Institute of Life Sciences, Hebrew University of Jerusalem, Jerusalem, Israel*
DOMINIC ESPOSITO • *Leidos Biomedical Research, Inc., Frederick National Laboratory for Cancer Research, Frederick, MD, USA*
JOHN FLANAGAN • *Division of Structural Biology, Henry Wellcome Building for Genomic Medicine, University of Oxford, Oxford, UK*
LESLIE A. GARVEY • *Protein Expression Laboratory, SAIC Frederick, Inc., Frederick National Laboratory for Cancer Research, Frederick, MD, USA*
ERIC R. GEERTSMA • *Institute of Biochemistry, Biocenter N220, Goethe-University Frankfurt, Frankfurt/M., Germany*
ROBERT J.C. GILBERT • *Division of Structural Biology, Henry Wellcome Building for Genomic Medicine, University of Oxford, Oxford, UK*
ANTONIO GRANELL • *Instituto de Biología Molecular y Celular de Plantas, Consejo Superior de Investigaciones Científicas (CSIC), Universidad Politécnica de Valencia, Valencia, Spain*
BARBARA A. HALKIER • *DynaMo Center of Excellence, Department of Plant and Environmental Sciences, Faculty of Science, University of Copenhagen, Copenhagen, Denmark*
RYAN E. HILL • *Department of Biochemistry, University of Otago, Dunedin, New Zealand*
NATHAN J. HILLSON • *Fuels Synthesis Division, Joint BioEnergy Institute, Emeryville, CA, USA; DOE Joint Genome Institute, Walnut Creek, CA, USA; Physical BioSciences Division, Lawrence Berkeley National Lab, Berkeley, CA, USA*
HISASHI HOSHIDA • *Department of Applied Molecular Bioscience, Yamaguchi University Graduate School of Medicine, Ube, Japan*

RUSLAN KALENDAR • *MTT/BI Plant Genomics Laboratory, Institute of Biotechnology, University of Helsinki, Helsinki, Finland; PrimerDigital Ltd, Helsinki, Finland*
EIRIN KORVALD • *Department of Biotechnology, Norwegian University of Science and Technology (NTU), Trondheim, Norway*
RAHMI LALE • *Department of Biotechnology, Norwegian University of Science and Technology (NTU), Trondheim, Norway*
DAVID LEE • *John Bingham Laboratory, National Institute of Agricultural Botany, Cambridge, UK*
SYLVESTRE MARILLONNET • *Department of Cell and Metabolic Biology, Leibniz-Institut für Pflanzenbiochemie, Halle, Germany*
JENNIFER L. MEHALKO • *Protein Expression Laboratory, SAIC Frederick, Inc., Frederick National Laboratory for Cancer Research, Frederick, MD, USA*
UFFE H. MORTENSEN • *Department of Systems Biology, Center for Microbial Biotechnology, Technical University of Denmark (DTU), Kongens Lyngby, Denmark*
MIKIKO NAKAMURA • *Innovation Center, Yamaguchi University, Ube, Japan*
DIEGO ORZAEZ • *Instituto de Biología Molecular y Celular de Plantas, Consejo Superior de Investigaciones Científicas (CSIC), Universidad Politécnica de Valencia, Valencia, Spain*
RAYMOND J. OWENS • *Oxford Protein Production Facility-UK, Oxfordshire, UK; Division of Structural Biology, Henry Wellcome Building for Genomic Medicine, University of Oxford, Oxford, UK*
OVE ØYÅS • *Department of Biotechnology, Norwegian University of Science and Technology, Trondheim, Norway*
JARLE PAHR • *Department of Biotechnology, Norwegian University of Science and Technology (NTU), Trondheim, Norway*
JORGE PALACI • *Instituto de Biología Molecular y Celular de Plantas, Consejo Superior de Investigaciones Científicas (CSIC), Universidad Politécnica de Valencia, Valencia, Spain*
YOAV PELEG • *Israel Structural Proteomics Center (ISPC), Faculty of Biochemistry, Weizmann Institute of Science, Rehovot, Israel*
LAUREN V. PROCTER • *Protein Expression Laboratory, SAIC Frederick, Inc., Frederick National Laboratory for Cancer Research, Frederick, MD, USA*
JIAYUAN QUAN • *Department of Biomedical Engineering and the Institute for Genome Sciences and Policy, Duke University, Durham, NC, USA; Tianjin Institute of Industrial Biotechnology, Chinese Academy of Sciences, Tianjin, China*
HEATHER RADA • *Oxford Protein Production Facility-UK, Oxfordshire, UK; Division of Structural Biology, Henry Wellcome Building for Genomic Medicine, University of Oxford, Oxford, UK*
GUNVOR RØKKE • *Department of Biotechnology, Norwegian University of Science and Technology (NTU), Trondheim, Norway*
BO SALOMONSEN • *DynaMo Center of Excellence, Department of Plant and Environmental Sciences, Faculty of Science, University of Copenhagen, Copenhagen, Denmark*
ALEJANDRO SARRION-PERDIGONES • *Instituto de Biología Molecular y Celular de Plantas, Consejo Superior de Investigaciones Científicas (CSIC), Universidad Politécnica de Valencia, Valencia, Spain*
JONATHAN L. SCHMID-BURGK • *Institute for Clinical Chemistry and Clinical Pharmacology, University Hospital, University of Bonn, Bonn, Germany*
ALAN H. SCHULMAN • *MTT/BI Plant Genomics Laboratory, Institute of Biotechnology, University of Helsinki, Helsinki, Finland; Biotechnology and Food Research, MTT Agrifood Research Finland, Jokioinen, Finland*

Julia M. Shifman • *Department of Biological Chemistry, The Alexander Silberman Institute of Life Sciences, Hebrew University of Jerusalem, Jerusalem, Israel*
Ayako Suzuki • *Department of Applied Molecular Bioscience, Yamaguchi University Graduate School of Medicine, Ube, Japan*
Frank Thieme • *Icon Genetic GmbH, Halle, Germany*
Jingdong Tian • *Department of Biomedical Engineering and the Institute for Genome Sciences and Policy, Duke University, Durham, NC, USA; Tianjin Institute of Industrial Biotechnology, Chinese Academy of Sciences, Tianjin, China*
Tamar Unger • *Israel Structural Proteomics Center (ISPC), Faculty of Biochemistry, Weizmann Institute of Science, Rehovot, Israel*
Vanessa E. Wall • *Protein Expression Laboratory, SAIC Frederick, Inc., Frederick National Laboratory for Cancer Research, Frederick, MD, USA*
Uwe Werling • *Department of Cell Biology, Albert Einstein College of Medicine, Bronx, NY, USA*
Zhen Xie • *Center for Synthetic and Systems Biology, TNLIST, Tsinghua University, Beijing, China*
Chun You • *Biological Systems Engineering Department, Virginia Tech, Blacksburg, VA, USA*
Yongwei Zhang • *Department of Cell Biology, Albert Einstein College of Medicine, Bronx, NY, USA*
Y.-H. Percival Zhang • *Biological Systems Engineering Department, Virginia Tech, Blacksburg, VA, USA; Institute for Critical Technology and Applied Science (ICTAS), Virginia Tech, Blacksburg, VA, USA; Gate Fuels Inc., Blacksburg, VA, USA*

Chapter 1

BioBrick Assembly Standards and Techniques and Associated Software Tools

Gunvor Røkke, Eirin Korvald, Jarle Pahr, Ove Øyås, and Rahmi Lale

Abstract

The BioBrick idea was developed to introduce the engineering principles of abstraction and standardization into synthetic biology. BioBricks are DNA sequences that serve a defined biological function and can be readily assembled with any other BioBrick parts to create new BioBricks with novel properties. In order to achieve this, several assembly standards can be used. Which assembly standards a BioBrick is compatible with, depends on the prefix and suffix sequences surrounding the part. In this chapter, five of the most common assembly standards will be described, as well as some of the most used assembly techniques, cloning procedures, and a presentation of the available software tools that can be used for deciding on the best method for assembling of different BioBricks, and searching for BioBrick parts in the Registry of Standard Biological Parts database.

Key words BioBrick, BioBrick standard assembly, BioBrick BB-2 assembly, BglBricks assembly, Silver assembly, Freiburg assembly, Fusion protein assembly, 3A Assembly, Amplified insert assembly, Gibson scarless assembly, The constructor, BioBrick search engine, Clotho, Gibthon

1 Introduction

The first BioBrick assembly standard was proposed by Tom Knight [1]. This very first assembly standard was called "Standard Sequence Assembly of BioBricks," and is no longer in use but since then many other assembly methods have been proposed.

All BioBrick parts [2] are flanked by a prefix and a suffix, and because of these sequences, it is possible to put together different BioBrick parts in order to create new parts with more complex functions than the constituents. What is different between the almost hundred proposed assembly standards are the prefix and suffix surrounding the BioBricks and the restriction sites contained in these sequences. Different assembly standards have been developed and optimized for different purposes, but the main goal of most of the assembly standards is to make fusion of BioBricks coding protein

Svein Valla and Rahmi Lale (eds.), *DNA Cloning and Assembly Methods*, Methods in Molecular Biology, vol. 1116, DOI 10.1007/978-1-62703-764-8_1, © Springer Science+Business Media New York 2014

Prefix for non-coding BioBrick parts: 5′ GAATTC GCGGCCGC T TCTAGAG 3′
EcoRI NotI XbaI

Prefix for protein-coding BioBrick parts: 5′ GAATTC GCGGCCGC T TCTAG (ATG) 3′
EcoRI NotI XbaI

Suffix: 5′ T ACTAGT A GCGGCCG CTGCAG 3′
SpeI NotI PstI

Fig. 1 BioBrick standard assembly prefix and suffix sequences

domains possible. Stability and degradation of the fusion proteins is also an issue that has led to new proposed assembly standards. In the following subchapters, the five assembly standards supported by the Registry of Standard Biological Parts, in addition to three of the most used assembly techniques will be explained.

1.1 Assembly Standards

1.1.1 BioBrick Standard Assembly

BBF RFC 10, called BioBrick standard assembly [3], was proposed by Tom Knight in May 2007 and is the BioBrick standard that is most commonly used. Most BioBricks listed on the website of the Registry of Standard Biological Parts (http://partsregistry.org/Main_Page) is compatible with this assembly method.

BioBricks following the BBF RFC 10 assembly standards have one standardized suffix, but the prefix may appear in one of the two ways, depending on whether or not the BioBrick in question is a protein-coding sequence or not. The two prefixes and the suffix used by bricks that are compatible with BioBrick standard assembly are given in Fig. 1.

In both prefixes, restriction sites for EcoRI (GAATTC) and NotI (GCGGCCGC) are present. Additionally, for noncoding BioBrick parts the prefix contains an XbaI (TCTAGA) restriction site. In the case of the prefix used for protein-coding parts, the XbaI restriction site appears only when the prefix sequence is fused with a protein-coding sequence starting with ATG. In the suffix, restriction sites for SpeI (ACTAGT), NotI (GCGGCCGC), and PstI (CTGCAG) are present.

The most common way of joining together two BioBrick parts following the BBF RFC 10 assembly standard is to digest the first BioBrick with EcoRI and SpeI, creating an insert, and the second BioBrick with EcoRI and XbaI, creating a backbone with an opening in front of the second BioBrick part. When the two digested fragments are joined together, the sticky ends created by EcoRI on the two DNA molecules will be ligated back together to restore the EcoRI restriction site. Both XbaI and SpeI create compatible overhangs, so the sticky ends created by digesting with XbaI and SpeI can be ligated together. When this is performed, the ligated sequence will be a mix of the two restriction sites, and the new

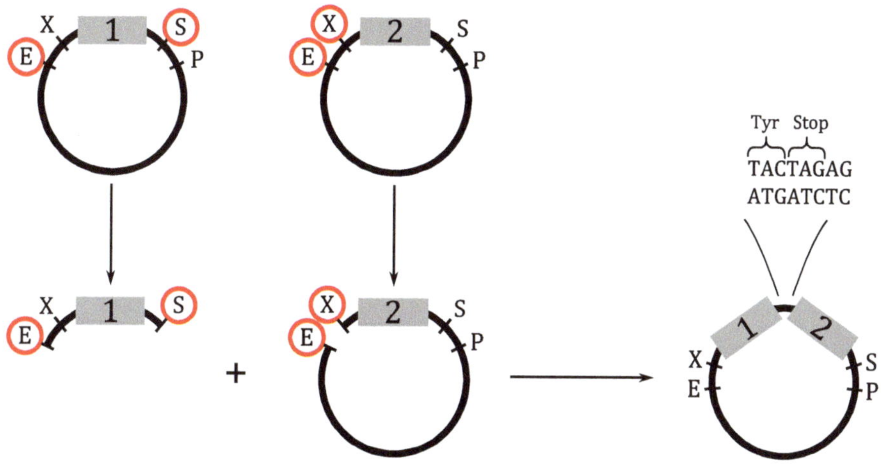

Fig. 2 Assembly by BBF RFC 10. Assembling two BioBrick parts using this assembly standard involves restriction digest of the first part with EcoRI and SpeI and of the second part with EcoRI and XbaI. The digested fragments are then ligated, yielding a new BioBrick made up by the two initial ones. Here E, X, S and P denotes EcoRI, XbaI, SpeI, and PstI, respectively

sequence is not recognized by any restriction enzyme. This means that when two BioBricks are joined together, the prefix in front of the first BioBrick is restored, and the suffix behind the second BioBrick is unchanged, while the sequence between the fused BioBricks, called a scar, will be TACTAG in the cases where the second BioBrick is a protein-coding part. Otherwise the scar will be TACTAGAG. The process of joining BioBricks together using the BBF RFC 10 standard is illustrated in Fig. 2. When translated, the scar sequence corresponds to a tyrosine and a stop codon.

When joining together two BioBricks following this standard, it is also possible to digest the first BioBrick with SpeI and PstI to create an opening behind the first brick, and the second BioBrick with XbaI and PstI. This will create the same scar as when digesting with EcoRI, XbaI, and SpeI.

For BioBricks to be compatible with the BBF RFC 10 standards, the part themselves must not contain EcoRI, XbaI, SpeI, PstI, or NotI restriction sites. According to the original draft standard by Tom Knight, restriction sites for PvuII, XhoI, AvrII, NheI, and SapI should also be avoided.

To make BBF RFC 10 compatible BioBricks by PCR, the following primer sequences should be used:

Forward primer for protein-coding parts: 5′ GTTTCTTCG AATTCGCGGCCGCTTCTAGAG 3′ <18–24 bp of matching primer>

Forward primer for other parts: 5′ GTTTCTTCGAATTCG CGGCCGCTTCTAG 3′ <18–24 bp of matching primer, beginning with ATG>

BB-2 prefix: 5′ GAATTC GCGGCCGC ACTAGT 3′
EcoRI NotI SpeI

BB-2 suffix: 5′ GCTAGC GCGGCCG CTGCAG 3′
NheI NotI PstI

Fig. 3 BioBrick BB-2 assembly prefix and suffix sequences

Reverse primer: 5′ GTTTCTTCCTGCAGCGGCCGCTACTAGTA 3′ <18–24 bp of matching primer (reverse complement)>

1.1.2 BioBrick BB-2 Assembly

The BBF RFC 12, or the BioBrick BB-2 assembly standard [4], was proposed by Tom Knight in November 2008. The assembly standard was made to tackle some of the problems associated with the original BioBrick assembly standard, BBF RFC 10. As explained in Subheading 1.1.1., the scar created when joining together two BioBricks using Standard Assembly is TACTAGAG (TACTAG when a protein-coding part is part number two). This corresponds to a tyrosine residue and a translation stop signal (Fig. 2). Normally, the scars are not being translated, but in the cases where the two BioBricks that are being joined together are both protein domains; the scar of the initial BioBrick standard will cause a potential problem, as the ribosome will receive a stop signal after having translated only one protein domain coded by the first BioBrick. Another problem with the initial BioBrick standard is that the scar consists of eight bases, which will yield an altered reading frame when joining protein domains.

For the BioBrick BB-2 standard, the enzymes used for digestion of the initial parts are almost the same as for the initial BioBrick standard. The prefix and suffix are, however, slightly modified compared to the initial standard (Fig. 3).

When joining two BB-2 parts, part 1 should be digested with EcoRI and NheI, to create an insert, while part 2 should be digested with EcoRI and SpeI in order to create a backbone with an opening in front of the part. The scar created in this process will be GCTAGT, which, when translated, corresponds to alanine and serine. It is recommended that BioBricks following the BB-2 standard should avoid restriction sites for PvuII, XhoI, AvrII, XbaI, and SapI inside the parts. When making BioBricks compatible with the BB-2 standard by PCR, the following set of primers should be used:

Forward primer: 5′ GTTTCTTCGAATTCGCGGCCGCACTAGA 3′ <18–24 bp of matching primer>

Reverse primer: 5′ GTTTCTTCCTGCAGCGGCCGCGCTAGC 3′ <18–24 bp of matching primer (reverse complement)>

BglBricks prefix: 5′ GAATTC ATG AGATCT 3′
EcoRI BglII

BglBricks suffix: 5′ GGATCC TTA CTCGAG 3′
BamHI XhoI

Fig. 4 BglBricks assembly prefix and suffix sequences

1.1.3 BglBricks Assembly

BBF RFC 21, more commonly known as the BglBricks assembly standard [5], was proposed by J. Christopher Anderson, John E. Dueber, Mariana Leguia, Gabriel C. Wu, Jonathan C. Goler, Adam P. Arkin, and Jay D. Keasling in September 2009, and in the same way as the BB-2 standard, the BglBricks standard was developed to make multiple fusion of protein domain BioBricks possible without altering the reading frame or introducing stop codons. The BglBricks standard uses the restriction enzymes BglII and BamHI [6], which cut with high efficiency and are unaffected by DNA methylation, as the inner restriction sites, and EcoRI and XhoI as the outer restriction sites in the prefix and suffix, respectively.

Officially, a BglBrick part is defined as a DNA sequence flanked by GATCT on the 5′ end and by G on the 3′ end, and lacking BglII, BamHI, EcoRI, and XhoI restriction sites. Additionally, a Bgl vector is defined as a DNA sequence flanked by GATCC on the 5′ end and A on the 3′ end. When ligated together, the BglBrick and the Bgl vector produce a BglII restriction site in front of the BglBrick and a BamHI restriction site after the brick. As previously stated, in addition to these, EcoRI and XhoI restriction sites are often included in the Bgl vector (Fig. 4).

When putting together two BglBricks, the principle is the same as for the two previous assembly standards described, but the enzymes combination is different. In order to put together two parts, EcoRI and BamHI can be used to digest the first part, while EcoRI and BglII can be used to digest the second part. As BamHI and BglII create compatible flanking ends, these two overhangs can be ligated together. The scar created by using this standard is GGATCT, which corresponds to a glycine and a serine residue. The process of putting together BioBricks using the BglBricks assembly method is given in Fig. 5.

As for BioBrick standard assembly, an additional method can be used to put together parts. Digesting the first BioBrick part with BglII and XhoI, and the second part with BamHI and XhoI prior to ligation will result in the same scar (GGATCT). For BioBricks to be compatible with the BglBricks assembly standard, the part sequences must not contain restriction sites for EcoRI, BglII, BamHI, or XhoI.

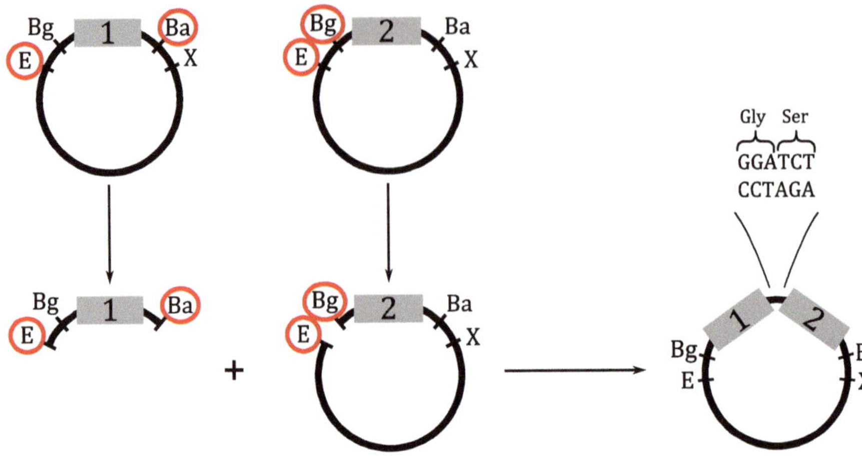

Fig. 5 Assembly of two BioBricks compatible with the BglBricks assembly standard. Digesting the first BioBrick with EcoRI and BamHI, and the second BioBrick with EcoRI and BglII, and ligating the two digested parts together, will result in a new BioBrick consisting of the two initial parts, and the scar GGATCT. E, Bg, Ba, and X denotes EcoRI, BglII, BamHI, and XbaI, respectively

Silver standard prefix: 5′ GAATTC GCGGCCGC T TCTAGA 3′
EcoRI XbaI

Silver standard suffix: 5′ ACTAGT A GCGGCCG CTGCAG 3′
SpeI PstI

Fig. 6 Silver assembly prefix and suffix sequences

1.1.4 *Silver Assembly*

The BBF RFC 23 standard, also called the Silver standard, was proposed by Ira Phillips and Pamela Silver in April 2006, and this assembly standard is a modified version of RFC 10. As with the RFC 12 and 21 standards, BBF RFC 23 (Fig. 6) also allows fusion of multiple BioBricks coding for protein sequences [7].

The restriction sites, enzymes, and assembly protocols used in the Silver standard are the same as for the RFC 10 standard. When joining two BioBricks using Silver assembly, the scar sequence will be ACTAGA, encoding the amino acids threonine and arginine. According to the N-end rule of protein stability, stating that the N-terminal amino acid of a protein determines its half-life, this scar sequence may accelerate protein degradation.

In the RFC 10 standard, spacer nucleotides are inserted between the part and the flanking XbaI and SpeI sites to prevent methylation of the DNA, which would inhibit restriction enzyme activity. In RFC 23, these spacer nucleotides are not present, and methylation is thus a possible problem. Especially the sequence TCn, where n could be any nucleotide, is problematic if present as the first codon of the part. The problem can be circumvented by

PhusionParts prefix: 5′ GAATTC GCGGCCGC T TCTAGA TG GCCGGC 3′
EcoRI NotI XbaI NgoMIV

PhusionParts suffix: 5′ ACCGGT TAAT ACTAGT A GCGGCCG CTGCAG 3′
AgeI SpeI NotI PstI

Fig. 7 Freiburg assembly prefix and suffix sequences. The original RFC 10 prefix and suffix sequences are underlined

replacing TCn with either AGT or AGC, as all these codons would encode for the same amino acid, serine.

For parts to be compatible with the Silver standard, it is important that they start and end in frame, and that no spacer nucleotides are present, as these would alter the reading frame of the protein.

1.1.5 Freiburg Assembly

RFC 25, called the Freiburg standard, is another assembly standard designed to facilitate protein fusion by shortening the spacer or "scar" sequence between parts [8]. It was proposed by the Freiburg iGEM 2007 team; it is backwards compatible with RFC 10 and is intended as an alternative to RFC 23. RFC 25 extends RFC 10 by allowing modular construction of fusion proteins by the use of BioBrick "FusionParts," defined in the standard (Fig. 7).

The underlined part of the sequence in Fig. 7 is the original RFC 10 prefix and suffix. It is important to note that the FusionParts prefix contains a start codon directly after the end of the RFC 10 prefix sequence. As a result, if a FusionPart is made from a complete protein-coding sequence, the amino acid sequence methionine–alanine–glycine will be added to the start (N-terminus) of the protein. This can be avoided by using a hybrid "N-part" format, which is done by using the RFC 10 prefix in place of the RFC 25 prefix.

FusionParts can be used as standard BioBrick Parts according to RFC 10. The scar sequence is longer than for RFC 10, as it contains the additional restriction sites AgeI or NgoMIV. Since the RFC 25 prefix and suffix contains the RFC 10 prefix and suffix, assembly can be performed like described in Fig. 1. Alternatively, the AgeI and NgoMIV sites can be used for assembly of fusion proteins, as depicted in Fig. 8. The AgeI and NgoMIV overhangs are compatible, with ligation resulting in the shortened scar sequence ACCGGC, coding threonine and glycine. As for the RFC 12, 21, and 23 standards, this scar contains neither a frameshift nor a stop codon. Compared to RFC 23, RFC 25 avoids potentially destabilizing changes to the N-terminal of the protein, and native protein start can be kept intact by using N-parts.

RFC 25 is compatible with Standard Assembly, 3A Assembly, and Gibson Scarless Assembly (see below), using the established protocols for RFC 10. With respect to fusion protein assembly,

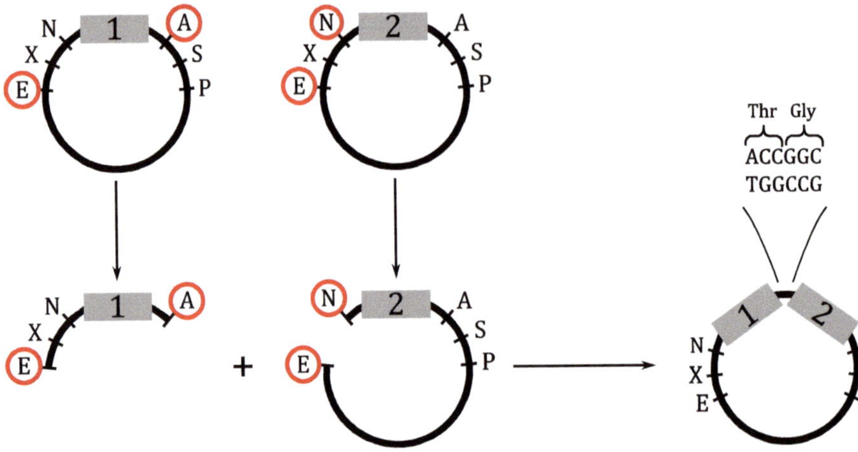

Fig. 8 Assembly of two FusionParts by digesting the first part with EcoRI and AgeI, and the second with EcoRI and NgoMIV. After ligation, the scar created between the parts will be ACCGGC, corresponding to Thr-Gly. E, X, N, A, S, and P denotes EcoRI, XbaI, NgoMIV, AgeI, SpeI, and PstI, respectively

RFC 25 is not directly compatible with RFC 23, meaning that fusion proteins cannot be made directly using two parts following the RFC 23 and RFC 25 standards, respectively.

When turning a DNA sequence into a FusionPart, it is important to keep in mind that restriction sites for EcoRI, XbaI, SpeI, PstI, and NotI must be avoided, like for the RFC 10 standard. In addition, FusionParts cannot contain restriction sites for the AgeI (ACCGGT) or NgoMIV (GCCGGC), as these are present in the suffix and prefix.

1.2 Assembly Techniques

The simplest way of performing assembly of two BioBricks is to use the enzyme combinations that are described for each assembly standard in the sections above. This is called standard assembly (not to be confused with BBF RFC 10, which is also called standard assembly, but referring to the assembly standard, and not the assembly technique), and requires that the backbone of one of the two BioBricks to be joined have to be used as backbone for the new composite BioBrick.

However, more assembly techniques exist, and most of them are compatible with several assembly standards. Three of the most used techniques are described in the following subchapters.

1.2.1 3A Assembly

3A assembly is short for "three antibiotic assembly" which allows cloning two parts together and selecting for correct assemblies through an antibiotic selection. This technique was developed by Reshma Shetty, Meagan Lizarazo, Randy Rettberg, and Thomas F. Knight Jr. in 2011 as an alternative to the standard assembly technique [9].

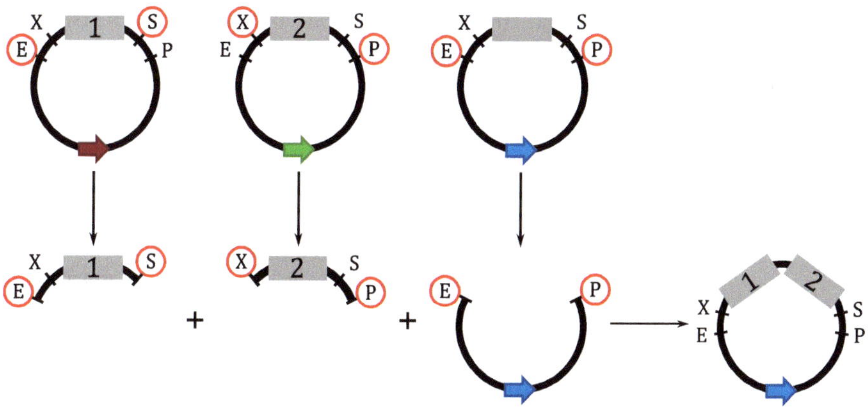

Fig. 9 The process of 3A assembly explained schematically. Here, E, X, S, and P denotes restriction sites for EcoRI, XbaI, SpeI, and PstI, respectively, and the colored arrows in the plasmid backbones indicate resistance genes for three different antibiotics. In this case, the "blue" antibiotic should be used when selecting for the correct ligation

3A assembly is similar to standard assembly in many ways, but the restriction digests are different between the two assembly standards. When two BioBricks are joined using the standard assembly technique, the backbone vector of one of the BioBricks will be used as backbone for the new composite brick. However, in 3A assembly, an independent vector is used as backbone, and the antibiotic resistance of the vector should be different from the ones contained in the backbones of the BioBricks. So in a 3A assembly, three restriction digests are carried out. The first BioBrick is digested with EcoRI and SpeI, the second BioBrick with XbaI and PstI, and the backbone to be used is digested with EcoRI and PstI. Then, the fragments are ligated, and in 3A assembly, this can be done directly without investigating the digested fragments on gel electrophoresis prior to ligation. Instead, the correct ligation is selected for using the antibiotic of the correct backbone vector in the medium when the ligated DNA is transformed. The entire process of 3A assembly is explained in Fig. 9.

Since the enzymes used for digestion in 3A assembly are the same as for the standard assembly, this indicates that BioBricks that are compatible with standard assembly, also are compatible with 3A assembly, and unlike the other assembly standards described in this chapter, 3A assembly does not require a prefix and suffix that is different from the standard prefix and suffix.

When using 3A assembly, one thing one needs to be aware of is that this assembly method yields less correct assembled plasmids compared to, for example, standard assembly, as 3A assembly uses a triple ligation process. The yield of a triple ligation process will be about 10 % of the assembled plasmids produced by a double ligation [10]. If running into this problem, a possible solution can be to assure that the competent cells are as effective as they should be. If they have been stored for a while, a good solution could be to make new, fresh competent cells.

1.2.2 Amplified Insert Assembly

Amplified insert assembly does not depend on a fixed prefix and suffix sequence. So in theory this method could be combined with standard assembly, BioBrick BB-2 assembly, BglBricks assembly, Silver assembly, and Freiburg assembly.

The assembly technique was proposed by Michael A. Speer and Tom L. Richard in 2011, and it solves the problem of low transformation rate in 3A assembly. Another advantage is that the assembly method does not require the involved plasmids to have different antibiotic resistance genes, like 3A assembly does [11].

The principle of amplified insert assembly is to eliminate the noise from uncut plasmids, and thus to decrease the possibility of creating plasmids with unwanted combinations of insert and backbone. The method achieves this by simply amplifying the insert using PCR prior to digestion, and also by treating the mixture with the restriction enzyme DpnI, which digests methylated DNA. Plasmids are often methylated, so DpnI will eliminate the template plasmids, and only the PCR amplified insert will be left.

To eliminate the possibility of religation of the backbone, the backbone sample is treated with phosphatase. For a detailed protocol describing amplified insert assembly, *see* Subheading 3.9.

1.2.3 Gibson Scarless Assembly

In contrast to most other assembly techniques, Gibson scarless assembly allows joining of multiple BioBricks simultaneously [12]. But in order to achieve this, the technique requires that the DNA sequences to be joined together should overlap with each other, 20–150 bp. This is, however, often not the case when assembling BioBricks. Therefore PCR primers are used to create overhangs between adjacent BioBricks, as shown in Fig. 10. The primers in question should be made so that they anneal to approximately 20 bp of the end of a certain BioBrick, in addition to approximately 20 bp of the start of the next BioBrick.

To achieve annealing of the overlapping sequences in practice, a T5 exonuclease is used. This enzyme chews back nucleotides from the 5′ end, and thus creates single-stranded DNA in the ends of all sequences, where the different components are designed to anneal. After annealing, a DNA polymerase fills the gaps between the different DNA parts, and finally; a *Taq* ligase seals the nicks.

When it comes to designing primers that overlaps with the BioBricks to be joined, several software tools have been developed. One of these is Gibthon, made by the iGEM 2010 Cambridge team, described in Subheading 4.5, and another software J5, described in Chap. 14.

Gibson assembly itself does not require a certain prefix and suffix as standard, BB-2, BglBricks, Silver and Freiburg assembly does.

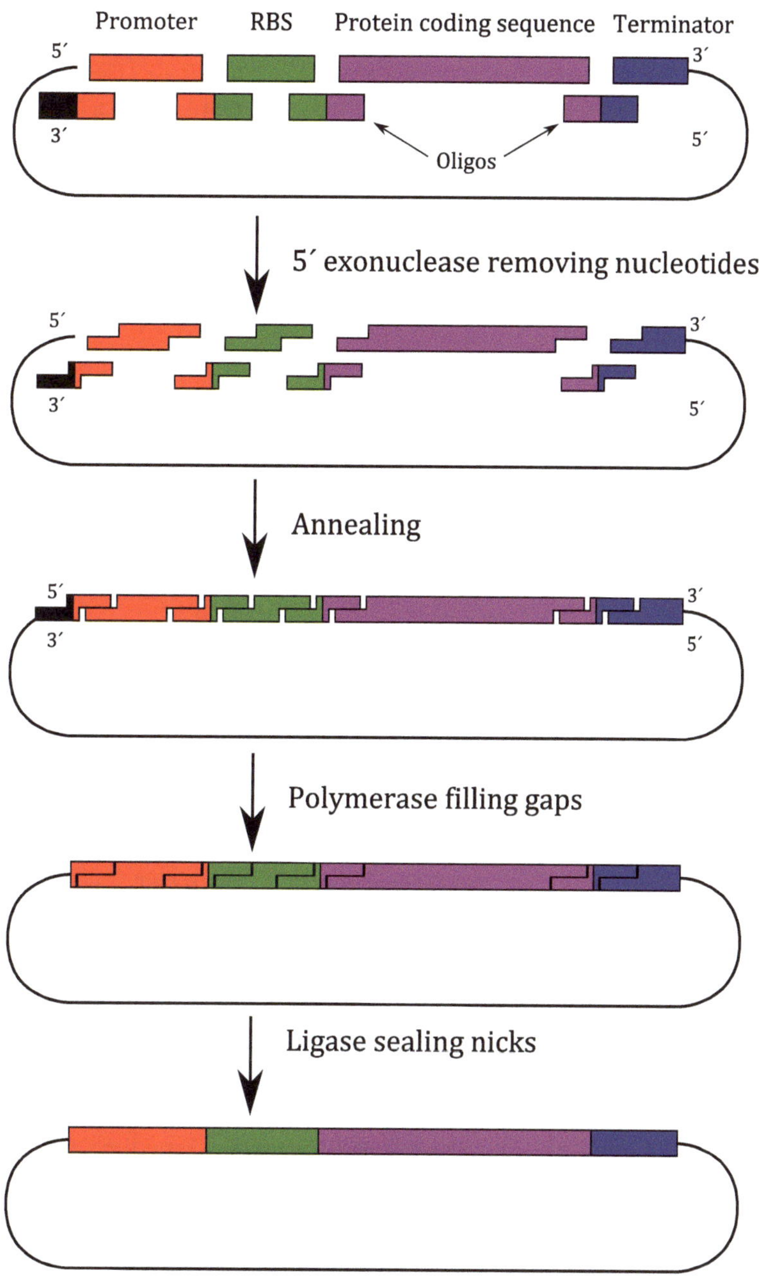

Fig. 10 Overview of the Gibson scarless assembly

However when joining BioBricks using this method, it is still recommended to keep the prefix and suffix associated with the composite BioBrick. This is to ensure that the part is compatible with one or more assembly standards, and it allows the use of restriction enzymes associated with a certain assembly standard possible.

2 Materials

2.1 Transformation

1. DNA to be used, resuspended.
2. Competent cells, 50 μL per transformation.
3. Ice in a container.
4. Water bath, 42 °C.
5. Glass spreader.
6. Clean SOC media.
   - SOC medium (1 L): 20 g bacto-tryptone, 5 g yeast extract, 0.5 g NaCl, 10 mL 250 mM KCl, 5 mL 2 M $MgCl_2$, 10 mL 1 M $MgSO_4$, dH_2O (adjust to 1 L).
   - Autoclave the medium.
   - Let the medium cool down, then add 20 mL 1 M sterile-filtered glucose.
7. 37 °C incubator, both shaking and normal.
8. Two petri dishes with Lysogeny Agar (LA) medium containing a suitable antibiotic.
   - LA medium: 10 g tryptone, 5 g yeast extract, 10 g NaCl, and 15 g agar in 1 L of water. Autoclave for 20 min, and let the medium cool down to approximately 5 °C before adding the suitable antibiotic.
   - After adding the suitable antibiotic (Ampicillin—100 μg/mL, Kanamycin—50 μg/mL, and Chloramphenicol—170 μg/mL), pour the mixture into petri dishes. Let them solidify, invert, and keep at 4 °C, in the dark.
9. Distilled H_2O.

2.2 Inoculation of a Colony

1. Autoclaved toothpicks.
2. Plastic tube, 15 mL.
3. Lysogeny Broth (LB) medium with suitable antibiotic.
   - LB medium: 10 g tryptone, 5 g yeast extract, and 10 g NaCl in 1 L of water. Autoclave for 20 min.
   - Since an antibiotic is needed, let the solution cool to approximately 55 °C before adding it.
   - Final concentrations of common antibiotics: Ampicillin—100 μg/mL, Kanamycin—50 μg/mL, and Chloramphenicol—170 μg/mL.
4. 37 °C incubator.

2.3 Plasmid DNA Isolation

1. Promega Wizard Plus SV Minipreps DNA Purification System A1460.
2. Ethanol 95 %.

3. Microcentrifuge capable of 14,000 × *g*.
4. Sterile 1.5 mL microcentrifuge tubes.
5. Centrifuge capable of 10,000 × *g*.

2.4 Restriction Digestion of BioBricks

1. Bucket with ice.
2. Microcentrifuge tubes, 1.5 mL.
3. Distilled H_2O.
4. NEB Buffer 1, 2, 3, and 4.
5. Bovine serum albumin (BSA).
6. Restriction enzymes: EcoRI, SpeI, XbaI, PstI (for BBF RFC 10, if another assembly standard is used, other enzymes might be required).
7. Water bath or thermal cycler, 37 and 80 °C.
8. Spectrophotometer (e.g., NanoDrop).

2.5 Gel Electrophoresis

1. Agarose gel (low EEO, for less smeared bands) with Gel Green™.
2. TBE buffer (Tris/Borate/EDTA).
   - Preparation of stock solution: 54 g tris base, 27.5 g borate, 20 mL 0.5 M EDTA, diluted up to 1 L with dH_2O.
   - For use: dilute 100 mL with 900 mL distilled H_2O.
3. Loading dye (0.2 volume of loading dye, e.g., 4–20 μL).
   - Example recipe: 25 mg bromophenol blue, 4 g sucrose, diluted with dH_2O to 10 mL. If using bromophenol blue, be aware that it migrates equally quickly as 200–400 bp sized DNA pieces, and will therefore hide those fragments. In cases where this applies, use another dye.
4. 1 kb DNA ladder.
5. Gel documentation unit.

2.6 Gel Purification

1. QIAquick Gel Extraction Kit (Qiagen).
2. Pieces of gel with DNA.
3. Ethanol (96–100 %).
4. Isopropanol (100 %).
5. Heating block or water bath, 50 °C.
6. Microcentrifuge capable of 14,000 × *g*.

2.7 Ligation

1. Digested BioBricks to be ligated (the amount of digested DNA is determined by expression 1).
2. T4 DNA Ligase Reaction Buffer.
3. T4 DNA Ligase.
4. Heating block, 16 and 80 °C.

2.8 Amplified Insert Assembly

1. Thermal cycler.
2. High-fidelity polymerase (e.g., Phusion, Pfu Turbo, or Vent).
3. Primers.
4. Antarctic Phospotase and Antarctic Phospotase Buffer.
5. Dpnl.

3 Methods

3.1 Transformation

The presented protocol is the official transformation protocol from the Registry of Standard Biological Parts [12] and is intended for inserting plasmid DNA into competent *E. coli* cells.

1. Thaw the competent cells by putting them on ice.
2. Add 1–2 μL of the DNA to the competent cells, and 1 μL dH_2O to the control (*see* **Note 1**).
3. Incubate the closed tubes on ice for 30 min.
4. Heat shock the competent cells by immersion in a 42 °C water bath for 60 s (important not to exceed that time).
5. Incubate on ice for 5 min.
6. Add 200 μL clean SOC media to each transformation.
7. Make sure the tubes are properly closed and incubate the cells for 1 h at 37 °C, with shaking (the time can be reduced, but this step is still very important for an efficient transformation).
8. Prepare two petri dishes of LA with the sufficient antibiotic(s), by labeling the petri dishes with part number and plasmid backbone (antibiotic resistance can be found in the name of the plasmid backbone). Plate out 20 μL and 200 μL onto the dishes.
9. Prepare two more petri dishes the same way for the control.
10. Incubate the plates at 37 °C for 12–14 h. *See* **Note 2**.
11. Pick a colony to store using a glycerol stock or inoculate a colony to be miniprepped.

3.2 Inoculation of a Colony

1. Prepare a 10 mL tube with 4 mL LB media and the appropriate antibiotics (for antibiotic concentrations, *see* Subheading 2).
2. Pick one colony from the petri dish with a sterilized toothpick and drop it into the 10 mL tube.
3. Incubate in a shaking incubator holding 37 °C for 12–16 h.

3.3 Plasmid DNA Isolation

The centrifugation protocol supplied with Promega Wizard Plus SV Minipreps DNA Purification System A1460 is presented [13].

Before starting, add 170 mL 95 % ethanol to dilute the Column Wash Solution, giving a final volume of 270 mL. Cell Resuspension

Solution, Cell Lysis Solution, Neutralization Solution, and Column Wash Solution are all found in the Promega kit. Perform all the steps at room temperature, and label the tubes used during the miniprep properly.

1. Centrifuge the 10 mL tube containing cells with DNA to be isolated at 10,000 × *g* for 5 min. Pour off supernatant and remove excess media on a paper towel by inverting the tube.
2. Add 250 μL of Cell Resuspension Solution, and vortex or pipette the solution until the pellet is completely resuspended. Transfer to a sterile 1.5 mL microcentrifuge tube. For the following steps, avoid vortexing and pipetting up and down, not to induce double strand breaks in the DNA.
3. Add 250 μL of Cell Lysis Solution and mix by inverting the tube four times. Incubate for 1–5 min, or until the cell suspension clears. Do not incubate for longer than 5 min.
4. Add 10 μL of Alkaline Protease Solution and mix by inverting the tube five times. Incubate for 5 min (do not exceed this time).
5. Add 350 μL of Neutralization Solution and immediately mix by inverting the tubes four times.
6. Centrifuge the bacterial lysate at 14,000 × *g* for 10 min, using a microcentrifuge.
7. Insert one Spin Column (found in the kit) into a 2 mL Collection Tube (also in the kit), for each sample.
8. Transfer the lysate to the Spin Column, avoid transferring any of the precipitate. If some of the precipitate is transferred, pour the contents into a new sterile 1.5 mL microcentrifuge tube and centrifuge again for 10 min. Transfer lysate.
9. Centrifuge for 1 min at maximum speed. Discard the flow through and reinsert the Spin Column into the Collection Tube.
10. To wash, add 750 μL of Column Wash Solution to the Spin Column.
11. Centrifuge for 1 min and discard flow through.
12. Repeat wash procedure, but use 250 μL of Column Wash Solution instead.
13. Centrifuge for 2 min at maximum speed.
14. Carefully transfer the Spin Column to a sterile 1.5 mL microcentrifuge tube, so no Column Wash Solution is transferred. Centrifuge for 1 min at maximum speed.
15. Transfer to a new sterile 1.5 mL microcentrifuge tube and elute the plasmid DNA (now attached to the filter) by adding 100 μL Nuclease-Free Water to the Spin Column. Centrifuge for 1 min at maximum speed.
16. Remove and discard the Spin Column. Close and label the tube and store at −20 °C or below.

Table 1
Enzymes, buffer needed and product given from a double restriction digest

Product	Enzyme 1	Enzyme 2	NEB buffer
Front insert	EcoRI	SpeI	2 or 4
Whole insert	EcoRI	PstI	3
Front vector	EcoRI	XbaI	2 or 4
Back vector	SpeI	PstI	1 or 2[a]
Back insert	XbaI	PstI	2 or 3[a]

[a]NEB buffers for SpeI + PstI and XbaI + PstI give only 75 % activity for the enzymes, the rest have 100 %

3.4 Restriction Digestion of BioBricks

To be sure to get enough of the substrates for the restriction digest, it is wise to measure the concentration of the DNA (*see* **Note 3**). The following protocol is a single reaction protocol from www.partsregistry.org [14].

1. Add 250 ng of DNA and the right amount of distilled water, for a total volume of 16 μL.
2. Add 2.5 μL of the appropriate NEB buffer (*see* Table 1)
3. Add 0.5 μL of BSA
4. Add 0.5 μL Enzyme 1
5. Add 0.5 μL Enzyme 2 (if there's only one enzyme, add 1 μL of the enzyme once).
6. This gives a total volume of 20 μL. Mix and spin down.
7. Incubate at 37 °C for 1 h.
8. Run a portion of the digest on a gel to check the lengths of the parts of interest.

All of the enzymes work optimally at 37 °C.

3.5 Gel Electrophoresis

1. Molding the gel:
   - Choose the appropriate comb based on the number of samples and the volume of each sample (*see* **Note 4**). Put the comb in place and pour the dissolved gel mixture slowly in the mold.
   - Let it harden for between half an hour and three quarters, and remove comb vertically.
   - Pour the diluted TBE buffer [15] into the tank, up to the point where it gets in contact with the electrodes and is completely covering the gel (*see* **Note 5**).
2. Preparation of samples:

- To the 20 μL of sample yielding from restriction digest, add 4 μL of 5× loading dye.
- Flick the sample or pipette up and down a couple of times to mix it.
- Load the samples in separate wells, be sure to write down which sample is where.
- Load the ladder in one or two wells.

3. Running gel electrophoresis:
   - Put the lid on and switch on the power, e.g., 80 V. To be sure it has started, look for bubbles near the electrodes.
   - Let it run for approximately 45 min (depending on the expected fragment size) and check if the separation is satisfactory. If not, let the gel run a little longer (*see* **Note 6**).
   - When the bands are well separated, cut out the bands of interest (compare to the ladder) using a clean scalpel.

3.6 Gel Purification

The protocol used here is given by QIAGEN [16], but several equivalent gel purification kit exists, and could as well be used.

1. Weight the gel piece. Add 3 volumes of Buffer QG to 1 volume gel (100 mg–100 μL). For >2 % agarose gels, add 6 volumes Buffer QG.
2. Dissolve the gel at 50 °C, vortexing the tube every 2–3 min (should take approximately 10 min).
3. If the color of the mixture is yellow like the Buffer QG (buffer QG is yellow at pH < 7.5), proceed. If it's orange or violet, add 10 μL 3 M sodium acetate, and mix. The solution will then turn yellow.
4. Add 1 gel volume of isopropanol, mix.
5. Place a QIAquick spin column in a 2 mL collection tube and apply the sample.
6. Centrifuge for 1 min to bind the DNA.
7. Wash by adding 0.75 mL Buffer PE to the column and centrifuge for 1 min. Discard flow through. Place the column back in the tube.
8. Centrifuge the column for 1 min to remove the rest of the wash buffer. Discard flow through and place the column in a clean 1.5 mL microcentrifuge tube.
9. Elution of DNA: add 50 μL Buffer EB or water to the center of the membrane and centrifuge the column for 1 min. For higher DNA concentration, add 30 μL Buffer EB to the center of the membrane, wait for 1 min and then centrifuge for 1 min.

After adding Buffer EB, increasing the incubation time up to 4 min can increase the DNA concentration.

10. Measure the concentration using a NanoDrop [17] and write the concentration on the tube. Store at −20 °C.

3.7 Ligation

The presented protocol is from partsregistry.org [18].

1. Use Eq. 1 to find volume of backbone and volume of insert to be used. Total volume (backbone + insert) should be 15 μL, and normally, the ratio 1:3 between backbone and insert can be used.

$$V_{insert} = \frac{size_{insert} \cdot C_{backbone} \cdot V_{backbone} \cdot n}{size_{backbone} \cdot C_{insert}} \quad (1)$$

2. Make two ligation samples in 1.5 mL tubes; one with the insert and one with dH_2O added instead of insert (the latter is called a religation and is used as a control experiment).
3. Add 2 μL of T4 DNA Ligase Reaction Buffer.
4. Add 1 μL of T4 DNA Ligase, mix, and spin down.
5. Incubate for 30 min at 16 °C and 20 min at 80 °C to heat kill the enzymes.
6. Store at −20 °C, or use 2 μL of the ligation mixture for transform to competent cells.

In this expression, the sizes of the insert and backbone are measured in base pairs, concentration is measured in ng/μL, and volume is measured in μL. n denotes the ratio between insert and backbone, with respect to the insert. For example; if three times as much insert as backbone is used, then $n = 3$.

3.8 Making a Construct of Two Parts

1. Obtain plasmids with the genes of interest, many can be found through the Registry of Standard Biological Parts (http://partsregistry.org/Catalog).
2. Decide which should be treated as insert and backbone (*see* **Note 7**), and perform a restriction digestion as described in Subheading 3.4. Enzymes and buffers to be used can be seen in Table 1.
3. To obtain the correct pieces of DNA, investigate the fragments resulting from the restriction digestion using gel electrophoresis (described in Subheading 3.5), followed by excision of the correct gel piece, and gel purification (described in Subheading 3.6).
4. Ligate the two pieces together, using the protocol for Ligation (described in Subheading 3.7) (*see* **Note 8**).
5. Use 2 μL of the ligation mix to perform a transformation (described in Subheading 3.1).

6. Transfer a colony to liquid medium (Subheading 3.2).
7. Miniprep the transformants (Subheading 3.3), to obtain plasmid DNA.

3.9 Amplified Insert Assembly Protocol

The protocol given here is taken from the OpenWetWare website [19].

1. Miniprep both insert and plasmid from their respective cultures (*see* Subheading 3.3).
2. Amplify the insert using PCR.
   - Use a high-fidelity polymerase
   - Primers:
     - The primers should flank the restriction sites by 100–150 bp.
     - The primers should have a T_m of 55–60 °C.
     - For most Biobrick applications, the primers VF2 and VR can be used.
   - Run 25–30 cycles, as this will help ensure high fidelity.
3. Start the vector digestion while the PCR is still running. Digest the vector with the appropriate restriction endonucleases for 2 h (*see* Subheading 3.4).
4. Purify the PCR product (*see* Subheading 3.6).
5. Digest the purified insert for 1 h with enzymes complementary to the vector digest. Include DpnI in the reaction mixture (*see* Subheading 3.4).
6. Add 6 μL Antarctic Phosphatase Buffer and 1 μL Antarctic Phosphatase to the vector digest, and incubate until the insert digest is done.
7. Heat inactivate all enzymes by incubating the reaction mixtures for 20 min at 80 °C.
8. Ligate at a molar ratio of 4:1 (insert:vector) (*see* Subheading 3.7).
9. Transform the ligation mixture to competent cells (*see* Subheading 3.1).
10. Plate out the transformed cells on agar plates containing a suitable antibiotic.
11. Celebrate!

4 Software Tools

4.1 Sources of Software Developed Within the iGEM Competition

Starting in 2008, awards for software tools were introduced in the iGEM competition [20]. The efforts of iGEM teams have since then produced a number of software tools performing a wide variety of tasks that are relevant when working with BioBrick parts from the Registry of Standard Biological Parts. The latest tools that have been created can be located through the iGEM repository at

GitHub, https://github.com/igemsoftware and a large selection of iGEM-developed software is available at http://igem.synbioreview.com/. The iGEM community pages could also be worth paying a visit.

The amount of programs available is vast and increases rapidly. In this section only a few notable examples will be listed. A point emphasizing the rapid advancement of this field is that, at the time of this being written, all the tools mentioned here have been around for less than 2 years.

4.2 The Constructor

The Constructor is created by iGEM teams from Wageningen UR. Its purpose is to optimize assembly of BioBrick parts given a list of parts that the user wishes to put together. It is available as an easy-to-use web application through http://www.systemsbiology.nl/the_constructor/. A screenshot of the tool is shown in Fig. 11. The Constructor also is the subject of an open access publication in the *Journal of Biological Engineering* entitled "The Constructor: a web application optimizing cloning strategies based on modules from the registry of standard biological parts" [21].

To use The Constructor, only an e-mail address is required, to which the results are sent, as well as a list of transcription units (TUs). Each TU should consist of one or more BioBrick parts from the Registry listed in the order that they are to be transcribed and separated by a space, e.g., "BBa_R0062 BBa_B0034 BBa_E1010 BBa_B0010 BBa_B0012." Some additional options are also available for filtering parts by their working status, previous experience, and availability. To generate a cloning procedure, click "go" and wait for the results to arrive by e-mail.

4.3 BioBrick Search Engine

The 2012 iGEM team from UT-Tokyo created a useful search engine that makes it easier to locate parts in the database of the Registry. It generally gives more relevant results than the Registry's own search engine, and it also offers some additional functionalities. The BioBrick search engine is located at http://igem-ut.net/bbsearch/.

This tool works like any other search engine: Enter a query, hit enter, and browse the list of search results. The query may be any text, such as "gfp," "strong promoter," or "E. coli," and the results are presented as a list of BioBrick parts sorted in descending order by a combination of reliability, relevance, and popularity. These three factors are each visualized as a scale of stars ranging from zero to five. Some other information from the Registry of Standard Biological Parts is also shown, such as the part type, the creator of the part, the year of submission, and a brief description of the part. It is also possible to filter the results by type or year, either directly in the search field or by using a panel on the left-hand side of the search results. A screenshot from the BioBrick Search Engine is shown in Fig. 12.

4.4 Clotho

Clotho differs fundamentally from the above-mentioned applications. First of all, it is not an application, but a platform, or converging point, for several different synthetic biology tools. In other words,

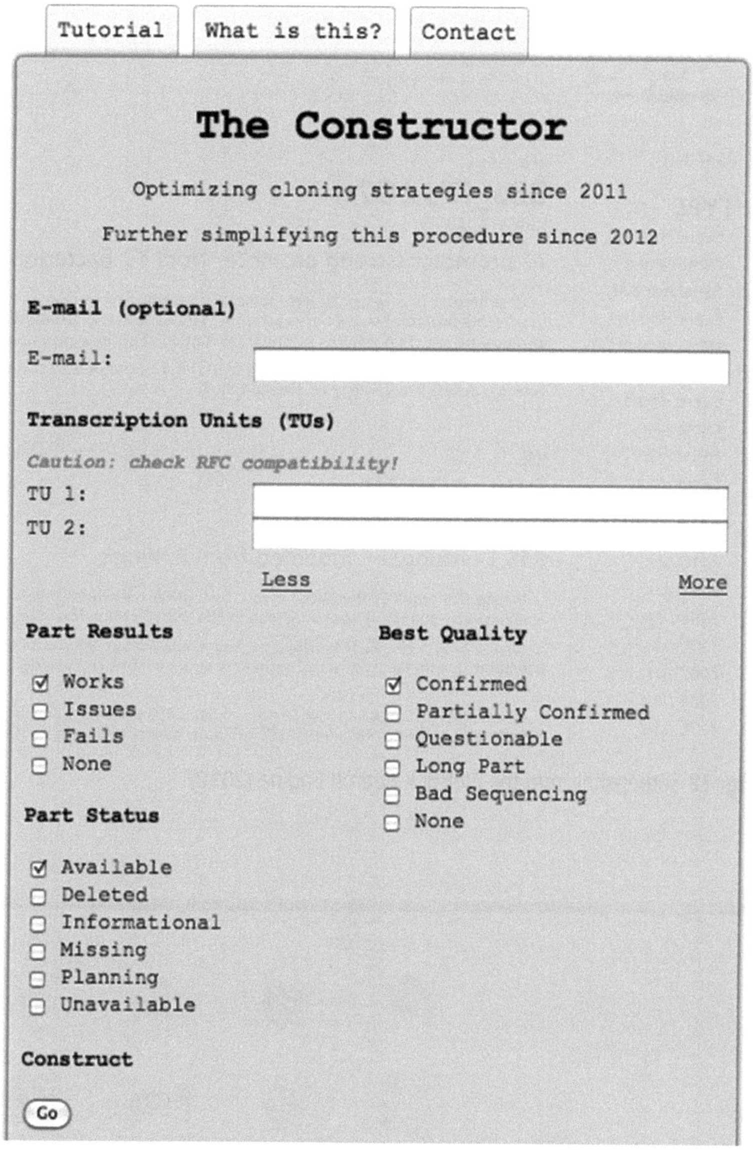

Fig. 11 Screenshot from the Constructor (Wageningen UR 2011)

Clotho itself does not do anything; it is more of an environment for running other programs. It is also downloadable rather than web-based.

In principle, Clotho is an app environment that is similar to a smart phone in the sense that anyone can create and share new tools using its platform. Using apps, it is possible for users to customize their own versions of Clotho to suit their needs. A notable advantage of running several synthetic biology tools on the Clotho platform is its standardized data model: All apps written for Clotho pass data seamlessly to one another and work together even when they were not originally designed to do so. A screenshot from Clotho is shown in Fig. 13.

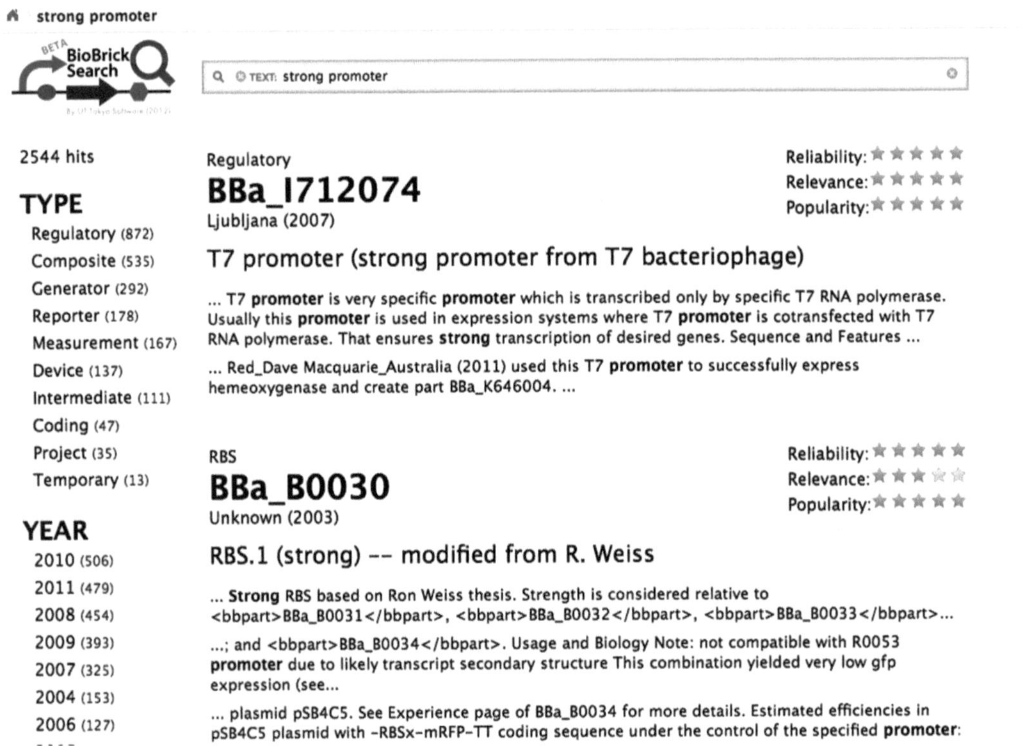

Fig. 12 Screenshot from the BioBrick Search Engine (2012)

Fig. 13 Screenshot from Clotho (2012)

The currently available Clotho applications perform several different tasks, from parsing parts from the Registry, to viewing and editing sequences, and calculating fragment sizes resulting from restriction digestion. New Clotho applications are continuously being developed, several through the iGEM competition. To read more or download Clotho, visit http://clothocad.org.

4.5 Gibthon

Gibthon is a suite of web-based tools to aid in the design and manufacture of synthetic parts and devices for biological systems. It was created by the 2010 iGEM team from Cambridge with the purpose of automating primer design for Gibson Assembly. Gibthon allows for seamless joining of an arbitrary number of sequences and facilitates the design of primers that are optimized for length and melting temperature.

As of today, the Gibthon project includes several functionalities that are structured as apps. In addition to the Construct Designer, which sets up the Gibson Assembly schemes, there are online calculators available for primer design, ligation ratios, solution molarity, and buffer choice for digestion reactions.

Gibthon is located at http://django.gibthon.org/. It is an open-source project and the source code can be found on http://www.github.com.

5 Notes

1. The DNA is fragile, so do not vortex to mix, instead carefully flick the tube with your finger.
2. Remember to put the agar side of the dishes up. Also, if incubated too long, the antibiotic(s) will break down, especially in transformations where ampicillin is used.
3. To measure the concentrations of the isolated DNA, a spectrophotometer can be used (e.g., the NanoDrop 1,000 Spectrophotometer from Thermo Scientific [17]).
4. A small comb will give smaller wells but also room for more samples. There should be at least one well for the DNA ladder, but since the DNA migration length could vary throughout the gel, so putting one ladder on each side is recommended [22].
5. When loading the gel, keep in mind that DNA migrates towards the plus pole.
6. Beware that the small DNA pieces requires less time to migrate and might move out of the gel if it is left for too long.
7. Smaller plasmids are good to use as backbone, and an insert should be no smaller than approximately 500 bp
8. Remember to make a religation test. A common source of background is either by religation of the vector, or simply by having uncut plasmid.

References

1. Knight T. Standard sequence assembly of BioBricks. http://openwetware.org/images/a/a3/BBFRFC8.pdf. Accessed 11 Dec 2012
2. Registry of Standard Biological Parts. Parts. http://partsregistry.org/Help:Parts. Accessed 11 Dec 2012
3. Knight T. (2007) Draft Standard for BioBrick Biological Parts. http://dspace.mit.edu/bitstream/handle/1721.1/45138/BBFRFC10.txt?sequence=1. Accessed 12 Dec 2012
4. Knight T. (2008) BBF RFC 12 Draft Standard for BioBrick BB-2 http://dspace.mit.edu/bitstream/handle/1721.1/45139/BBFRFC12.txt?sequence=1. Accessed 12 Dec 2012
5. Anderson JC, Dueber JE, Leguia M. et al. (2009) BBF RFC 21: BglBricks Assembly Standard http://dspace.mit.edu/bitstream/handle/1721.1/46747/BBFRFC21.pdf?sequence=1. Accessed 12 Dec 2012
6. Anderson JC, Dueber JE, Leguia M et al (2010) BglBricks: a flexible standard for biological part assembly. J Biol Eng 4:1
7. Phillips IE, Silver PA (2006) A new BioBrick assembly strategy designed for facile protein engineering. http://hdl.handle.net/1721.1/32535. Accessed 27 Dec 2012
8. Müller KM, Arndt KM, iGEM 2007 Team Freiburg et al. (2009) BBF RFC 25: Fusion Protein (Freiburg) BioBrick assembly standard. http://hdl.handle.net/1721.1/45140. Accessed 5 Nov 2012
9. Shetty R, Lizarazo M, Rettberg R et al (2011) Assembly of BioBrick standard biological parts using three antibiotic assembly. Methods Enzymol 498:311–26
10. The Registry of Standard Biological Parts. 3A Assembly. http://partsregistry.org/Assembly:3A_Assembly. Accessed 24 Jan 2013
11. Speer MA, Richard TL (2011) Amplified insert assembly: an optimized approach to standard assembly of BioBrick genetic circuits. J Biol Eng 5:17
12. Registry of Standard Biological Parts. Transformation. http://partsregistry.org/Help: Protocols/Transformation. Accessed 5 November 2012
13. Promega (2010) Wizard® *Plus* SV Minipreps DNA Purification System. http://tinyurl.com/promega-miniprep
14. Registry of Standard Biological Parts. Restriction Digests. http://partsregistry.org/Help:Protocols/Restriction_Digest. Accessed 6 Nov 2012
15. About.com. Make TBE Buffer. http://biotech.about.com/od/buffersandmedia/ht/MakeTBE.htm. Accessed 6 Nov 2012
16. QIAGEN (2010) QIAquick® Gel Extraction Kit. http://www.qiagen.com/literature/render.aspx?id=201083. Accessed 7 Nov 2012
17. Thermo Scientific (2008) NanoDrop 1000 spectrophotometer V3.7 user's manual http://www.nanodrop.com/library/nd-1000-v3.7-users-manual-8.5x11.pdf. Accessed 6 Nov 2012
18. Registry of Standard Biological Parts. Ligation. http://partsregistry.org/Help:Protocols/Ligation. Accessed 7 Nov 2012
19. OpenWetWare. Amplified Insert Assembly. http://openwetware.org/wiki/Amplified_insert_assembly. Accessed 25 Jan 2013
20. iGEM.org. About iGEM. http://igem.org/About. Accessed 8 Jan 2013
21. Hesselman MC, Koehorst JJ, Slijkhuis T et al (2012) The constructor: a web application optimizing cloning strategies based on modules from the registry of standard biological parts. J Biol Eng 6:14
22. Lewis M. Agarose gel electrophoresis (basic method) http://www.methodbook.net/dna/agarogel.html. Accessed 6 Nov 2012

Chapter 2

Plasmid Construction by SLIC or Sequence and Ligation-Independent Cloning

Ryan E. Hill and Julian J. Eaton-Rye

Abstract

Sequence and ligation-independent cloning (Nat Methods 4:251–256, 2007) is a powerful tool for the construction of multi-fragment complex plasmids in a simple and efficient manner. Plasmids consisting of 6–7 DNA fragments can be assembled in a single day, with additional 2 days for screening and extraction. SLIC requires PCR products with overlapping regions of 30–40 bp at the 5′ and 3′ ends, T4 DNA polymerase, and an optional RecA protein for construction.

Key words Synthetic biology, SLIC, Plasmid construction, Molecular biology, Cloning

1 Introduction

Plasmid construction and cloning have been a mainstay of molecular biology for several decades [1, 2]. Until recently, cloning and plasmid construction have made use of restriction enzymes and ligases to cleave and join desired fragments of DNA. While cloning of DNA fragments using restriction enzymes is suitable for two-fragment cloning (i.e., a DNA insert into a plasmid), this method becomes inefficient upon increasing the number of DNA fragments to be assembled. This inefficiency is largely attributed to the short 2–4 bp single-stranded overhangs created by restriction enzyme digestion, the so-called sticky-ends. Such short sequences of DNA do not provide sufficient stability or specificity for efficient construction of plasmids comprising three or more fragments. Additionally, as the number of fragments increases, it becomes difficult to find unique restriction sites for assembling the plasmid. Construction of multi-fragment plasmids via conventional cloning involves a large amount of sub-cloning steps that in some cases can take from weeks to even months of plasmid construction to reach the final desired plasmid.

Svein Valla and Rahmi Lale (eds.), *DNA Cloning and Assembly Methods*, Methods in Molecular Biology, vol. 1116, DOI 10.1007/978-1-62703-764-8_2, © Springer Science+Business Media New York 2014

With the emergence of the field of synthetic biology, a number of techniques have been developed for construction of multi-part plasmids in as little time as 3 days. Sequence and ligation-independent cloning (SLIC) [3] is one such protocol and is the focus of this chapter. Similar protocols include circular polymerase extension cloning (CPEC, Chap. 8) [4] and the Gibson method (Chap. 1) [5]. The potential of these methods has been realized with the advent of large-scale commercial DNA oligomer synthesis, which provides cheap primers of lengths up to 100 bp and delivered in as little as 24 h. The development of fast, high fidelity polymerase systems such as Phusion Hot-Start II polymerase [6] and PCR protocols including Touch-Down PCR (TD-PCR) [7, 8] has also increased the efficiency of these molecular biology techniques.

The SLIC method involves designing custom, highly specific, sticky-ends of 20–40 bp through careful primer design and subsequent digestion of the 3′ ends by T4 DNA polymerase. The T4 DNA polymerase exhibits 3′–5′ exonuclease activity in the absence of dNTPs, facilitating the formation of long sticky-ends. The final plasmid is assembled from DNA fragments (PCR products) where each fragment's 3′ sticky-end is a reverse complement to the adjacent fragment's 5′ sticky-end (Fig. 1). The sticky-ends are stable at <40 °C and so upon mixing of the fragments together the plasmid self-assembles without the need for an in vitro ligase enzyme to stabilize the construction process. Upon transformation, the natural repair processes of *Escherichia coli* will repair the nicks (single-stranded breaks) in the plasmid at the overlaps between the fragments [9–11]. This repair process is efficient enough for rapid construction of plasmids consisting of up to six fragments; however, increasing the number of fragments beyond six does significantly decrease the efficiency of successful construction [3]. For construction of plasmids with greater than six fragments, CPEC [4], the Gibson method [5], or the DNA assembler method [12] is recommended.

In this chapter we describe a series of routine steps that enable successful cloning of multiple DNA fragments using the SLIC method.

2 Materials

All solutions should be made using ultra-pure water (prepared by purifying deionized water to attain a sensitivity of 18 MΩ cm at 25 °C) and analytical grade reagents. All chemicals were purchased from Sigma-Aldrich Corporation, St. Louis, MO, USA, unless otherwise stated.

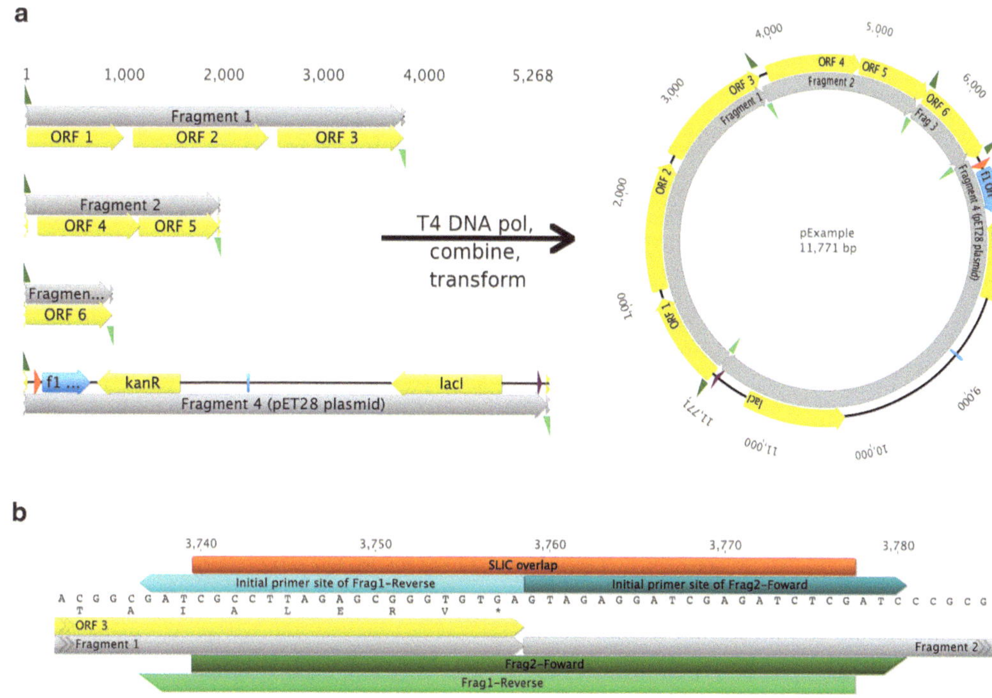

Fig. 1 (**a**) An example plasmid, pExample based on pET28a (Novagen), consisting of four PCR fragments (*grey*) which when digested with T4 DNA polymerase and combined together create the 11,771 bp plasmid. The 3′ end of each fragment matches the 5′ end of the next fragment. Primer binding sites for each fragment are indicated (forward, *dark green*; reverse, *light green*). Annotations: open reading frames (*yellow*), origins of replication (*blue*), T7 terminator (*orange triangle*), T7 promoter (*purple triangle*). (**b**) Detailed example of the overlap between Fragment 1 and Fragment 2. Primers (*light/dark green*) were designed such that a 38 bp overlap (*orange*) is created between the two fragments after PCR and T4 DNA polymerase digest. This overlap has a T_m of 77 °C and is stable during "ligation" at 37 °C. The forward primer binds the template with an initial T_m of 57 °C (initial primer site, *dark blue*), while the reverse primer has an initial T_m of 65 °C (initial primer site, *light blue*)

2.1 Sodium Borate Gel Electrophoresis

1. 20× Sodium Borate (SB) buffer: 0.2 M NaOH, 0.73 M boric acid, pH 8.0 [13]. Dissolve 8 g of NaOH in 500 mL of water followed by 45 g of boric acid. Make up to 1 L with water (*see* **Note 1**).
2. 1× SB: 10 mM NaOH, 36.5 mM boric acid, pH 8.0. Mix 50 mL of 20× SB buffer with 950 mL of water. Store at room temperature (*see* **Note 1**).
3. 10× sample buffer: 0.25 % (w/v) bromophenol blue, 0.25 % (w/v) xylene cyanol FF, 30 % (w/v) glycerol. Dissolve 0.05 g bromophenol blue and 0.05 g xylene cyanol FF in 10 mL of water. Add 6 mL of glycerol and make up to a final volume of 20 mL with water, and store at 4 °C.

4. Ethidium bromide solution 0.01 % w/v.
5. UV gel imager, e.g., Gel-Doc UV imager (Bio-Rad Laboratories, USA).
6. UltraPure agarose (Invitrogen, USA).

2.2 Touch-Down PCR

1. Phusion Hot-Start II High-Fidelity DNA Polymerase Kit (Finzymes, Finland).
2. 20 mM deoxyribonucleotide triphosphate (dNTP) stock: 20 mM dATP, 20 mM dTTP, 20 mM dCTP, 20 mM dGTP. Using a 100 mM dNTP Set PCR Grade (Roche Diagnostics GmbH, Germany), add 20 μL of each dNTP to a tube containing 20 μL of water to give a final volume of 100 μL. Store at −20 °C (*see* **Note 2**).
3. Primers: 10 μM working stock for each primer. Assuming a stock concentration of 100 μM, mix 5 μL of primer with 45 μL of water. Store at −20 °C (*see* **Note 2**).
4. PureLink PCR Purification Kit (Invitrogen).

2.3 SLIC–T4 Digestion and Ligation

1. DpnI restriction enzyme (Roche Diagnostics GmbH, Germany).
2. T4 DNA polymerase kit: T4 DNA polymerase, 10× NEBuffer2, BSA 10 mg/L (New England BioLabs, USA).
3. RecA Protein (Epicentre Biotechnologies, USA).
4. Ligation buffer (Roche Diagnostics GmbH, Germany).
5. NanoDrop spectrophotometer (NanoDrop Products, USA).
6. Heat-shock or electroporation-competent *E. coli* cells.
7. Agar selection plates, 1 % w/v peptone, 0.5 % w/v yeast extract, 0.5 % NaCl, 1.5 % w/v agar + appropriate antibiotic.

3 Methods

3.1 Primer Design

1. Construct a virtual plasmid of the intended final plasmid with a molecular biology software suite such as Geneious [14, 15], and ensure to annotate clearly the regions of each fragment so that primers can be designed to overlap each fragment (*see* **Note 3**; Fig. 1).
2. For each boarder, create a forward and a reverse primer of 40–45 bp. Ensure at least 30 bp of overlapping sequence between the two primers, with initial melting temperature (T_m) of >55 °C (typically 20–25 bp), and ensure the 3′ terminal base of each primer is a G or a C (*see* **Note 4**; Fig. 1).
3. For each primer, check that no predicted secondary structures are present with a T_m equal to or greater than the initial T_m of

Table 1
Example TD-PCR protocol

Pre-phase 1	Time
98 °C	30 s
Phase 1 (10 cycles, 65–55 °C)	
98 °C	10 s
65 °C (−1 °C/cycle)	20 s
72 °C	30 s[a]
Phase 2 (15 cycles)	
98 °C	10 s
65 °C	20 s
72 °C	30 s[a]
Post-phase 2	
72 °C	30 s[a]

[a]Refer to **Note 8** for additional information

the primer. A useful online tool for oligomer analysis is OligoAnalyzer [16], by Integrated DNA Technologies, BVBA, Leuvan, Belgium (*see* **Note 5**).

3.2 Touch-Down PCR of SLIC Fragments

1. Prepare a 50 μL Phusion PCR for each fragment of the final plasmid. Each reaction has 34 μL of water, 10 μL of 5× HF Phusion buffer, 0.5 μL of 20 mM dNTP, and 0.5 μL of Phusion polymerase (0.25 U). Prepare a base mix by scaling up for the number of PCRs and mix thoroughly by pipetting (avoid bubbles/foaming). Aliquot 45 μL of base mix into 0.2 mL thin-walled PCR tubes, and add 1 μL of template (*see* **Note 6**) and 2 μL of each 10 μM primer stock (two per reaction), bringing the final volume to 50 μL.
2. Due to the nature of TD-PCR, a single PCR protocol should suffice for most, if not all, of the fragment reactions. The plasmid backbone, however, typically requires a longer extension time (*see* **Note 7**). The protocol presented in Table 1 should be a good starting point, *see* **Note 8** for further information.
3. While the TD-PCRs are running, prepare a 0.8 % SB gel of sufficient volume for the gel tank system of choice. For a gel of 85 × 60 × 5 mm, mix 0.2 g of agarose with 25 mL of 1× SB buffer in a 100 mL conical flask, microwave for 30 s, mix by swirling, and microwave for a further 20 s. Pour the gel, insert the comb, and allow to set for at least 15 min (*see* **Note 9**). Once the gel is set, remove the comb and gel dams (if applicable), and fill the tank with 1× SB buffer such that it covers the gel with approximately 5 mm of buffer.

4. On completion of TD-PCR, place the reactions on ice while running the SB gel. For each reaction, set up a tube containing 7 μL water, 2 μL of PCR mix, and 1 μL of 10× sample buffer. In addition, set up one tube in a similar manner but replace the PCR mix with 1 μL of 1 kb Plus DNA marker (Invitrogen) (*see* **Note 10**). Load each sample into a separate well and run the gel for 15 min at 200 V (25 V/cm gel length). Remove gel and soak in a solution of 0.01 % (w/v) ethidium bromide for 10 min to stain the DNA.
5. Observe the PCR products in a Gel-Doc UV imager (Bio-Rad Laboratories, USA) or other similar UV gel viewing system. For each reaction ensure that only a single band of the correct size is present (*see* **Note 11**).

3.3 T4 DNA Polymerase Digestion and Ligation

1. Following confirmation that the correct PCR products have been obtained, add 0.5 μL of DpnI to each PCR in which the DNA template was a plasmid. Incubate for 30–60 min at 37 °C (*see* **Note 12**).
2. After 1 h, purify the PCR product with a PCR purification kit; the Invitrogen PureLink Purification Kit is recommended. Follow the manufacturer's instructions but ensure to keep the final elution volume small (20–30 μL) to keep the DNA as concentrated as possible for further reactions (*see* **Note 13**).
3. For each purified PCR fragment, determine its concentration (e.g., via a NanoDrop spectrophotometer (NanoDrop Products, USA)) using 1–2 μL of the sample. Calculate the total amount of DNA required for each fragment, and determine how much T4 DNA polymerase is needed to achieve 0.5 U/μg DNA. Add the necessary amount of T4 DNA polymerase 10× buffer, taking into account the volume of T4 DNA polymerase. Typically this equates to 3 μL of buffer, with 0.5–1 μL of T4 DNA polymerase at 0.5 U/μL. Incubate the reactions for 30 min at 37 °C. Following incubation, immediately place on ice and add one-tenth volume of 10 mM dCTP to stop the reaction (*see* **Note 14**). Calculate the final concentration for each fragment, remembering to adjust for the added volumes (*see* Table 2 for an example).
4. An SLIC "ligation" is achieved by combining all the fragments in a new 0.2 mL thin-walled PCR tube (*see* **Note 15**). Each fragment should be added in sequential order, starting with the most 5′ fragment, to the most 3′ fragment of the total insert. Add the plasmid last (150 ng). Ensure that you properly mix the ligation each time a fragment is added. The total amount (ng) of each fragment to be added is calculated on a 1:1 (insert-to-plasmid) ratio. Divide the length of the fragment (in kilobases, kb) by the length of the plasmid fragment

Table 2
T4 DNA polymerase digestion and ligation

T4 DNA polymerase digest							
Fragment (28 μL)	**Conc. (ng/μL)**	**Total DNA (ng)**	**T4 Pol (0.5 U/μL) (μL)**	**Buffer (μL)**	**dCTP (10 mM) (μL)**	**Final vol. (μL)**	**Final conc. (ng/μL)**
1	54	1,512	1.5	3.3	3.6	36	42
2	45	1,260	1.3	3.3	3.6	36	35
3	83	2,324	2.3	3.4	3.7	37	62
4 (plasmid)	55	1,540	1.5	3.3	3.6	36	42

Ligation				
Fragment	**Conc. (ng/μL)**	**Size (kb)**	**Amount (ng)**	**Volume (μL)**
1	42	3.80	108	2.6
2	35	1.96	56	1.6
3	62	0.89	25	0.4
4 (plasmid)	42	5.27	150	3.5
Lig. Buffer				0.9
RecA (optional)				0.1
(200 ng/μL)			Total	9.1

(kb). This is then multiplied by 150 ng to give the required amount (ng) of insert (*see* Table 2 for an example). Finally add one-tenth volume 10× ligation buffer. Incubate at 37 °C for 30 min (*see* **Note 15**).

5. Place the "ligation" on ice for 5 min and then transform the entire reaction into competent *E. coli* cells, which have a transformation efficiency of >10^6 colonies per μg of DNA. Follow the transformation procedure provided with the competent cells (be it heat-shock, electroporation, etc.). Plate 150 μL of cells onto one agar selection plate, concentrate the remaining cells to about 150 μL, and plate onto a second selection plate (*see* **Note 16**).

3.4 Screening for Correct Assembly

1. Assuming transformation was successful, screen 5–15 colonies via colony PCR utilizing primers used in construction of the plasmid. Use a forward primer from one fragment and a reverse primer from a 3′ adjacent fragment. Colony PCR uses the same reaction mix as described in Subheading 3.2; however,

use 10 μL reactions. For the template, use 1 μL of desired cells resuspended in 3 μL water. Plate the remaining 2 μL onto an appropriate selection plate and incubate for no more than 8 h at 37 °C (*see* **Note 17**). For the TD-PCR, use a similar protocol to that described in Subheading 3.2 except extend the pre-phase by one 98 °C step from 30 s to 12 min to allow inactivation of all cellular metabolism and lysis of the cells. Additionally, use a higher phase one annealing temperature range of 75–65 °C. Adjust the extension time as necessary to match the expected product size (*see* **Note 7**).

2. Once the colony PCR is completed, add 1 μL of 10× sample buffer to each reaction, and run on an SB gel as described in Subheading 3.2.
3. Set up 5 mL overnight cultures for three to four positive colonies and extract the plasmids via alkaline lysis or through the use of a commercial plasmid extraction kit. Confirm correct assemblage of the SLIC plasmids via restriction digest. Expect a success rate of 75–90 % from positively identified colonies from colony PCR, depending on the complexity of the SLIC plasmid (*see* **Note 18**).
4. Finally, sequence the extracted plasmid to confirm seamless construction of the SLIC plasmid and that no mutations are present. Ensure the sequencing covers the connections between each fragment, as there is a chance of misaligned overhangs, which introduce or remove additional bases.

4 Notes

1. When dissolving NaOH and boric acid, dissolve slowly as heat will be produced as the two compounds mix. Once fully dissolved the solution should be at pH 8.0 without further adjustment. The 20× solution can be stored indefinitely at room temperature. 1× SB can be used repeatedly so do not discard until it takes on a blue tint (from the sample buffer) or if resolution or performance is impaired.
2. Small volume working stock solutions for primers and dNTPs are necessary because constant freeze-thawing will gradually degrade them. Contamination/degradation of a small working stock solution is also of no great loss as a new stock can be easily prepared.
3. The use of software designed for molecular biology and bioinformatics is recommended. Geneious by Biomatters Ltd., Auckland, New Zealand, comes highly recommended as a software package. Geneious includes tools for sequence manipulation and annotation, primer design, restriction enzyme

analysis, sequencing data analysis and contig assembly, virtual plasmid construction, direct access to NCBI databases, and is compatible with multiple file-types. It is a multi-platform program (Java-based), and the user interface is intuitively designed.

4. Ending a primer on the 3′ end with one to two G/C pairs (GC-clamp) will increase PCR yield and performance as GC hydrogen bonding is stronger than for AT pairs.
5. A strong (T_m >50 °C) secondary structure, especially if residing at the 3′ end of the primer, will greatly reduce the performance of PCR leading to low or no product amplification, and primer replication.
6. If the template is a plasmid, linearize the plasmid if possible by restriction digest to reduce the number of colonies after transformation harboring non-SLIC plasmids. Final template concentrations should be approximately 20 ng/50 μL reaction for plasmids and 100 ng/reaction for genomic DNA.
7. Phusion polymerase has a flexible extension time of 15–30 s/1 kb of DNA. As a general rule, reactions with products equal to or less than 2 kb can be completed with a 30 s extension time. It is advised, however, to add an additional 30 s for every 2 kb of product. For DNA fragments greater than 4 kb, use an extension time of 20 s/1 kb. In our lab, we have replicated up to 11.7 kb, with extension times of 3 min, and regularly replicate 6 kb with extension times of 2 min.
8. A SLIC primer at ~40 bp will have an average T_m of about 70 °C and an initial binding temperature of 55–65 °C (phase one), because only half of the primer will match the initial template. The protocol provided is a variant of TD-PCR in that it will slowly drop annealing temperatures for 10 cycles to allow for initial binding, then return to a higher annealing temperature (phase two) for 15 cycles. If the PCR is unsuccessful, try adjusting the initial binding range (phase one) from 65–55 °C to 60–50 °C. If additional bands are observed, i.e., nonspecific amplification, try increasing to 70–60 °C. The initial annealing temperature range should encompass both primers' initial T_m values. Phase two annealing temperature should be adjusted such that it is 1–2 °C below the lowest T_m.
9. If the gel is not set, then the wells can be damaged when the comb is removed. Additionally, the separation of product can be impaired to the point where the bands are smeared beyond recognition. To ensure the gel is set in a timely manner, it is suggested that the flask of molten agarose be cooled by briefly running under a cold-water tap (1–2 min). Alternatively, simply pour the gel at the beginning of the TD-PCR protocol, as the TD-PCR will take a minimum of 45 min to complete, which is ample time for the gel to set.

10. Depending on the number of wells per comb, prepare one marker for every 7–9 samples, or a minimum of one marker per comb.
11. While it is possible to continue with the presence of contaminating bands, doing so will reduce the overall efficiency of the SLIC reaction. It is generally best to avoid this, especially if the number of SLIC fragments is greater than four. The desired product can be extracted by gel extraction; however, this is not recommended as the yield is usually not sufficient for SLIC reactions. It is preferable to simply repeat the PCR with slightly higher annealing temperatures (*see* **Note 8**).
12. DpnI has a recognition sequence of GATC and is methylation-sensitive. This enzyme should digest the plasmid template into a large number of fragments (typically more than 20 fragments). DpnI will not digest PCR products, as they are not methylated. Digestion with DpnI should drastically reduce the number of colonies after transformation harboring non-SLIC plasmids (background). If, however, there is still a large background, try amplifying the fragment again using 1 μL of the previous DpnI-digested reaction (purified) as template.
13. It is important for the success of SLIC to keep the DNA fragments as concentrated and in as small a volume as possible. Our lab uses the Invitrogen PureLink PCR Purification Kit; however, the following procedures can be applied to other kits to increase yield. Pass the sample, in binding buffer, through the column twice by pipetting the eluate back into the column reservoir. When eluting the washed DNA from the column (i.e., the final step), warm the elution buffer briefly in a microwave for 5–10 s, add 15 μL to the column, wait 1 min, elute, and then repeat, pooling the two elutions.
14. The dCTP can be substituted with dATP or dTTP or dGTP, but only one dNTP can be present. Having only a single dNTP causes T4 DNA polymerase to stall, rendering it incapable of both polymerase and exonuclease activity. Once the dCTP has been added, the reactions can be stored at −20 °C if desired.
15. Fragments of <200 bp should be added in a 5:1 (insert-to-plasmid) ratio; however, do not exceed a final volume of 20 μL (if possible, keep below 15 μL). Note that while it is called a ligation, no ligase is present. The newly created overhangs are stable at <40 °C, and often as high as 45 °C. When transformed into *E. coli* the nicks in the plasmid will be repaired by the cell's endogenous enzymes. To boost efficiency, 20 ng of RecA protein can be added to the ligation reaction. However, in our experience this is usually unnecessary unless the plasmid concentration is low (<20 ng/μL; also *see* ref. 3). When using RecA, only 15 ng of plasmid is required (remember to adjust

other fragments accordingly), and if the final volume is less than 10 μL make up the difference with water.

16. The efficiency of SLIC can vary considerably, depending on the complexity, number of fragments, the length of overlapping regions, and the concentration and volume of the "ligation" reaction. Plating of the concentrated cells is necessary to obtain as many colonies as possible for screening.
17. The remaining 2 μL of resuspended cells should be set out in a grid-like manner on a new plate, with each 2 μL drop numbered to match the corresponding PCR. The reason for this is to correlate the positive reactions with the source colony for extraction of the plasmid and further analysis.
18. The more fragments that comprise the final SLIC plasmid, the higher the chance of a mistake occurring. With 35–40 bp overlaps, however, plasmids consisting of 6–7 fragments and a final plasmid size of up to 12 kb are achievable in our experience.

Acknowledgement

REH was supported by a Todd Foundation Postgraduate Scholarship in Energy Research and JER acknowledges the support of New Zealand Marsden Contract 08-UOO-043.

References

1. Kawakami B (1994) Enzymes used for recombinant DNA technology produced by recombinant microbes. Bioprocess Technol 19:311–323
2. Tolmachov O (2009) Designing plasmid vectors. In: Walther W, Stein US (eds) Gene therapy of cancer. Humana, New York, NY, pp 117–129
3. Li MZ, Elledge SJ (2007) Harnessing homologous recombination *in vitro* to generate recombinant DNA via SLIC. Nat Methods 4:251–256
4. Quan J, Tian J (2011) Circular polymerase extension cloning for high-throughput cloning of complex and combinatorial DNA libraries. Nat Protoc 6:242–251
5. Gibson D, Young L, Chuang R (2009) Enzymatic assembly of DNA molecules up to several hundred kilobases. Nat Methods 6:343–345
6. Wang Y, Prosen DE, Mei L et al (2004) A novel strategy to engineer DNA polymerases for enhanced processivity and improved performance *in vitro*. Nucleic Acids Res 32:1197–1207
7. Hecker KH, Roux KH (1996) High and low annealing temperatures increase both specificity and yield in touchdown and stepdown PCR. Biotechniques 20:478–485
8. Korbie DJ, Mattick JS (2008) Touchdown PCR for increased specificity and sensitivity in PCR amplification. Nat Protoc 3:1452–1456
9. Heitman J, Zinder ND, Model P (1989) Repair of the *Escherichia coli* chromosome after *in vivo* scission by the EcoRI endonuclease. Proc Natl Acad Sci USA 86: 2281–2285
10. Kuzminov A (1999) Recombinational repair of DNA damage in *Escherichia coli* and bacteriophage λ. Microbiol Mol Biol Rev 63:751–813
11. Heitman J, Ivanenko T, Kiss A (1999) DNA nicks inflicted by restriction endonucleases are repaired by a RecA- and RecB-dependent pathway in *Escherichia coli*. Mol Microbiol 33:1141–1151
12. Shao Z, Zhao H, Zhao H (2009) DNA assembler, an *in vivo* genetic method for rapid construction of biochemical pathways. Nucleic Acids Res 37:e16
13. Brody JR, Kern SE (2004) Sodium boric acid: a tris-free, cooler conductive medium for DNA electrophoresis. Biotechniques 36:214–216

14. Drummond A, Ashton B, Buxton S et al (2011) Geneious v5.4, Available from http://www.geneious.com.
15. Kearse M, Moir R, Wilson A et al (2012) Geneious basic: an integrated and extendable desktop software platform for the organization and analysis of sequence data. Bioinformatics 28:1647–1649
16. Owczarzy R, Tataurov AV, Wu Y et al (2008) IDT SciTools: a suite for analysis and design of nucleic acid oligomers. Nucleic Acids Res 36:W163–W169

Chapter 3

Quick and Clean Cloning

Frank Thieme and Sylvestre Marillonnet

Abstract

Identification of unknown sequences that flank known sequences of interest requires PCR amplification of DNA fragments that contain the junction between the known and unknown flanking sequences. Since amplified products often contain a mixture of specific and nonspecific products, the quick and clean (QC) cloning procedure was developed to clone specific products only. QC cloning is a ligation-independent cloning procedure that relies on the exonuclease activity of T4 DNA polymerase to generate single-stranded extensions at the ends of the vector and insert. A specific feature of QC cloning is the use of vectors that contain a sequence called catching sequence that allows cloning specific products only. QC cloning is performed by a one-pot incubation of insert and vector in the presence of T4 DNA polymerase at room temperature for 10 min followed by direct transformation of the incubation mix in chemo-competent *Escherichia coli* cells.

Key words Ligation-independent cloning, Flanking sequences, Specific products, Genome walking, Exonuclease digestion

1 Introduction

Many protocols have been developed to identify unknown sequences that flank a known sequence of interest [1–6]. In all cases, a fragment containing part of the known sequence and part of the unknown flanking sequence is amplified by PCR. The unknown sequence can usually be identified by directly sequencing the PCR product using a primer that binds a region of the known sequence. However, if a library of flanking sequences is amplified rather than a single sequence (for example, when amplifying sequences flanking transposons or transgenes present in multiple copies in a genome, or when amplifying the variable regions of immunoglobulins from a B-cell population), the amplified PCR product needs to be cloned and some of the clones obtained sequenced individually.

Amplification of the fragment containing known and unknown sequences requires attaching adaptor sequences to the fragment of unknown sequence. Since adaptor sequences are not attached

Svein Valla and Rahmi Lale (eds.), *DNA Cloning and Assembly Methods*, Methods in Molecular Biology, vol. 1116, DOI 10.1007/978-1-62703-764-8_3, © Springer Science+Business Media New York 2014

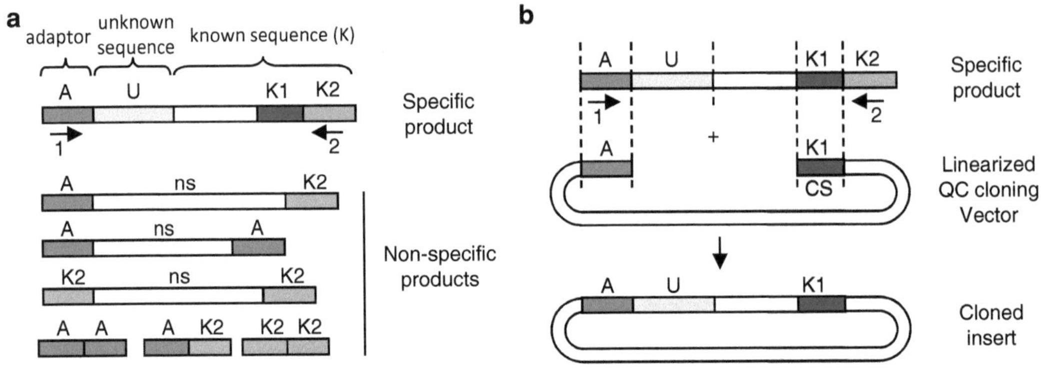

Fig. 1 Principle of the QC cloning strategy. (**a**) DNA fragments containing known and unknown flanking sequences are amplified by PCR using a primer homologous to an adaptor sequence (primer 1 and sequence A) attached to the end of the unknown sequence (U) and a primer homologous to a region of the known sequence (primer 2 and region K2). In addition to specific products, PCR amplification can yield nonspecific products (ns) and primer dimers. (**b**) Fragments are cloned via homology between sequences present in both the insert and the vector: sequences A and sequence K1 (also called the catching sequence, CS). Since primer 2 used for PCR amplification of the insert does not overlap with region K1, nonspecific products and primer dimers cannot be cloned

exclusively to the sequences flanking the known sequence of interest, a mixture of specific and unspecific products can be amplified. This is in addition to other nonspecific products such as primer dimers that can be amplified as by-products in all PCR amplifications (Fig. 1a). The QC cloning strategy was designed to clone exclusively the specific products from a mix that contains both specific and unspecific products. QC cloning is a ligation-independent cloning strategy that uses homology between sequences present in both the vector and the insert. One end of the linearized QC cloning vector is homologous to the adaptor sequence used for amplification of the insert (Fig. 1b). In contrast, the other end of the linearized vector does not have homology to the other primer used for amplification of the insert (binding to region K2), but rather has homology to a sequence of the known region nested and non-overlapping with the primer used for amplification (region K1). Since this sequence is only present in specific products, only these can be cloned.

Cloning of specific sequences using QC cloning therefore requires preparing a specific cloning vector for each new sequence that needs to be cloned. The insert is cloned in the QC vector using a one-pot incubation with T4 DNA polymerase followed by direct transformation of the incubation mix in competent cells. Individual clones can then be sequenced with a primer in the known region or with vector primers.

2 Materials

2.1 PCR

1. Novagen KOD Hot Start DNA polymerase, supplied with 10× buffer, 25 mM $MgSO_4$, and 2 mM dNTPs, or any other high fidelity DNA polymerase.
2. Custom-made primers (can be ordered from many commercial vendors).
3. Thermal Cycler.
4. PCR tubes (0.2 mL, sterile, DNase-free).
5. Macherey-Nagel NucleoSpin Extract II kit, or any other kit for purification of PCR products.

2.2 Cloning

1. T4 DNA polymerase.
2. Restriction endonucleases PstI, supplied with 10× Buffer 3 (500 mM Tris–HCl pH 7.9, 100 mM $MgCl_2$, 1 M NaCl, 10 mM DTT), and BpiI (10 U/μL), supplied with 10× Buffer G (100 mM Tris–HCl pH 7.5, 100 mM $MgCl_2$, 500 mM NaCl, 1 mg/mL BSA).
3. T4 DNA Ligase 3 U/μL or T4 DNA Ligase (HC) 20 U/μL, both supplied with 10× ligation buffer (300 mM Tris–HCl pH 7.8, 100 mM $MgCl_2$, 100 mM DTT, 10 mM ATP).
4. For measuring DNA concentration, we use a NanoDrop ND1000.
5. Lysogeny Broth (LB) Medium: 1 % bacto-tryptone, 0.5 % yeast extract, and 1 % NaCl in deionized water, adjusted to pH 7.0 with 5 N NaOH. For plates, 1.5 % agar is added.
6. Antibiotic carbenicillin (used instead of ampicillin): filter-sterilized stocks of 50 mg/mL in H_2O (stored in aliquots at −20 °C) are diluted to a final concentration of 100 μg/mL in an appropriate amount of medium after the medium has been autoclaved and cooled down.
7. 1.5 mL tubes (sterile, DNase-free).
8. A 25 °C incubator is not required, since incubation is carried out in a PCR block, which is missing in Materials but described in Methods.
9. Chemo-competent *E. coli* DH10B cells.
10. LB agar plates supplemented with antibiotics, e.g., carbenicillin (100 μg/mL) and X-Gal (40 μg/mL) for blue–white selection.
11. LB medium.
12. 5-Bromo-4-chloro-3-indolyl-β-D-galactopyranoside (X-gal): stock solution of 20 mg/mL in dimethylformamide (DMF). For preparation of plates, the stock is diluted 1:500 (final concentration: 40 μg/mL) in an appropriate amount of LB agar after autoclaving/melting and cooling down.

2.3 Screening of Colonies

1. Macherey-Nagel NucleoSpin Plasmid Quick Pure, or any other kit for isolation of plasmid DNA at Miniprep scale.
2. DNA ladder: The Thermo Scientific GeneRuler 1 kb DNA Ladder Plus is used as marker for gel electrophoresis.
3. 50× TAE buffer: 242.0 g Tris, 57.1 mL acetic acid, and 100 mL 0.5 M EDTA, pH 8.0, in 1 L of deionized water.
4. For preparation of gels for electrophoresis, agarose (0.7–1.5 %) in 1× TAE is melted in a microwave oven and one drop of a 0.025 % ethidium bromide solution is added per 100 mL of melted agarose.
5. Running buffer for agarose gels is 1× TAE.
6. Gels are checked visually using a Syngene GelVue transilluminator, and pictures are taken by using a Quantity one® gel analysis software (Biorad).
7. DNA maps of plasmids are made by using the Vector NTI software (Invitrogen).

3 Methods

3.1 Preparation of the QC Cloning Vector

Unlike other ligation-independent cloning strategies where the insert is designed to fit a chosen cloning vector, for QC cloning, a vector is made to fit the specific product that needs to be cloned. Therefore, a specific vector has to be prepared for each new type of insert. The vector contains a sequence homologous to the adapter sequence that is found at one end of the insert (sequence A, Fig. 2), a selectable marker such as *lacZα*, and a sequence homologous to part of the known region (the catching sequence, CS). The vector also contains two restriction sites for linearization of the vector; in this case, two restriction sites for the enzyme PstI. Digestion with PstI will remove the *lacZα* marker. There are many strategies for making a QC cloning vector, including a strategy that we described earlier [7]. We provide here a strategy that only requires pUC19 as starting material. A PCR fragment containing *lacZα* flanked by the adaptor sequence and the catching sequence is amplified from pUC19 by PCR. The purified fragment is cloned using BpiI in a vector backbone fragment that is also amplified from pUC19 (Fig. 2). An alternative strategy that only requires PCR amplification of a vector backbone fragment is also provided in **Note 1**.

1. Design primers 1 and 2. Primers 1 and 2 are required to amplify the *lacZα* fragment. Primer 1 (ttt **gaagac aa catt** *gtccagagccgtccagcaac* <u>tgcag</u> gcagctggcacgacaggtttc) contains the sequence of a BpiI site (bold) followed by a sequence homologous to the adaptor sequence (italics), then by a PstI site (underlined), and finally by sequences of pUC19. The adaptor sequence has to

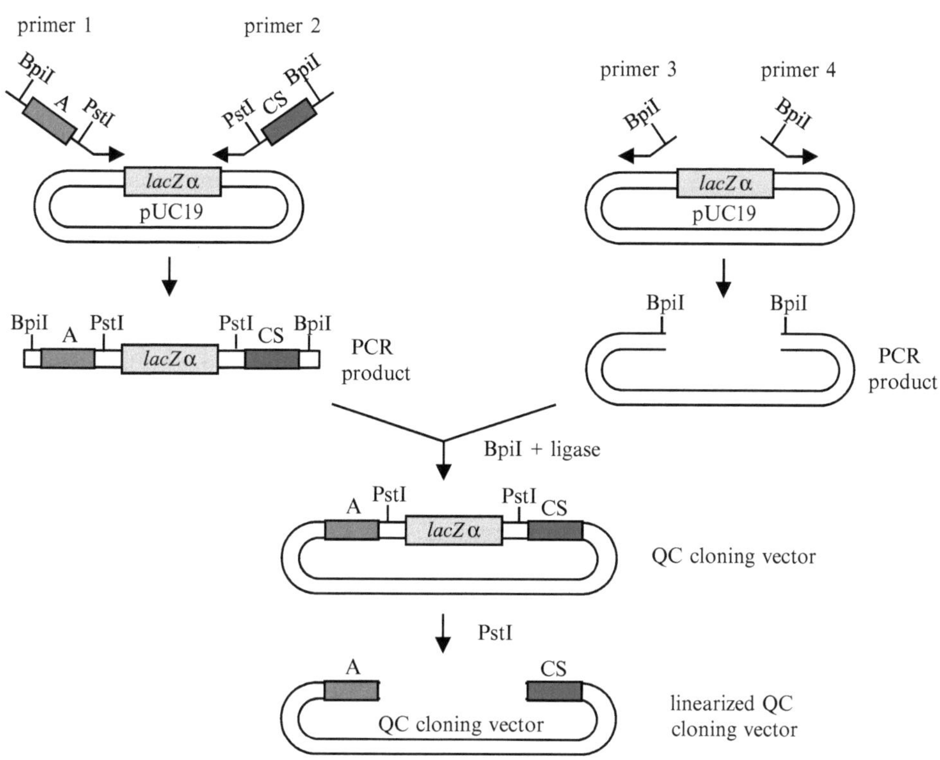

Fig. 2 Preparation of QC cloning vectors. QC cloning vectors can be made by amplifying a *lacZα* fragment and a vector backbone fragment from pUC19, and assembling the two fragments using a restriction–ligation with the restriction enzyme BpiI and T4 DNA ligase. The QC cloning vector is linearized by restriction digest with PstI before its use for QC cloning

be replaced by the user to fit the adaptor sequence that was used for amplification of the DNA insert of interest. Since the same adaptor can be used to amplify different sequences of interest, primer 1 can be reused for many different cloning experiments. Primer 2 (ttt **gaagac aa gacg** *ggggaagtagtccttgac caggcagcccaggg*c tgcag tcacagcttgtctgtaagcggatg) has a similar structure as primer 1 except that the region in italics consists of a sequence complementary to a fragment of the known region (sequence K1, Fig. 1b; in this example, it consists of a fragment of a human immunoglobulin constant region). This sequence has to be changed by the user to fit the known sequence of interest. The size of the catching sequence can vary from 20 to 50 nucleotides (*see* **Note 2**).

2. Amplify the *lacZα* fragment from pUC19 with primers 1 and 2 using a proof-reading polymerase.
3. Amplify a pUC19 backbone fragment with primers 3 (ttt gaagac aa cgtc tccgggagctgcatgtg) and 4 (ttt gaagac aa aatg

aatcggccaacgcgcgg) using a proof-reading polymerase. Primers 3 and 4 can be reused for construction of other QC cloning vectors without modification.

4. Check the PCR products by agarose gel electrophoresis.
5. Column-purify the fragments using a commercial kit to remove remaining polymerase and dNTPs.
6. Assemble both purified fragments using a restriction–ligation reaction containing 100 ng *lacZα* PCR product (2 μL), 20–50 ng pUC19 backbone PCR product (1 μL), 2 μL 10× T4 DNA ligase buffer, 14.5 μL water, 1 μL BpiI, and 0.5 μL T4 DNA ligase (final volume 20 μL). The reaction mix is incubated for 1 h at 37 °C and inactivated for 20 min at 65 °C (*see* **Note 3**).
7. Pipet the 20 μL reaction mix into a tube containing 100 μL chemo-competent *E. coli* DH10B cells, and mix briefly by flicking.
8. Incubate the transformation mix for 10 min on ice (incubation of up to 30 min is also fine).
9. Incubate the tubes at 42 °C in a water bath for 90 s.
10. Cool down the tubes on ice for 2 min.
11. Add 1 mL liquid LB medium. Incubate at 37 °C for 45 min.
12. Spread 100 μL of the transformation on an LB plate containing 100 μg/mL carbenicillin and 40 μg/mL X-Gal. Spin down the rest of the transformation. Resuspend the cells in 100 μL of LB and plate everything on a second plate.
13. Incubate the plates at 37 °C overnight.
14. Inoculate 2–4 blue colonies in culture vials containing 5 mL LB medium supplemented with carbenicillin (50 μg/mL) (*see* **Note 4**). Incubate overnight at 37 °C in a shaker-incubator (approximately 170 rotations per minute).
15. Purify plasmid DNA using a DNA extraction kit.
16. Screen the clones by digestion of miniprep DNA with PstI. Correct constructs should display three fragments of 2.1 kb, 383 bp, and 200 bp.
17. Measure the DNA concentration of the selected clone using NanoDrop.
18. Confirm the sequence of the vector (adaptor and catching sequence) by sequencing with primers seqf (TGG AAA AAC GCC AGC AAC GC) and seqr (TGT CTC ATG AGC GGA TAC AT) (*see* **Note 5**).
19. Digest the QC vector using PstI: pipet into a PCR tube 3 μL QC cloning vector (0.15–0.6 μg), 3 μL 10× NEBuffer 3, 0.3 μL 100× BSA (10 mg/mL), 1 μL PstI (20 U/μL), and

22.7 µL water (final volume of 30 µL). Incubate the digest at 37 °C for 2 h in a PCR block, followed by incubation at 80 °C for 20 min to inactivate the PstI enzyme (*see* **Note 6**).

20. Check for proper digestion of the vector by agarose gel electrophoresis.

The digested QC cloning vectors can be used directly for cloning or can be stored at −20 °C. There is no need to purify the digested DNA by ethanol precipitation or with a kit.

3.2 Preparing the Insert

Many protocols are available for amplifying unknown sequences that flank known sequences of interest [6]. The protocols also vary depending on whether the starting material is RNA or DNA. Therefore a protocol for amplifying flanking sequences is not provided here. Whatever procedure has been used for amplification of the insert, the PCR product needs to be purified using a column to remove remaining polymerase and dNTPs. Removing dNTPs is necessary for Subheading 3.3, **step 2** described below to prevent the DNA ends made single-stranded by the exonuclease activity of T4 DNA polymerase to be filled again due to its polymerase activity.

1. Analyze the PCR products by agarose gel electrophoresis. Depending on the type of starting material and the protocol used for amplification, the PCR product may consist of a fragment of defined size but may also consist of a variable number of fragments of different sizes or even of a smear (e.g., amplification of flanking sequences with TAIL PCR [5]). Therefore evaluation of the quality of PCR product will depend on the nature of amplified products.
2. Column-purify the PCR products using a kit.
3. Quantify the amount of purified DNA using NanoDrop device (*see* **Note 7**).

3.3 QC Cloning Procedure

QC cloning is a ligation-independent cloning strategy. As many other ligation-independent cloning strategies, it relies on the 3′–5′ exonuclease activity of T4 DNA ligase to generate single-stranded DNA at the ends of the insert and vector [8–10]. Unlike other ligation-independent cloning protocols, exonuclease digestion of the insert and of the vector is performed at the same time in a single tube. The single-stranded complementary sequences generated at the end of the vector and insert are then able to anneal. The mix is directly transformed in chemo-competent DH10B cells where the annealed vector–insert complex is repaired by the *E. coli* DNA repair machinery.

1. Pipet the following components in a PCR tube: 14.5 µL water, 2 µL 10× T4 ligase buffer (*see* **Note 8**), 0.5 µL T4 DNA polymerase (5 U/µL) (*see* **Note 9**), 1 µL PstI-digested T4 cloning vector (~5–20 ng/µL), and 2 µL column-purified PCR product (~10–100 ng/µL) (*see* **Notes 10** and **11**).

2. Incubate the reaction(s) in a PCR block at 25 °C for 10 min (*see* **Note 12**). Incubation is followed by a 4 °C incubation step to stop the reaction (*see* **Note 13**).
3. Add the entire 20 μL reaction mixture to 50–100 μL chemo-competent *E. coli* DH10B cells and mix briefly by flicking (*see* **Note 14**).
4. Incubate the transformation mix 10 min on ice (incubation of up to 30 min is also fine).
5. Incubate at 42 °C in a water bath for 90 s.
6. Cool down the tubes on ice for 2 min.
7. Add 1 mL liquid LB medium and incubate at 37 °C for 45 min (*see* **Note 15**).
8. Spread 100 μL of the transformation on an LB plate containing 100 μg/mL carbenicillin and 40 μg/mL X-Gal. Spin down the rest of the transformation and resuspend the cells in 100 μL of LB. Plate everything on a second plate.
9. Incubate the plates at 37 °C overnight. Usually, a few hundred white colonies should be obtained. However, this number can vary depending on the quality of the insert (the ratio of specific to nonspecific fragments), the quality of the enzymes used, and the quality of the competent cells. Most colonies should be white, but occasionally a few blue colonies containing undigested vector can be found.
10. Clones can be screened by colony PCR using vector primers seqf (TGG AAA AAC GCC AGC AAC GC) and seqr (TGT CTC ATG AGC GGA TAC AT) (example shown in Fig. 3). The PCR products can be sequenced using seqf or seqr primers, or using a primer in the known region of the insert.

4 Notes

1. Making a QC cloning vector requires some work, but this is worth it when the same vector needs to be reused several times, for example, for cloning the variable regions of immunoglobulins from a large number of samples [11]. However, making a QC cloning vector might seem quite time-consuming if it is not reused many times. We provide here a solution that eliminates the need for constructing a specific plasmid for each new cloning experiment. Instead a vector backbone is amplified from pUC19 using two primers that add extensions homologous to the adaptor sequence on one end and to region K1 on the other end. The first primer, primer 1 (*gttgctggacggctctggac* attaatgaatcggccaacgcgcggggagag), contains a 5′ extension that is complementary to the adaptor sequence (italics) followed by

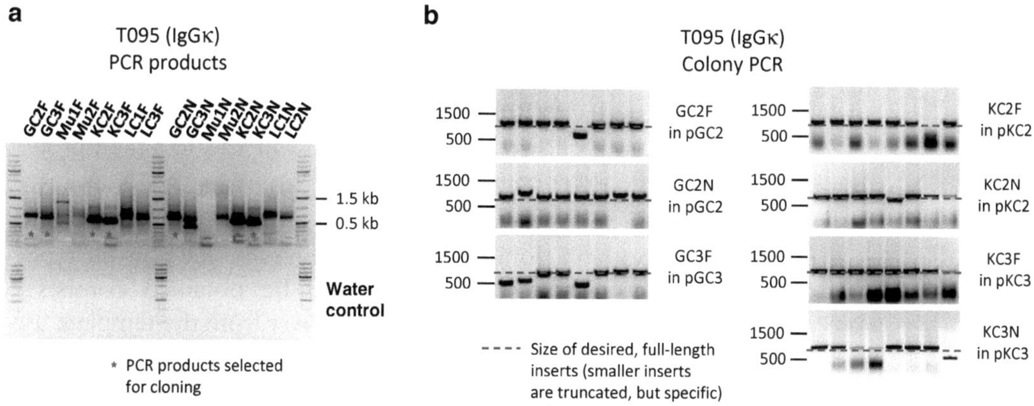

Fig. 3 QC cloning of immunoglobulin fragments from patient sample T095. (**a**) PCR products containing the variable region (unknown region) and a fragment of the constant region (the known region) of immunoglobulin Kappa light chains amplified from non-Hodgkin lymphoma biopsy sample T095. The products were amplified using a G-tail anchor primer (binds the adaptor sequence) and immunoglobulin constant region-specific primers (GC2F, GC3F, etc.). The PCR products indicated by an *asterisk* were cloned in corresponding QC cloning vectors. (**b**) Randomly chosen clones from each cloning experiment were analyzed by colony PCR using vector primers. PCR products were separated on a 1 % agarose gel supplemented with ethidium bromide and visualized under UV light. The expected insert size is indicated by a *dashed line*

pUC19 sequences. Since the same adaptor sequence can be used for amplification of different sequences of interest, this primer can be reused for different cloning experiments. The second primer, primer 2 (*gccctgggctgcctggtcaaggactacttcccc* cgtctccgggagctgcatgtg), contains an extension that is homologous to region K1 (italics). Since this region is specific to each sequence of interest, primer 2 needs to be modified for each new cloning project. PCR amplification performed using primers 1 and 2 and pUC19 as a template results in a 2.7 kb backbone fragment. The PCR product just needs to be column-purified and can be used directly for QC cloning. Since it is a linear fragment, the PstI digestion step described for linearization of QC cloning vectors is not needed anymore.

2. We have investigated the length of catching sequences suitable for QC cloning. A CS of 12–15 nucleotides is sufficient for QC cloning (as is also known for other LIC strategies [12]). However, a polymorphism of one nucleotide in the region of the amplified insert homologous to a 12-nucleotide CS is likely to prevent cloning of this insert. In contrast, using a 50-nucleotide CS will allow homologous sequences containing one or a few mismatches to be cloned (as long as a stretch of 12–15 consecutive nucleotides is present in both the insert and the catching sequence).
3. BpiI is a type IIS restriction enzyme whose cleavage site is located outside of its DNA recognition site. As a result, ligation

of both PCR products will result in a construct lacking BpiI restriction sites. Restriction and ligation can therefore be performed at the same time since the ligated product will not be redigested with BpiI. Restriction–ligation is more efficient than restriction followed by ligation, and will lead to a high proportion of correct constructs [13].

4. Using type IIS restriction enzymes and restriction–ligation will lead to a high cloning efficiency. However, one possible type of non-correct construct that could be obtained consists of pUC19 vector that may be carried over from the template used for PCR amplification in Subheading 3.1, **steps 2** and **3**. This will produce blue colonies, as with the desired QC cloning vector. One solution to avoid these clones is to amplify a *lacZα* fragment from a vector carrying a different antibiotic marker (for Subheading 3.1, **step 2**), and to amplify the backbone fragment from a pUC19-derived plasmid already containing an insert, and therefore with a nonfunctional *lacZα* gene (for Subheading 3.1, **step 3**).
5. It is useful to sequence the vector in the region corresponding to the adaptor and the catching sequences, since both could contain mutations derived from the primer. Sequencing the rest of the plasmid is not important since the plasmid is replicating and is therefore functional, and the *lacZα* fragment provides a blue color and is therefore also functional.
6. Inactivation of PstI is necessary, as incubation of the digested vector with insert (in subsequent Subheading 3.3, **step 2**) might lead to digestion of inserts that would contain a PstI site in the unknown region. Such digestion would of course prevent cloning of these inserts. PstI digestion of the vector and the following heat inactivation step could be performed in a water bath, but are more conveniently performed in a thermal cycler.
7. Even though the amount of DNA has been quantified using a NanoDrop device or any other method, it is always a good idea to confirm visually the amount and the quality of purified PCR product by agarose gel electrophoresis.
8. We use T4 DNA ligase buffer, but since the reaction does not require ATP, other buffers can be used instead; e.g., NEBuffer 2 is suitable for QC cloning.
9. The quality of T4 DNA polymerase is critical. It is important to check that the enzyme lot for the T4 DNA polymerase is not expired as cloning might become less efficient. If cloning has failed several times for unknown reasons, one should try another lot of enzyme from the same manufacturer, or enzyme from another manufacturer. In practice, we obtained good results using the T4 DNA polymerase supplied by Fermentas (cat. no. EP0061).

10. In case a PCR product contains a low concentration of DNA, the volume of product added to the reaction mix can be increased up to 16.5 μL (for a 20 μL reaction).
11. Try to keep the reaction mix on ice between pipetting steps, especially when preparing several samples. This is important, since the exonuclease activity of T4 DNA polymerase will start to digest DNA ends at temperatures higher than 4 °C. To synchronize digestion duration for all samples, the vector(s) and the various inserts are first pipetted in all tubes. A master mix of enzyme, buffer, and water is then quickly added to all tubes.
12. The reaction is temperature- and time-dependent. The incubation conditions given in the method section are for a 50-nucleotide catching sequence. For a 20-nucleotide catching sequence an incubation of 20 min at 15 °C is optimal [7]. Temperatures >25 °C are not recommended, since the reaction is then too fast and difficult to control.
13. When the program is finished, place the QC cloning reactions on ice and proceed immediately. If the mix is not kept at low temperature, exonuclease digestion will continue and may decrease cloning efficiency.
14. Since repair and ligation of the vector takes place in *E. coli* cells, the choice of strain is important for QC cloning. We recommend using the *E. coli* strain DH10B since it gave the best results for our experiments. The original DH10B strain should be used, because some derivatives gave only poor results; e.g., One Shot® *ccdB* survival cells (Invitrogen) were not suitable for QC cloning.
15. Please keep in mind that the *E. coli* need to repair and ligate the insert and vector DNA. Therefore, the incubation prior to plating on selective medium should not be too short. We recommend at least 30 min.

References

1. Frohman MA, Dush MK, Martin GR (1988) Rapid production of full-length cDNAs from rare transcripts: amplification using a single gene-specific oligonucleotide primer. Proc Natl Acad Sci USA 85:8998–9002
2. Mueller PR, Wold B (1989) In vivo footprinting of a muscle specific enhancer by ligation mediated PCR. Science 246:780–786
3. Riley J, Butler R, Ogilvie D et al (1990) A novel, rapid method for the isolation of terminal sequences from yeast artificial chromosome (YAC) clones. Nucleic Acids Res 18:2887–2890
4. Rosenthal A, Jones DS (1990) Genomic walking and sequencing by oligo-cassette mediated polymerase chain reaction. Nucleic Acids Res 18:3095–3096
5. Liu YG, Whittier RF (1995) Thermal asymmetric interlaced PCR: automatable amplification and sequencing of insert end fragments from P1 and YAC clones for chromosome walking. Genomics 25:674–681

6. Tonooka Y, Fujishima M (2009) Comparison and critical evaluation of PCR-mediated methods to walk along the sequence of genomic DNA. Appl Microbiol Biotechnol 85:37–43
7. Thieme F, Engler C, Kandzia R et al (2011) Quick and clean cloning: a ligation-independent cloning strategy for selective cloning of specific PCR products from non-specific mixes. PLoS One 6:e20556
8. Aslanidis C, de Jong PJ (1990) Ligation-independent cloning of PCR products (LIC-PCR). Nucleic Acids Res 18:6069–6074
9. Yang YS, Watson WJ, Tucker PW et al (1993) Construction of recombinant DNA by exonuclease recession. Nucleic Acids Res 21:1889–1893
10. Li MZ, Elledge SJ (2007) Harnessing homologous recombination in vitro to generate recombinant DNA via SLIC. Nat Methods 4:251–256
11. Bendandi M, Marillonnet S, Kandzia R et al (2010) Rapid, high-yield production in plants of individualized idiotype vaccines for non-Hodgkin's lymphoma. Ann Oncol 21:2420–2427
12. Aslanidis C, de Jong PJ, Schmitz G (1994) Minimal length requirement of the single-stranded tails for ligation-independent cloning (LIC) of PCR products. PCR Methods Appl 4:172–177
13. Engler C, Kandzia R, Marillonnet S (2008) A one pot, one step, precision cloning method with high throughput capability. PLoS One 3:e3647

Chapter 4

Hierarchical Ligation-Independent Assembly of PCR Fragments

Jonathan L. Schmid-Burgk, Zhen Xie, and Yaakov Benenson

Abstract

The emerging field of synthetic biology requires novel cloning techniques that allow the rapid assembly of multiple expression units to build artificial genetic circuits. Here, we describe a rapid, flexible, and cost-efficient cloning method that requires only standard laboratory equipment and skills. Our technique relies on the 3′–5′ exonuclease activity of T4 DNA polymerase to generate 20 nt single-stranded DNA overhangs that allow annealing and ligation-independent cloning (LIC) of four DNA fragments in one tube. The resulting intermediate-size constructs can be reused to hierarchically assemble constructs of more than 24 kb by the same method.

Key words Ligation-independent cloning, Hierarchical assembly, DNA fragments

1 Introduction

The traditional restriction–ligation cloning method performs poorly when assembling large artificial genetic circuits because the sequential digestion–ligation–transformation–preparation steps are time-consuming and the number of available restriction enzyme cleavage sites decreases with the increasing number and size of DNA sequences to be assembled. Ligation-independent cloning (LIC) has been described more than 20 years ago, relying on the annealing of DNA fragments that bear complementary single-stranded 5′-overhangs longer than the 4-nt overhangs generated by common restriction enzymes [1]. A convenient way to generate such long overhangs is to include base-restricted terminal sequences in PCR primers and to make the termini of the resulting PCR products single-stranded by using the controllable 3′-exonuclease activity of T4 DNA polymerase when provided with one "missing" dNTP during the reaction [1]. Using LIC, not only two DNA fragments were assembled in one tube [1–4], but Donahue et al.

Svein Valla and Rahmi Lale (eds.), *DNA Cloning and Assembly Methods*, Methods in Molecular Biology, vol. 1116, DOI 10.1007/978-1-62703-764-8_4, © Springer Science+Business Media New York 2014

described the direct assembly of three fragments in a single tube [5], where the single-stranded overhangs were generated by incorporating ribose residues into the PCR primers to stop the polymerase from terminally extending the 3′ ends. Furthermore, Geu-Flores et al. performed a 4-parts assembly after generating ssDNA overhangs using enzymatic nicking at uracil residues introduced by the PCR primer as well [6].

The aforementioned methods for multi-part assemblies are not well suited to hierarchically assemble larger constructs because these methods require PCR amplification of all assembly intermediates, making it increasingly difficult to obtain the mutation-free PCR product with increasing length of the intended product. Gibson et al. introduced a hierarchical LIC assembly method that uses either T4 DNA polymerase or T5 exonuclease for unspecific DNA recession in order to generate overlapping ssDNA regions between DNA fragments to be assembled [7, 8]. Because of the unspecific chew-back it is necessary to repair gaps in the annealed junctions of DNA fragments by a DNA polymerase and to ligate them by a DNA ligase in vitro. While this technique has proven very powerful for the assembly of large DNA constructs starting from synthetic sequences in the range of kilobases, it remains unclear whether a short DNA fragment would withstand the unspecific chew-back rather than being completely degraded. More recently, Gibson et al. further showed how to assemble 25 overlapping DNA fragments at once in yeast, providing a valuable route for the assembly of large DNA fragments in the scale of whole bacterial genomes [9]. However, this approach may not be suitable for the rapid generation of custom gene circuits with many repetitive regions such as common promoters, genes, or terminators, since those can lead to undesired recombination events in yeast cells. This poses an obstacle especially when well-characterized, standardized genetic parts are used for circuit engineering.

We have described a method that utilizes the nucleotide-specific chew-back capacity of T4 DNA polymerase in the presence of a stop nucleotide to generate 20-nt ssDNA overhangs (Fig. 1), which enables reliable assembly of four DNA fragments of lengths ranging from 200 bp to 9 kb in a single tube [10]. The DNA fragments can be produced either by PCR amplification or by enzymatic digestion of intermediate plasmid DNA constructs. Therefore it is possible to assemble ten gene-size fragments in two hierarchical steps of assembly (Fig. 2), providing a rapid route to the construction of genetic circuitry of high complexity.

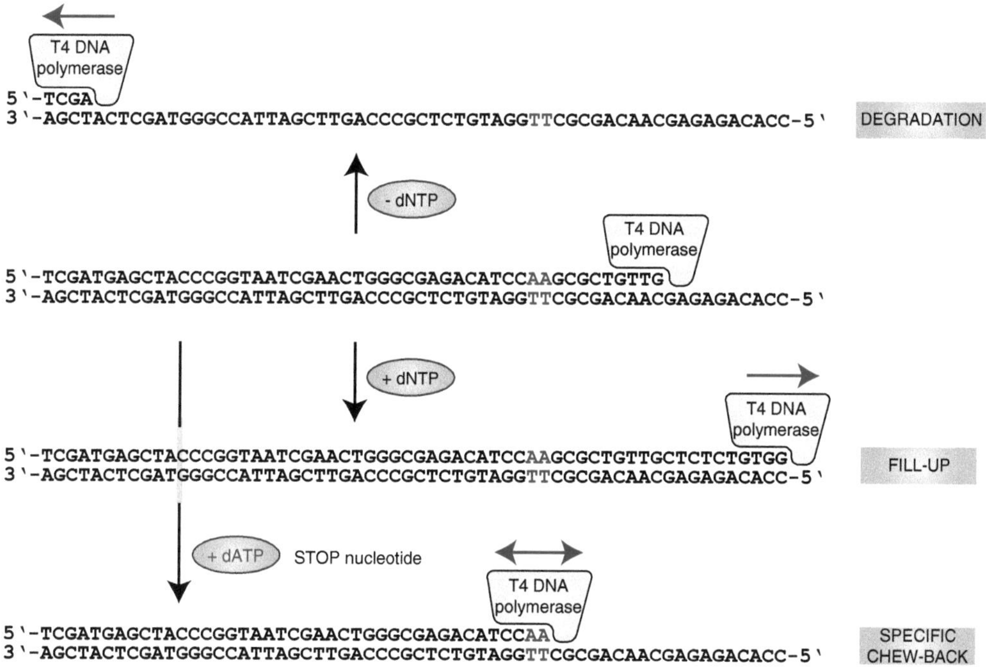

Fig. 1 Principle of sequence-specific chew-back executed by T4 DNA polymerase. In the absence of dNTPs, T4 DNA polymerase processively degrades DNA beginning from the 3′ end. In contrast, when providing dNTPs the polymerase activity of the enzyme outcompetes its exonuclease activity, thus reversing degradation to a fill-in reaction. By providing only one dNTP (dATP in this example) the 3′ exonuclease activity of T4 DNA polymerase removes bases from a dsDNA molecule until the dNTP can be incorporated, whereupon the enzyme stalls at that position. Thereby, well-defined 5′ overhangs of DNA molecules can be generated

2 Materials

2.1 Oligonucleotide Primers for Backbone Amplification

Primer	Sequence
BB1.1_fwd	GAGAGGCAGCAAGCAACGAATTTTGCAGGTG CTTCTGAGGCGGAAAGAACCAG
BB1.1_rev	CGAACACCAGCAAGAGACAATTTTGCAGGTG GCGTATATCTGGCCCGTACATC
BB1.2_fwd	GGTTCTTTTTCGTTGGGCGTAAAAGCAGGTG CTTCTGAGGCGGAAAGAACCAG
BB1.2_rev	TTCGTTGCTTGCTGCCTCTCAAAAGCAGGTG GCGTATATCTGGCCCGTACATC
BB1.3_fwd	AACACCGGAACAAGAAAGGCTTTTGCAGGTG CTTCTGAGGCGGAAAGAACCAG
BB1.3_rev	ACGCCCAACGAAAAAGAACCTTTTGCAGGTG GCGTATATCTGGCCCGTACATC
BB2_fwd	AACACCGGAACAAGAAAGGCTTTTGCAGGTG CTTCTGAGGCGGAAAGAACCAG
BB2_rev	CGAACACCAGCAAGAGACAATTTTGCAGGTG GCGTATATCTGGCCCGTACATC

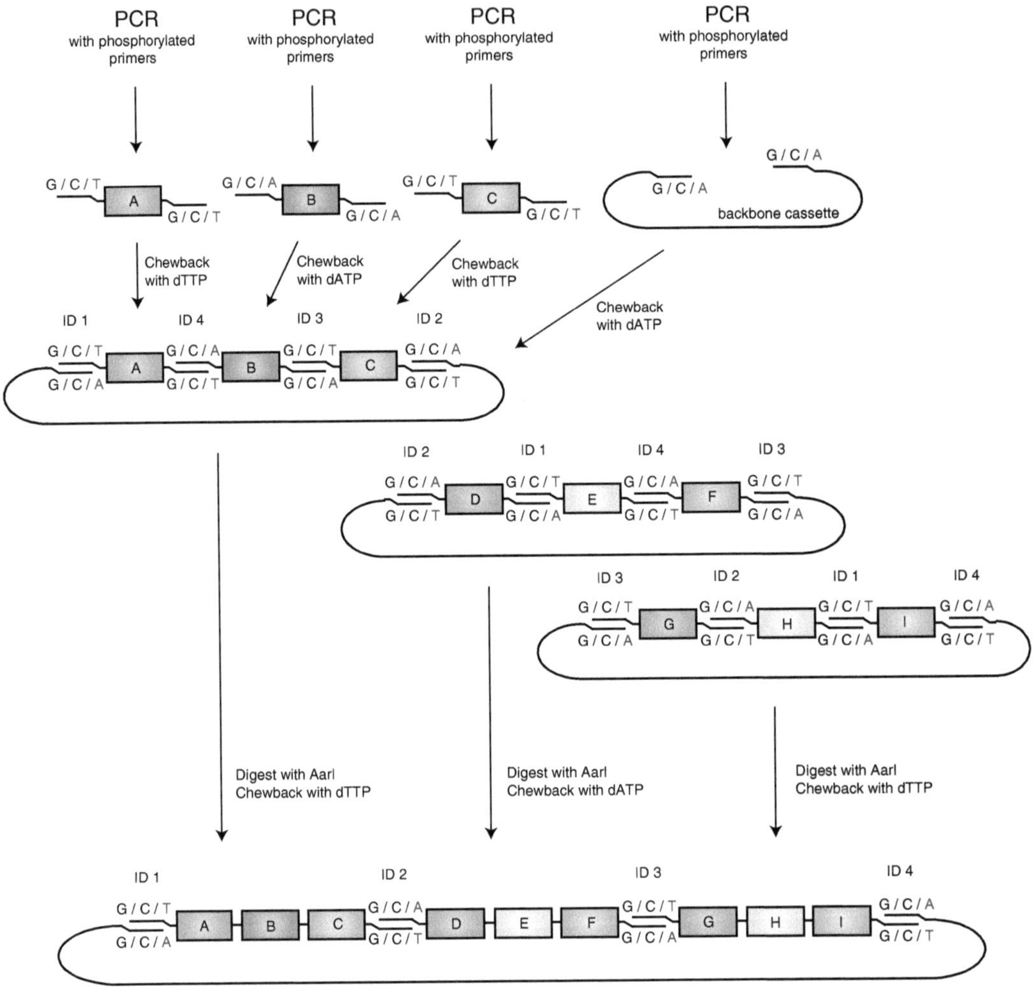

Fig. 2 Schematic overview of the hierarchical assembly of ten PCR products. In the first round of assembly, four PCR products are generated using 5′ phosphorylated primers. One of these PCR products contains a backbone sequence portion for bacterial amplification of the resulting plasmids. After chew-back of the dsDNA fragments using T4 DNA polymerase and one specific dNTP, the four pieces are annealed and transformed into *E. coli*. The first-round assembly process is performed three times in parallel using different PCR products. The three resulting intermediate assembly plasmids can be digested by AarI to release only the PCR fragments of interest, discarding all backbone portions. Together with a new PCR backbone fragment, the three excised composite fragments are assembled the same way as during the first round to form a final 10-parts assembly construct

2.2 Polynucleotide Kinase Phosphorylation Reagents

1. 10× T4 polynucleotide kinase (PNK) buffer (NEB): 700 mM Tris–HCl, 100 mM $MgCl_2$, 50 mM Dithiothreitol (DTT), pH 7.6.
2. 10 mM ATP.
3. 10 U/μL T4 PNK (NEB).

2.3 PCR Reagents

1. 10× Pfu Ultra II PCR buffer (Stratagene).
2. 10 mM dNTP mixture.
3. Pfu Ultra II Fusion HS (Stratagene).

2.4 Gel Electrophoresis Components

1. UltraPure Agarose (Life technologies).
2. Agarose gel running chamber and power supply.
3. 25 mg/mL Ethidium Bromide stock solution.
4. 10× agarose gel loading buffer (2.5 % Ficoll 400, 11 mM EDTA, 3.3 mM Tris–HCl, 0.017 % SDS, 0.015 % Bromophenol Blue, pH 8.0 @ 25 °C).
5. GeneRuler DNA Ladder Mix (Fermentas).
6. UV illumination and camera system.
7. Agarose gel purification kit (Qiagen).

2.5 Chew-Back Assembly Reagents

1. 10× T4 DNA polymerase buffer: 100 mM Tris–HCl, 500 mM NaCl, 100 mM $MgCl_2$, 10 mM DTT, pH 7.9.
2. 1 mg/mL bovine serum albumin (NEB).
3. 20 mM dATP dissolved in H_2O.
4. 20 mM dTTP dissolved in H_2O.
5. 3 U/μL T4 DNA polymerase (NEB).

2.6 Bacterial Components

1. Chemo-competent *E. coli* in aliquots of 50 μL, stored at −80 °C, more than 10^7 cfu/μg.
2. Lysogeny Broth (LB) medium (25 g powder dissolved in 1 L water; autoclaved).
3. 100 mg/mL Ampicillin stock solution.
4. QIAprep Spin Miniprep Kit (Qiagen).

2.7 Enzymatic Restriction Reagents

1. 10× AarI buffer (Fermentas).
2. 2 U/μL AarI (Fermentas).
3. 25 μM AarI supportive oligonucleotide (Fermentas).
4. DpnI (Fermentas).

2.8 Plasmids

1. pcDNA3.1 (Life technologies).

2.9 Software

1. Ape Plasmid Editor (http://biologylabs.utah.edu/jorgensen/wayned/ape/).

2.10 Devices

1. PCR thermal cycler.
2. Spectrophotometer (NanoDrop 1000).
3. Thermoblock (Eppendorf Thermomixer comfort).

3 Methods

Carry out all procedures on ice unless otherwise specified.

3.1 Primer Design

1. Design primer pairs for nine amplicons to be assembled, naming them according to their position in the final assembly (1_fwd, 1_rev, 2_fwd, …).
2. Choose the ID for each specific primer according to the following table:

1_fwd	1_rev	2_fwd	2_rev	3_fwd	3_rev	4_fwd	4_rev	5_fwd
ID 1	ID 4	ID 4	ID 3	ID 3	ID 2	ID 2	ID 1	ID 1
5_rev	6_fwd	6_rev	7_fwd	7_rev	8_fwd	8_rev	9_fwd	9_rev
ID 4	ID 4	ID 3	ID 3	ID 2	ID 2	ID 1	ID 1	ID 4

3. According to the ID number you have chosen, add the ID and termination sequence given by the following table to the 5′ end of the primer sequence:

ID 1 fwd	TTGTCTCTTGCTGGTGTTCGAAAAA
ID 1 rev	CGAACACCAGCAAGAGACAATTTTT
ID 2 fwd	GAGAGGCAGCAAGCAACGAATTTTT
ID 2 rev	TTCGTTGCTTGCTGCCTCTCAAAAA
ID 3 fwd	GGTTCTTTTTCGTTGGGCGTAAAAA
ID 3 rev	ACGCCCAACGAAAAAGAACCTTTTT
ID 4 fwd	AACACCGGAACAAGAAAGGCTTTTT
ID 4 rev	GCCTTTCTTGTTCCGGTGTTAAAAA

4. Buy the primers along with the backbone amplification primers (*see* Subheading 2) from a DNA synthesis company as desalted DNA oligonucleotides.

For cases where a different number of amplicons has to be assembled, **Note 1** describes a way of increasing the flexibility of this method.

3.2 Primer Phosphorylation

Phosphorylate all primers using the following protocol:

1. Mix 2 μL 10× T4 PNK buffer, 2 μL 10 mM ATP, 1 μL 100 μM primer solution, 14.2 μL H_2O, and 0.8 μL T4 PNK enzyme.
2. Incubate the reaction at 37 °C for 1 h.
3. Heat-inactivate the enzyme by incubating the reaction at 65 °C for 20 min.

3.3 Polymerase Chain Reaction

Perform nine cassette-specific PCRs and four backbone PCRs. Use the plasmid pcDNA3.1 as template for all backbone PCRs:

1. Mix 2.5 μL 10× Pfu Ultra II PCR buffer, 20 ng of template plasmid DNA, and 1 μL of each crude primer phosphorylation reaction with 0.625 μL of 10 mM dNTP mixture and fill up to 24 μL with H_2O. Add 1 μL of Pfu Ultra II Fusion HS polymerase and mix.
2. Perform the following cycling temperature protocol in a PCR thermal cycler:
   (a) 2 min at 95 °C.
   (b) 20 s at 95 °C.
   (c) 20 s at 68 °C.
   (d) 15 s/kb at 72 °C.
   (e) 3 min at 72 °C.
   (f) Hold at 4 °C.

   Steps b–d are repeated 30 times.

   In order to decrease the cloning background of the protocol, digest the PCR products with DpnI as discussed in **Note 2**.

3.4 Gel Purification

Purify all PCR products using the following protocol:

1. Mix 25 μL PCR with 2.5 μL 10× gel loading buffer and load a 1 % agarose gel containing 0.5 μg/mL Ethidium Bromide. Load 3 μL of loading-ready DNA ladder into a separate lane. Run the gel for 40 min at 100 V.
2. Visualize DNA bands under UV light and cut out the bands of the expected size.
3. Purify the DNA from the agarose gel pieces using a gel purification kit according to the manufacturer's instructions.

3.5 Chew-Back Assembly

1. Measure the concentration of the purified DNA fragments using a NanoDrop 1000 spectrophotometer.
2. Prepare six chew-back reactions, each containing 2 μL 10× T4 DNA polymerase buffer, 2 μL 1 mg/mL BSA, and, according to the following table, 1 μL 20 mM STOP dNTP and 40 fmol of each indicated DNA fragment:

Reaction Nr.	1	2	3	4	5	6
STOP dNTP	dTTP	dATP	dTTP	dATP	dTTP	dATP
DNA fragments	cassette 1 cassette 3	cassette 2 BB 1.1	cassette 5 BB 1.2	cassette 4 cassette 6	cassette 7 cassette 9	cassette 8 BB 1.3

3. Fill up the reactions to 19.6 μL with H_2O, add 0.4 μL T4 DNA polymerase, and mix.

4. Incubate the reactions at 27 °C for 5 min, chill on ice until the thermal cycler has reached 75 °C, and incubate the reactions at 75 °C for 20 min. During the incubation time, mix reaction 1 with 2, 3 with 4, and 5 with 6 while still being kept at 75 °C.
5. Cool down the reactions to 55 °C at a ramp rate of 1 °C/min. Incubate at 55 °C for 30 min and then lower to room temperature at a ramp rate of 0.4 °C/min.

3.6 Bacterial Propagation

Transform the three remaining assembly mixes into bacteria using the following protocol:

1. Mix 4 μL of each assembly mixture with 50 μL of chemocompetent *E. coli* cells. Incubate at 42 °C for 60 s and chill on ice for 2 min.
2. Reconstitute the bacteria by adding 500 μL of LB medium and by shaking at 37 °C for 30 min.
3. Spin down the bacteria at 3,000 rcf for 2 min. Resuspend the pellets in 50 μL LB medium and plate on agar plates containing 100 μg/mL Ampicillin. Incubate the plates at 37 °C overnight.
4. Inoculate 1.5 mL liquid cultures in LB medium each supplemented with 1.5 μL Ampicillin stock solution by picking single bacterial colonies from the plates using a sterile pipette tip. The cultures can be conveniently grown in 2 mL reaction tubes with a hole punched into the lid using a hot needle. Let the cultures grow for 12–16 h.
5. Use a miniprep kit to extract the plasmid DNA from the liquid cultures. Elute the DNA from the silica columns in 50 μL H_2O.

3.7 AarI Digestion

1. Mix 2 μL of 10× AarI buffer with 1 μg of plasmid DNA and 0.4 μL of the supportive oligonucleotide solution. Fill up to 19 μL with H_2O. Add 1 μL AarI enzyme solution, mix, and incubate at 37 °C for 3 h.
2. Mix the AarI restriction reactions with 2 μL of the 10× gel loading buffer and run them on a 1 % agarose gel as described above. Cut the appropriate bands and purify the DNA using a gel purification kit. Elute the DNA from the column in 20 μL H_2O.

   As described in **Note 3**, alternative restriction enzymes can be employed instead of AarI.

3.8 Second-Level Chew-Back Assembly

1. Use the AarI-digested and gel-purified fragments for a second-level chew-back assembly using the same protocol as described for the first-level assemblies.

Reaction Nr.	1	2
STOP dNTP	dTTP	dATP
DNA fragments	composite fragment 1 composite fragment 3	composite fragment 2 BB 2

2. Transform the reaction mix into chemo-competent *E. coli* and inoculate liquid cultures as described.

   Note 4 provides additional information for the design of the second-level backbone in order to maximize the fidelity of the cloning process.

3. Prepare the plasmid DNA using a miniprep kit.

3.9 Quality Control

1. Check the second-level assembly clones for being correct by mixing 5 μL plasmid DNA solution with 1 μL 10× FastDigest green buffer, 3 μL H_2O, and 2× 0.5 μL FastDigest restriction enzyme of choice. Incubate for 30 min at 37 °C and analyze on a 1 % agarose gel for the correct band pattern as predicted using the software “Ape Plasmid Editor”.

4 Notes

1. Due to the requirement for an alternating use of the extension nucleotides, only even numbers of DNA fragments can be assembled in a single tube. This requirement can be omitted by introducing short paired oligonucleotide fragments bearing appropriate overhangs. Two oligonucleotides should be designed such that their 3′ ends are reverse complementary for at least 20 bases and their 5′ ends are complementary to the overhangs generated by T4 DNA polymerase chew-back of PCR products. The oligonucleotides should be added at a similar molarity to the PCR fragments after the chew-back has been heat-inactivated, since otherwise they could be subject to rapid degradation.
2. By digesting the PCR products with DpnI prior to the assembly, the cloning background due to template plasmid DNA contamination bearing the same resistance gene as the final assembly construct can be reduced. DpnI digests only methylated plasmid DNA of bacterial origin, while PCR products are left intact.
3. The type-IIs restriction enzyme AarI can be replaced by other enzymes. If no outside cutter is used, the sequence portion of the enzyme’s recognition site which is retained after cutting and 3′ chew-back determines parts of the ID sequences to be used. Importantly, restriction enzymes leaving 3′ overhangs generally predetermine less of the ID sequences than enzymes leaving 5′ overhangs, since 3′ overhangs are removed at the time of chew-back irrespective of their base composition. Suitable alternatives to AarI are PacI, BsmBI, and BsmI.

4. By alternating the antibiotic resistance gene between the hierarchical levels of assembly (e.g., Ampicillin versus Kanamycin), the cloning background resulting from undigested first-level plasmid DNA can be significantly reduced.

Acknowledgement

The authors thank Veit Hornung for his support. This work is financially supported by the Bauer Fellows Program, SFB 670, ETH Zurich, ERC starting grant, National Institutes of Health (1R01CA155320-01), National Institute of General Medical Sciences Grant for Centers of Systems Biology, and the German National Academic Foundation.

References

1. Aslanidis C, Dejong PJ (1990) Ligation-independent cloning of PCR products (LIC-PCR). Nucleic Acids Res 18:6069–6074
2. Kodumal SJ, Patel KG, Reid R et al (2004) Total synthesis of long DNA sequences: synthesis of a contiguous 32-kb polyketide synthase gene cluster. Proc Natl Acad Sci U S A 101:15573–15578
3. Reisinger SJ, Patel KG, Santi DV (2006) Total synthesis of multi-kilobase DNA sequences from oligonucleotides. Nat Protoc 1:2596–2603
4. Aslanidis C, Dejong PJ, Schmitz G (1994) Minimal length requirement of the single-stranded tails for ligation-independent cloning (LIC) of PCR products. PCR Methods Appl 4:172–177
5. Donahue WF, Turczyk BM, Jarrell KA (2002) Rapid gene cloning using terminator primers and modular vectors. Nucleic Acids Res 30:e95
6. Geu-Flores F, Nour-Eldin HH, Nielsen MT et al (2007) USER fusion: a rapid and efficient method for simultaneous fusion and cloning of multiple PCR products. Nucleic Acids Res 35:e55
7. Gibson DG, Benders GA, Andrews-Pfannkoch C et al (2008) Complete chemical synthesis, assembly, and cloning of a *Mycoplasma genitalium* genome. Science 319:1215–1220
8. Gibson DG, Young L, Chuang RY et al (2009) Enzymatic assembly of DNA molecules up to several hundred kilobases. Nat Methods 6:343–345
9. Gibson DG, Benders GA, Axelrod KC et al (2008) One-step assembly in yeast of 25 overlapping DNA fragments to form a complete synthetic *Mycoplasma genitalium* genome. Proc Natl Acad Sci U S A 105:20404–20409
10. Schmid-Burgk JL, Xie Z, Frank S et al (2012) Rapid hierarchical assembly of medium-size DNA cassettes. Nucleic Acids Res 40:e92

Chapter 5

USER-Derived Cloning Methods and Their Primer Design

Bo Salomonsen, Uffe H. Mortensen, and Barbara A. Halkier

Abstract

Uracil excision-based cloning through USER™ (Uracil-Specific Excision Reagent) is an efficient ligase-free cloning technique that comprises USER cloning, USER fusion, and USER cassette-free (UCF) USER fusion. These USER-derived cloning techniques enable seamless assembly of multiple DNA fragments in one construct. Though governed by a few simple rules primer design for USER-based fusion of PCR fragments can prove time-consuming for inexperienced users. The Primer Help for USER (PHUSER) software is an easy-to-use primer design tool for USER-based methods. In this chapter, we present a PHUSER software protocol for designing primers for USER-derived cloning techniques.

Key words Primer design, USER fusion, PCR-based cloning, Seamless DNA fusion, Ligation-free cloning, PHUSER

1 Introduction

The invention of tools allowing DNA fragments to be combined in vitro has served as a cornerstone in both molecular biology and biotechnology. The original technology, discovered by Mertz and Davis in 1972 [1], combines DNA ligation of any two complementary cohesive ends [2, 3] formed through digestion with restriction endonucleases [4]. Since then other discoveries have made cloning a routine operation within molecular biology. Most noteworthy was the implementation of strategies based on the polymerase chain reaction (PCR) [5, 6] that allows for incorporation of restriction enzyme recognition sites in the primers for direct cloning of any DNA fragment [7]. Moreover, PCR gave rise to a plethora of new methods for DNA manipulation including, e.g., site-directed mutagenesis and assembly of multiple DNA fragments by overlap extension [8–11]. Importantly, PCR released cloning from the limitations of the traditional restriction digest and ligation cloning method [12–17] as it enabled introduction of complementary bases in the ends of one or more linear DNA fragments. These ends can then be joined by creating single-stranded

Svein Valla and Rahmi Lale (eds.), *DNA Cloning and Assembly Methods*, Methods in Molecular Biology, vol. 1116, DOI 10.1007/978-1-62703-764-8_5, © Springer Science+Business Media New York 2014

DNA overhangs through PCR [13, 14], exonuclease activity [12, 17–20], or by excision of uracil [15, 16, 21–23]. These methods possess a high degree of freedom as they allow for seamless assembly of any DNA stretch and the most well-known commercial names within this category are In-Fusion® (Chapter 15), Gibson Assembly™ (Chapter 1), and USER™ (Uracil-Specific Excision Reagent). Common for these methods is that they require only a single-step process of approximately 1 h to assemble the DNA fragments. Also, they allow for both cloning into plasmids and joining of PCR fragments, which are features that make these ideal cloning methods for construction of new vectors; for example, generating donor plasmids for recombinational systems as Gateway® (Chapter 14). Although similar possibilities are available when using methods based on hybridization of ssDNA overhangs, the methods differ with respect to the way the overhang is generated and in the recommended length of the overhang. In-Fusion® recommends 15 bp sequence overlap and relies on inherent 3′ exonuclease function of the vaccinia virus DNA polymerase [19], whereas the Gibson Assembly™ recommends 40 bp overlaps and includes joint action of a 5′ exonuclease, a DNA polymerase, and a DNA ligase [18]. In the USER cloning method, the USER primer used to generate the DNA fragment has uracil incorporated usually 8 nucleotides (nt) from the 5′ end of the PCR primer. Upon excision of the uracil, an 8 nt overhang is generated as the remaining 7 nt fragment of the 5′ end of the USER primer (referred to as USER tail) is lost from the PCR fragment by diffusion as the few hydrogen bonds between the USER tail and the PCR fragment melt at room temperature.. Though coined in the early 1990s [15, 16], the use of uracil excision-based cloning was uncommon mainly because it was not compatible with high fidelity PCR polymerases. This changed when Nour-Eldin and colleagues [23] reported the application of uracil-insensitive proof-reading polymerases as well USER cassettes based on restriction enzymes with 8 bp recognition site flanked by two recognition sites for a nicking enzyme (Fig. 1). Uni-directional cloning is obtained by the "directional bases" placed between the recognition sites of restriction and nicking enzymes. The overhang created through uracil excision is the same for a given USER cassette and it is independent of intrinsic restriction sites in the amplified DNA. The improved USER technique [23] was further developed to include site-directed mutagenesis [21] and fusion of several PCR fragments into a vector USER cassette through USER fusion [22]. Plasmids carrying USER cassettes have been generated for diverse organisms such as plants [24], yeast [25], filamentous fungi [26, 27], bacteria [28], and mammalian cell lines [29]. As an additional development to USER cloning, the overhangs generated in PCR products can be used to construct almost any DNA plasmid from

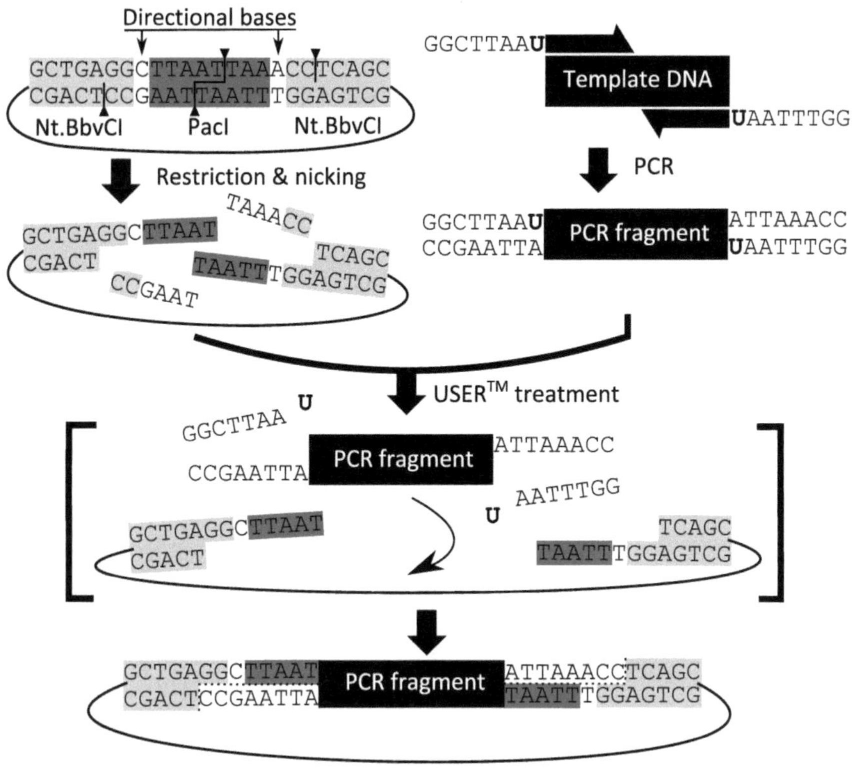

Fig. 1 Schematic illustration of a USER cloning. The plasmid with a USER cassette, here the PacI/Nt.BbvCI cassette, is prepared for cloning by restriction digest and nicking (*upper left corner*) to generate 8 nt single-stranded overhang. PCR amplification of the template DNA with primers including uracil allows for generation of compatible single-stranded overhangs on the PCR product after treatment with USER™ enzyme. Subsequently, the complementary overhangs will hybridize and the product transformed into chemically competent *E. coli* without ligation. Recognition site of Nt.BbvCI is marked with *light grey*. The *PacI* recognition site is marked with *dark grey*. The "directional bases" that ensure uni-directional cloning are not colored

PCR fragments, including PCR-generated plasmid backbone [30, 31]. The degree of freedom in this USER cassette-free (UCF) USER fusion cloning method is ideal for assembly of chimeric plasmids by combining properties from other plasmids. However, though the USER primer design is governed by a few simple rules it can prove difficult for inexperienced users. The Primer Help for USER (PHUSER, http://www.cbs.dtu.dk/services/PHUSER/) software is programmed to make primer design a fast and straightforward procedure for not only USER cloning and USER fusion [32] but also UCF-USER fusion. We here describe primer design for all USER methods, using the PHUSER software. The procedure is illustrated with assembly of a new chimeric yeast expression plasmid utilizing the UCF-USER fusion cloning method.

2 Materials

Unless stated otherwise use 18 mΩ cm water in all buffers and media and perform the procedures at room temperature.

2.1 PCRs

1. 10 mM dNTPs mixture (2.5 mM each of dATP, dCTP, dGTP, and dTTP).
2. *PfuTurbo*® Cx Hotstart DNA Polymerase (Agilent) or preferably the proof-reading X-7 polymerase [30].
3. 10× CxL polymerase buffer (200 mM Tris–HCl, 100 mM KCl, 60 mM $(NH_4)_2SO_2$, 20 mM $MgSO_4$, 1 mg/mL, 1 % Triton X-100): Prepare 1 M stock concentrations of Tris–HCl base, KCl, $(NH_4)_2SO_4$, and $MgSO_4$ in the following way. Dissolve 15.756 g Tris–HCl base in 100 mL ddH_2O and adjust pH to 8.8 with NaOH (*see* **Note 1**). Prepare three 100 mL Blue cap bottles, one with 7.46 g of KCl, the second with 13.2 g of $(NH_4)_2SO_4$, and 22.8 g of $MgSO_4 \cdot 6H_2O$ in the last and fill to 100 mL with ddH_2O. Mix 20 mL 1 M Tris–HCl pH 8.8 with 10 mL of 1 M KCl, 6 mL of 1 M $(NH_4)_2SO_4$, and 2 mL of 1 M $MgSO_4$. Add 10 mL of 100× BSA stock solution (10 mg/mL, New England Biolabs) and 1 mL Triton X-100. Fill to 100 mL with ddH_2O, sterile filter, and store at −20 °C.
4. Template DNA.
5. 10 μM primer solutions.
6. Enzyme (New England Biolabs): DpnI (20,000 units/mL).

2.2 Preparation of Plasmid with USER Cassette

This step is not necessary when performing UCF-USER fusion.

1. Appropriate restriction- and nicking enzyme with buffers for the digestion of the USER compatible vector (*see* **Note 2**). For example, the USER cassette in Fig. 1 should be digested with PacI and Nt.BbvCI (New England Biolabs).

2.3 USER Cloning

1. Enzyme (New England Biolabs): USER™ enzyme (1,000 units/mL).
2. Solid Lysogeny Broth (LB) media: Dissolve 5 g of yeast extract, 10 g of sodium chloride (NaCl), 10 g of tryptone, and 20 g of agar in 1,000 mL ddH_2O. Autoclave and add antibiotics for selection.
3. Liquid LB media: Autoclave 1,000 mL ddH_2O containing yeast extract (5 g/L), sodium chloride (NaCl 10 g/L), and tryptone (10 g/L). Add selective antibiotics after autoclavation.
4. Chemically competent *E. coli* cells, e.g., TOP10 or DH10B.
5. Plasmid isolation: GenElute Plasmid Miniprep Kit (Sigma-Aldrich).

2.4 DNA gel Electrophoresis

1. SeaKem® LE agarose (Lonza).
2. 1 L Blue cap bottles, 100 mL Blue cap bottles.
3. Magnetic stirrer and Teflon®-coated magnetic stir bar.
4. Buffer for gel electrophoresis: Prepare a 0.5 M solution of EDTA by dissolving 186.1 g EDTA disodium salt in 1 L H_2O and adjust pH to 8.0 using NaOH (*see* **Note 3**). Prepare 1 L 50× TAE buffer as stock solution by dissolving 242 g Tris–HCl base in 750 mL H_2O (*see* **Note 1**). Add 100 mL of 0.5 M EDTA solution (pH 8.0) and 57.1 g of glacial acetic acid, before adjusting the final volume to 1 L. Autoclave the mixture and store at room temperature. Dilute to a 1× TAE working solution.
5. GelRed™ Nucleic acid gel stain or similar. Prepare a 5× working solution of GelRed™ by dissolving 10 g of sucrose together with 50 mg of Orange G (Sigma-Aldrich, O-7252) in 40 mL H_2O. Add 1,250 μL GelRed™ (Biotium) and fill to 50 mL.
6. 1 kb Plus DNA Ladder (Invitrogen).
7. Use an E-C Apparatus VWR 105 Compact Power Supply or similar combined with appropriate chamber for DNA gel electrophoresis and a gel imaging apparatus.
8. Millipore DNA Gel Extraction Kit (Millipore).

3 Methods

The following section describes primer design using the PHUSER program for USER cloning and UCF-USER fusion. The UCF-USER fusion technique is illustrated through construction of a plasmid for yeast expression named pBOSAL-2, corresponding to a low copy number plasmid.

3.1 Primer Design

Designing primers for USER cloning and USER fusion with the PHUSER program includes three fast steps (Fig. 2). (A) Submit your sequence(s) target for USER cloning in FASTA format. (B) Decide the maximum length of finished PCR product (*see* **Note 4**) and (C) choose the correct target USER cassette with the same directional bases as your target plasmid (Figs. 1 and 2) (*see* **Notes 5** and **6**).

3.1.1 Traditional USER Cloning into Existing Plasmid

1. Go to http://www.cbs.dtu.dk/services/PHUSER/.
2. Insert the DNA fragment(s) of interest in FASTA format and in the desired order of assembly or upload them as a text file (*.txt) containing the sequence(s) in FASTA format in the box marked with A (Fig. 2).

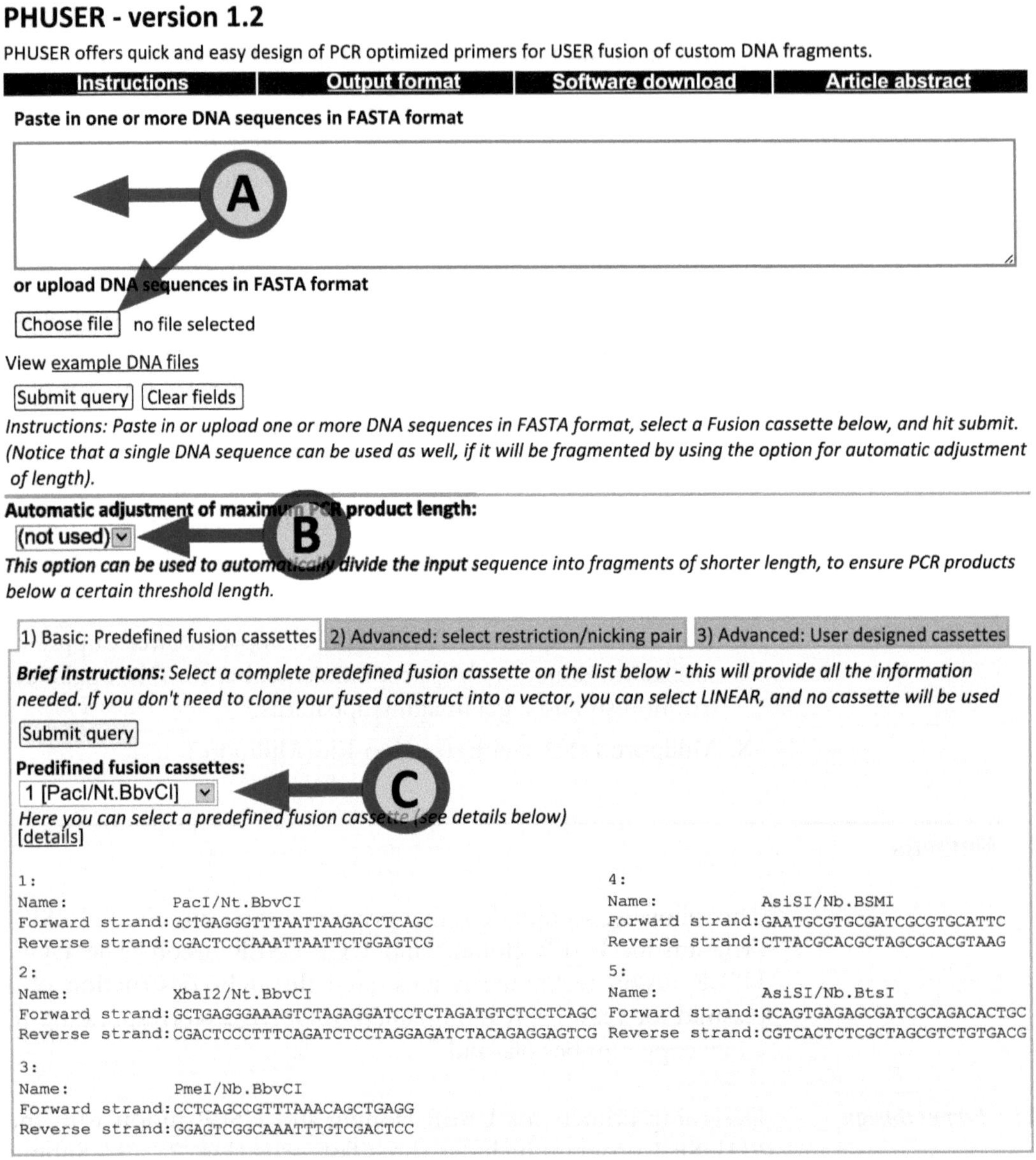

Fig. 2 Screenshot of the Primer Help tool for USER (http://www.cbs.dtu.dk/services/PHUSER/). Three important areas are *highlighted*: (A) where you insert your sequence(s); (B) where you specify the maximum desired length of PCR products; and (C) where you select the target USER cassette

3. Decide the maximum PCR product length using the drop-down menu marked with B (Fig. 2) (*see* **Note 4**).
4. Choose the USER cassette corresponding to one of your plasmid, i.e., having the same directional bases, and restriction- and nicking-sites by using the drop-down menu identified by C (Fig. 2) (*see* **Notes 5** and **6**). Press "submit query." The program returns the primers needed for cloning fusion into the

```
>fragment1v1
CCACTCGTCAGGTGGCACTTTTCGGGGAAATGTGCGCGGAACCCCTATTTGTTTATTTTTCTAAATACATTCAAATATGTATC
CGCTCATGAGACAATAACCCTGATAAA..2000 bp..GCCGCAGCCGAACGACCGAGCGCAGCGAGTCAGTGAGCGAGGA
AGCGGAAGAGCGCCCAATACGCAAACCGCCTCTCCCCGCGCGTTGGCCGATTCATTAATGCAGCTGGCACGACAAAGCA

>fragment2
GCTGAGCGACCTCATGCTATACCTGAGAAAGCAACCTGACCTACAGGAAAGAGTTACTCAAGAATAAGAATTTTCGTTTTAAA
ACCTAAGAGTCACTTTAAAATTTGTAT..1750 bp..TTCATCTTACAAAAAGATCACGTGATCTGTTGTATTGAACTGA
AAATTTTTTGTTTGCTTCTCTCTCTCTCTTTCATTATGTGAGATTTAAAAACCAGAAACTACATCATCGAATTCCAG

>fragment3
GCCAATAATAGGGGTTCCGCGCACATTTCCCCGAAAAGTGCCACCTGGGTCCTTTTCATCACGTGCTATAAAAATAATTATAA
TTTAAATTTTTTAATATAAATATATAA.. 600 bp ..CCCAAATTATTCCATCATTAAAAGATACGAGGCGCGTGTAAGT
TACAGGCAAGCGATCCGTCCTAAGAAACCATTATTATCATGACATTAACCTATAAAAATAGGCGTATCACGAGGCATCC

>fragment1v2
CCACTCGTCAGGTGGCACTTTTCGGGGAAATGTGCGCGGAACCCCTATTTGTTTATTTTTCTAAATACATTCAAATATGTATC
CGCTCATGAGACAATAACCCTGATAAA..2000 bp..GCCGCAGCCGAACGACCGAGCGCAGCGAGTCAGTGAGCGAGGA
AGCGGAAGAGCGCCCAATACGCAAACCGCCTCTCCCCGCGCGTTGGCCGATTCATTAATGCAGCTGGCACGACAAAGCA
```

Fig. 3 Example of inserted sequences to obtain primers for USER cassette-free USER fusion. The inserted fragments correspond to those used to produce pBOSAL-2: "fragment 1v1" and "fragment1v2" are from pUC19 and carry origin and ampicillin resistance; "fragment2" carries the yeast replication origin CEN6ARSH4 from pRS415; and "fragment3" carries a *URA3* yeast marker and an AsiSI/Nb.BsmI USER cassette flanked by the two terminators from CYC1 and ADH1. The 3′- and 5′-end of the fragments are visualized and the approximate size is written on the fragments

USER cassette. In addition to the returned primers, the program also returns an overview of how each of the primers aligns to the desired sequence along with valuable data on expected primer properties as melting temperature, Tm (*see* **Note 7**).

3.1.2 Plasmid Assembly by UCF-USER Fusion

For UCF-USER fusion of PCR products (*see* **Note 8**), do as follows.

1. Go to http://www.cbs.dtu.dk/services/PHUSER/.
2. Insert the DNA fragment sequences for the new plasmid in the box marked with A (Fig. 2), in FASTA format. For construction of pBOSAL-2 this included insertion of a total of three DNA fragments: "fragment 1" from pUC19, "fragment 2" from pXI-1, and "fragment 3" from pRS415 having approximate sizes of 2,000, 1,750, and 600 bp, respectively (Figs. 3 and 4) (*see* **Note 9**).
3. Insert "fragment 1" a second time, so it is represented as the first sequence and as the last sequence. Remember to give it two different names, e.g., fragment1v1 and fragment1v2, and press "submit query" (Fig. 3). When introducing the DNA fragments in this way, the program returns primers for seamless (*see* **Note 10**) fusion of the four inserted fragments into a circular plasmid by traditional USER cloning:

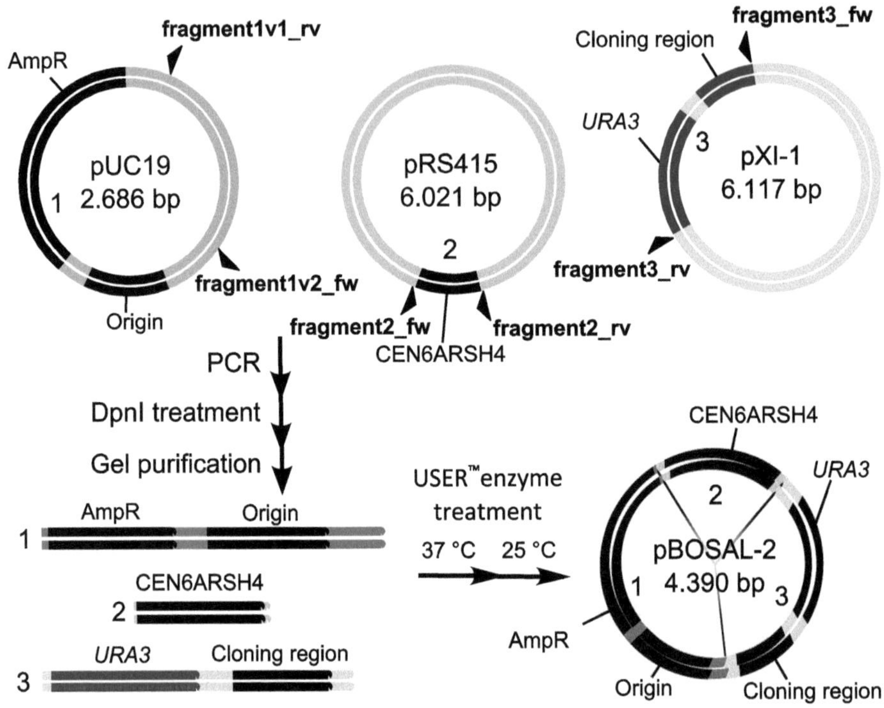

Fig. 4 Schematic drawing of pBOSAL-2 construction. Primers for amplifying the fragments are illustrated with an *arrow* and correspond to the primers returned by the PHUSER program. The three amplified sections are nominated from one to three and correspond to the fragments inserted in PHUSER

```
overview of the needed primers (5'-3'):
---------------------------------------------------------
forward primer for fragment1v1: GGGTTTAAUCCACTCGTCAGGTGGCAC
reverse primer for fragment1v1: AGCTGCUTTGTCGTGCCAGCTGCAT

forward primer for fragment2: AGCAGCUGAGCGACCTCATGCTATA
reverse primer for fragment2: ATTGGCCUGGAATTCGATGATGTAGTTTCTGG

forward primer for fragment3: AGGCCAAUAATAGGGGTTCCGCGCAC
reverse primer for fragment3: AGTGGGGAUGCCTCGTGATACGCCTATTTTTAT

forward primer for fragment1v2: ATCCCCACUCGTCAGGTGGCACTTTTCG
reverse primer for fragment1v2: GGTCTTAAUTGCTTTGTCGTGCCAGCT
```

To perform UCF-USER fusion, use the "reverse primer for fragment1v1" and "forward primer for fragment1v2" to amplify "fragment 1." The other primers correspond to the remaining fragments. "Forward primer for fragment1v1" and "reverse primer for fragment1v2" are not needed for UCF-USER fusion (Fig. 4).

Table 1
PCR mix

Template DNA	0.5 μL (1–5 ng)
10× CxL polymerase buffer	5 μL
dNTP mix (2.5 mM each)	4 μL
Primer-1 (10 μM)	4 μL
Primer-2 (10 μM)	4 μL
Proof-reading X-7 polymerase [30]	1 μL
ddH_2O	31.5 μL
Total	50 μL

Table 2
PCR program

98 °C	2 min	
98 °C 57 °C 72 °C	15 s 15 s 2 min (1 kb/30 s)	30 cycles
72 °C	3 min	
4 °C	∞	

3.2 Creating DNA Fragments Through PCR

1. PCR amplify (*see* Tables 1 and 2), with primers containing a uracil base, the desired feature from plasmids or other DNA (*see* **Note 11**). This enables creation of long single-stranded 3′ ends for ligation-independent cloning. Use a proof-reading polymerase compatible with uracil in the primers, e.g., the commercial *PfuTurbo*® Cx Hotstart DNA Polymerase (Agilent) or preferably the proof-reading X-7 polymerase [30] (*see* **Note 12**).
2. Add 1 μL of DpnI to each of the PCRs and incubate for 1 h at 37 °C, followed by 20 min of heat inactivation at 80 °C (*see* **Note 13**).
3. Prepare a 1 % agarose gel in 1× TAE buffer for gel electrophoresis (*see* **Note 14**). Pour the gel in a gel tray prepared with comb. The gel should be hardened within 30 min and the comb can be removed. Place the gel tray with gel in a gel chamber and over with 1× TAE buffer.
4. Mix 2 μL of each DpnI-treated PCR product with 2 μL ddH_2O and 1 μL 5× GelRed™ and load into the wells of the prepared gel. Load 5 μL of 1 kb Plus DNA Ladder into a neighboring well and perform the gel electrophoresis at 100 V for 25 min.

5. Visualize the bands on the gel using a gel imaging system. If desired bands are present, gel purify using Millipore DNA Gel Extraction Kit (Millipore) according to the manufacturer's instructions (*see* **Note 15**).

3.3 Preparing Plasmid with USER Cassette for Cloning

This step is not necessary when performing UCF-USER fusion.

1. Perform an overnight digest of approximately 20 μg plasmid with 40 units of the appropriate restriction enzyme for your USER cassette in a final volume of 200 μL, according to the manufacturer's instructions.
2. The following day, add 20 units of nicking enzyme and incubate for an additional hour according to the manufacturer's instructions.
3. Visualize the digested plasmid through gel electrophoresis to make sure that it is linearized. Perform gel purification using Millipore DNA Gel Extraction Kit (Millipore) according to the manufacturer's instructions.

3.4 USER Cloning

If preparing a UCF-USER fusion start at **step 2**.

1. Aliquot 2 μL of vector prepared for USER cloning into an empty 1.5 mL Eppendorf™ tube and add the PCR products as described in **step 2**.
2. Determine the ratio between PCR products based on the intensity of the bands from the gel. Make an equimolar mixture of the PCR products in a final volume of 18 μL (*see* **Note 16**). Divide into two 1.5 mL Eppendorf™ tubes before adding 1 μL of USER enzyme mix to one of the Eppendorf™ tubes and 1 μL of ddH_2O to the other as a negative control. Incubate in a 1.5 mL Eppendorf™ tube for 30 min at 37 °C followed by 30 min at 20 °C.
3. Transform chemically competent cells (*see* **Note 17**) with the USER-treated DNA fragments according to the manufacturer's instructions or do as follows: Thaw chemically competent cells on ice. Add 100 μL of chemically competent cells to the USER reaction mixtures and incubate on ice for 20–30 min before applying a 90 s 42 °C heat shock using a water bath. Keep the cells on ice for 5 min, then add 1 mL LB media, and incubate at 37 °C for 1 h with continuous shaking.
4. Spin down the cells at 12,000 × *g* for 30 s. Remove the supernatant and resuspend in 100 μL LB media (*see* **Note 18**). Plate the resuspended cells on LB media containing appropriate selective antibiotics. Incubate at 37 °C O/N.
5. Use liquid LB media containing the appropriate selective antibiotics to grow 3 mL of O/N cultures by inoculating with colonies from the cloning plate (*see* **Note 19**).

6. Isolate cloned plasmids from the O/N cultures using the GenElute™ Plasmid Miniprep Kit (Sigma-Aldrich) according to the manufacturer's instructions. Verify correct assembly of PCR product by restriction analysis.
7. Send a plasmid verified by restriction analysis for sequencing.

4 Notes

1. Add the Tris–HCl base to a beaker with water and a magnetic stir bar, to allow for stirring during addition.
2. When performing a UCF-USER fusion reaction this step is not needed, as the technique is solely based on assembly of PCR products.
3. EDTA will not go into solution unless the pH is adjusted to 8.0.
4. The standard setting is "(not used)" which usually works. However, when amplifying large genes or backbone fragments, it can be advantageous to limit the PCR products to a certain threshold length. As PHUSER designs primers to create seamless fusions this will not change the final outcome.
5. As USER cassettes are named according to the restriction- and nicking enzymes without listing the directional bases there are different cassettes carrying the same name. Example is the predefined cassette 1 PacI/Nt.BbvCI differing from the PacI/Nt.BbvCI found in [23, 33], as the program does not support less than two directional bases (*see* **Note 6**).
6. In case of using an USER cassette that has less than two directional bases, which is not supported by PHUSER, please replace the USER cassette-specific primer tails with tails for your specific cassette. Always replace the primer tails located on the forward primer of the first fragment and the reverse primer of the last fragment. The remaining USER fusion primers are independent of the USER cassette.
7. Tm of the designed primers does not always comply with the optimal temperature region of 55–72 °C; be sure to check if your primers are reasonable for the intended program.
8. Primer design for direct assembly of PCR products may be featured on a later version of PHUSER (Rasmus Wernersson, personal communication).
9. Fragment 1 was a part of pUC19 carrying the ampicillin resistance gene and a bacterial origin corresponding to approximately 2,000 bp. Fragment 2 was a DNA stretch from pXI-1 containing the URA3 marker of *Kluyveromyces lactis* and an AsiSI/Nb.BsmI USER cassette flanked by the two terminators

ADH1(T1) and CYC1(T2) corresponding to 1,750 bp [25]. Fragment 3 was from pRS415 containing the CEN6ARSH4 low copy yeast origin (Fig. 3).

10. The PHUSER website is capable of including linkers in junctions between DNA fragments. To do so merely include the desired linker sequence in FASTA format between the two fragments where the linker is needed and name it ">linker."
11. Remember that your finished plasmids must contain origin and a selection marker when performing UCF-USER fusion.
12. The PCR program described is optimized for the proof-reading X-7 polymerase [30]. When using the commercial *PfuTurbo*® Cx Hotstart DNA Polymerase (Agilent) follow the manufacturer's instructions.
13. DpnI is a 4 bp restriction enzyme acting on methylated target DNA. As plasmid DNA isolated from bacteria is methylated, residual plasmid DNA can be removed by DpnI digestion leaving only non-methylated PCR product. DpnI is functional in the buffer for proof-reading X-7 polymerase and the buffer for the *PfuTurbo*® Cx Hotstart DNA Polymerase.
14. Place the 1 L Blue cap flask on the magnetic stirrer and pour in 500 mL 1× TAE buffer along with a Teflon®-coated magnetic stir bar. Add 10 g SeaKem® LE agarose while stirring. Remove the lid and microwave the flask at max effect until the gel material is dissolved. Fill the bottle to the 1 L mark with 1× TAE buffer. Caution: The liquid can become superheated and flash boil upon agitation. The dissolved gel can be stored at 60 °C for a couple of days.
15. When constructing pBOSAL-2, the DNA fragments 1 and 3 were contaminated with by-products from the PCR. To save time in verification of correct clone in subsequent steps, it was chosen to perform a gel purification to remove these impurities before proceeding. This step is rarely necessary after DpnI treatment and can often be disregarded.
16. PCR products have a tendency to be very concentrated which can inhibit the USER reaction. Try adding approximately 50–100 ng/kb per PCR product and fill with water to 9 μL.
17. Do not perform transformation by electroporation, as the electroshock causes the hybridized tails to dissociate thus ruining the circular product.
18. If you remove the supernatant by pouring, the residual liquid is sufficient for resuspension.
19. Ideally, the negative control plate should not have any colonies. However, even in instances where the colonies on the

negative control plate easily outnumber the colonies on the cloning plate, we have found that the majority of the colonies on the cloning plate to be correct. Thus these should not be disposed without being examined.

Acknowledgement

This work was supported by a strategic PhD stipend from the Faculty of LIFE Sciences.

References

1. Mertz JE, Davis RW (1972) Cleavage of DNA by RI restriction endonuclease generates cohesive ends. Proc Natl Acad Sci USA 69:3370–3374
2. Gellert M (1967) Formation of covalent circles of lambda DNA by *E. coli* extracts. Proc Natl Acad Sci USA 57:148–155
3. Weiss B, Richardson CC (1967) Enzymatic breakage and joining of deoxyribonucleic acid, I. Repair of single-strand breaks in DNA by an enzyme system from *Escherichia coli* infected with T4 bacteriophage. Proc Natl Acad Sci USA 57:1021–1028
4. Smith HO, Welcox KW (1970) A restriction enzyme from *Hemophilus influenzae*: I. Purification and general properties. J Mol Biol 51:379–391
5. Mullis K, Faloona F, Scharf S et al (1986) Specific enzymatic amplification of DNA in vitro: the polymerase chain reaction. Cold Spring Harb Symp Quant Biol L1:263–273
6. Mullis KB, Faloona FA (1987) Specific synthesis of DNA in vitro via a polymerase-catalyzed chain reaction. Methods Enzymol 155:335–350
7. Scharf SJ, Horn GT, Erlich HA (1986) Direct cloning and sequence analysis of enzymatically amplified genomic sequences. Science 233:1076–1078
8. Ho SN, Hunt HD, Horton RM et al (1989) Site-directed mutagenesis by overlap extension using the polymerase chain reaction. Gene 77:51–59
9. Horton RM, Hunt HD, Ho SN et al (1989) Engineering hybrid genes without the use of restriction enzymes: gene splicing by overlap extension. Gene 77:61–68
10. Vallette F, Mege E, Reiss A et al (1989) Construction of mutant and chimeric genes using the polymerase chain reaction. Nucleic Acids Res 17:723–733
11. Yon J, Fried M (1989) Precise gene fusion by PCR. Nucleic Acids Res 17:4895
12. Aslanidis C, De Jong PJ (1990) Ligation-independent cloning of PCR products (LIC-PCR). Nucleic Acids Res 18:6069–6074
13. Jones D, Sakamoto K, Vorce R et al (1990) DMA mutagenesis and recombination. Nature 344:793–794
14. Shuldiner AR, Scott LA, Roth J (1990) PCR-induced (ligase-free) subcloning: a rapid reliable method to subclone polymerase chain reaction (PCR) products. Nucleic Acids Res 18:1920
15. Nisson PE, Rashtchian A, Watkins PC (1991) Rapid and efficient cloning of Alu-PCR products using uracil DNA glycosylase. Genome Res 1:120–123
16. Smith C, Day P, Walker MR (1993) Generation of cohesive ends on PCR products by UDG-mediated excision of dU, and application for cloning into restriction digest-linearized vectors. Genome Res 2:328–332
17. Yang YS, Watson WJ, Tucker PW et al (1993) Construction of recombinant DNA by exonuclease recession. Nucleic Acids Res 21:1889–1893
18. Gibson DG, Young L, Chuang RY et al (2009) Enzymatic assembly of DNA molecules up to several hundred kilobases. Nat Methods 6:343–345
19. Evans DH, Willer DH, Yao XD (2003) DNA joiningmethod. USPatentUS/2003/0162265 A1
20. Li MZ, Elledge SJ (2007) Harnessing homologous recombination in vitro to generate recombinant DNA via SLIC. Nat Methods 4:251–256
21. Bitinaite J, Rubino M, Varma KH et al (2007) USER™ friendly DNA engineering and cloning method by uracil excision. Nucleic Acids Res 35:1992–2002

22. Geu-Flores F, Nour-Eldin HH, Nielsen MT et al (2007) USER fusion: a rapid and efficient method for simultaneous fusion and cloning of multiple PCR products. Nucleic Acids Res 35:e55
23. Nour-Eldin HH, Hansen BG, Nørholm MHH et al (2006) Advancing uracil-excision based cloning towards an ideal technique for cloning PCR fragments. Nucleic Acids Res 34:e122
24. Geu-Flores F, Nielsen MT, Nafisi M et al (2009) Glucosinolate engineering identifies a γ-glutamyl peptidase. Nat Chem Biol 5:575–577
25. Mikkelsen MD, Buron LD, Salomonsen B et al (2012) Microbial production of indolylglucosinolate through engineering of a multi-gene pathway in a versatile yeast expression platform. Metab Eng 14:104–111
26. Hansen BG, Salomonsen B, Nielsen MT et al (2011) Versatile enzyme expression and characterization system for *Aspergillus nidulans*, with the *Penicillium brevicompactum* polyketide synthase gene from the mycophenolic acid gene cluster as a test case. Appl Environ Microbiol 77:3044–3051
27. Bjarne GH, Xin ES, Hans JG et al (2012) Adaptive evolution of drug targets in producer and non-producer organisms. Biochem J 441:219–226
28. Burow M, Losansky A, Müller R et al (2009) The genetic basis of constitutive and herbivore-induced ESP-independent nitrile formation in Arabidopsis. Plant Physiol 149:561–574
29. Wang Z (2009) Epitope tagging of endogenous proteins for genome-wide chromatin immunoprecipitation analysis. Methods Mol Biol 567:87–98
30. Nørholm M (2010) A mutant Pfu DNA polymerase designed for advanced uracil-excision DNA engineering. BMC Biotechnol 10:21
31. Annaluru A, Muller H, Ramalingam S et al (2011) Assembling DNA fragments by USER fusion. Methods Mol Biol 852:77–95
32. Olsen LR, Hansen NB, Bonde MT et al (2011) PHUSER (Primer Help for USER): a novel tool for USER fusion primer design. Nucleic Acids Res 39:W61–W67
33. Nour-Eldin HH, Geu-Flores F, Halkier BA (2010) USER cloning and USER fusion: the ideal cloning techniques for small and big laboratories. Methods Mol Biol 643:185–200

Chapter 6

Application of the Restriction-Free (RF) Cloning for Multicomponents Assembly

Yoav Peleg and Tamar Unger

Abstract

Molecular manipulations, including DNA cloning and mutagenesis, are currently employed on a routine basis in all life science disciplines. Over the last decade new methodologies have emerged that expanded and facilitated the applications for DNA cloning. The classical Ligation-Dependent Cloning (LDC) is gradually replaced by Ligation-Independent Cloning (LIC) techniques. The Restriction-Free (RF) cloning was originally developed for introduction of a foreign DNA into a plasmid at any desired position. The RF methodology is based on generation of a PCR product, which serves as a set of mega-primers for subsequent incorporation into any desired position within a circular plasmid. We have expanded the applications of the RF methodology for multiple simultaneous alterations of a target DNA and for multicomponents assembly. In the current manuscript we describe a step-by-step protocol for application of the RF methodology for simultaneous multiple DNA fragments assembly in tandem and at distinct positions within an expression vector.

Key words DNA cloning, Restriction-Free (RF) cloning, Ligation-Independent Cloning (LIC), Multicomponents assembly

1 Introduction

Assembly of DNA fragments is a basic and essential tool, currently used on a routine basis in many applications in biology and medical research. The Ligation-Dependent Cloning (LDC) approach using restriction enzymes digestion followed by a ligation step is currently being partially replaced by more sophisticated methods. These methods are primarily Ligation-Independent Cloning (LIC) techniques. Numerous such approaches have been described in recent years. Some of these methods rely on recombination between the DNA insert and the destination vector [1, 2]. The recombination approach is being used in commercial kits such as the Gateway® (Invitrogen) (Chapter 14) and In-Fusion™ (Clontech) (Chapter 15). Other LIC approaches

Svein Valla and Rahmi Lale (eds.), *DNA Cloning and Assembly Methods*, Methods in Molecular Biology, vol. 1116, DOI 10.1007/978-1-62703-764-8_6, © Springer Science+Business Media New York 2014

designated LIC-PCR [3], LIC [4], and enzyme-free cloning [5] use complementary single-strand overhangs to combine the insert and the linearized vector.

In recent years along with the growing interest in synthetic biology several tools were developed for integration of multiple DNA fragments. The Sequence and Ligation-Independent Cloning (SLIC, Chapter 2) is based on homologous in vitro recombination and single-strand assembly [6, 7]. The SLIC was used for tandem assembly of up to ten DNA fragments, simultaneously. Additional approach is the Gibson Assembly™ (Chapter 1) (New England Biolabs), which joins together multiple 3′ overhangs DNA fragments. The DNA assembly method was used to join and clone DNA fragments up to several hundred kilobases in *E. coli* [8].

In this chapter on the Restriction-Free (RF) methodology and in the following chapter on the Transfer-PCR (TPCR) platform we describe DNA assembly methods based on whole plasmid amplification of the vector and DNA insert(s). In both approaches gene amplification is carried out by two primers each containing at the 3′ a target-gene-specific sequence and at the 5′ an extension, overlapping with the sites of integration in the destination vector [9, 10]. Whereas, the RF methodology requires purification of the intermediate PCR product generated in the first amplification stage for integration into the destination vector [9, 11], in the TPCR approach the synthesis of the PCR product and its assembly into the destination vector occurs in a single-tube [10]. A useful web tool was developed to facilitate the design of the primers for cloning using the RF and TPCR methods [12]. Both methods were used for various DNA manipulation applications, including cloning and single- and multiple-site-targeted mutagenesis.

We have demonstrated the strength of the RF cloning not only for single DNA fragment integration but also for simultaneous multiple DNA fragments [9]. In the current protocol we describe a detailed procedure for implementation of the RF cloning for simultaneous multicomponents DNA assembly at distinct positions or in tandem within a circular expression vector of choice.

2 Materials

2.1 Bacterial Strains

For DNA cloning and plasmid preparation procedures, *E. coli* DH5α (Agilent Technology, Stratagene division, Santa Clara, CA) was used. For protein expression *E. coli* BL21(DE3) (Novagen/EMD Millipore Chemicals, Darmstadt, Germany) was employed.

2.2 Vectors

For cloning experiments described in the methods section the expression vectors pET21a and pACYCDuet-1 (Novagen/EMD Millipore Chemicals, Darmstadt, Germany) were used as destination vectors.

2.3 RF Cloning Reaction and Analysis

2.3.1 RF Cloning Reaction

1. Phusion high fidelity thermo-stable DNA polymerase, 5× Phusion HF buffer, and dimethyl sulfoxide (DMSO) (New England Biolabs, Ipswich, MA; Catalog # M0530S).
2. dNTPs set (Fermentas/Thermo Fisher Scientific, Glen-Burnie, MD; Catalog # R0181). Prepare 10 mM dNTPs stock solution in ultrapure water.
3. Synthetic primers (forward and reverse) for RF cloning. Both primers are diluted with TE buffer (10 mM Tris–HCl, 1 mM EDTA, pH 8) or ultrapure water, to 25 μM stock solution (*see* **Note 1** and Table 1 for primer design).
4. PCR product purification kit (MEGAquick-spin Total, Intron Biotechnology, Daejoen, South Korea; Catalog # 17286) or equivalent.
5. Plasmid DNA purification kit (QIAprep spin mini prep kit, Qiagen, Hilden, Germany; Catalog # 27106) or equivalent.
6. Template DNA. If plasmid is used as a template dilute DNA with TE buffer (10 mM Tris–HCl, 1 mM EDTA, pH 8) or ultrapure water, to 10 ng/μL stock solution.
7. Destination plasmid is diluted with TE buffer (10 mM Tris–HCl, 1 mM EDTA, pH 8) or ultrapure water, to 10 ng/μL stock solution.
8. Spectrophotometer (NanoDrop/Thermo Fisher Scientific, Wilmington, DE; Catalog # ND-1000) or equivalent.
9. Thermocycler machine.
10. DpnI (New England Biolabs, Ipswich, MA; Catalog # R0176S) or equivalent.
11. High efficiency *E. coli* DH5α competent cells for heat shock transformation (*see* **Note 2**).
12. Sterile spreaders.
13. Ultrapure water purified using Barnstead Nanopure ultrapure water purification system (Thermo Fisher Scientific Inc., Waltham, MA, USA) or equivalent.
14. Lysogeny Broth (LB) agar plates (10 g/L tryptone, 5 g/L yeast extract, 5 g/L NaCl, and 12 g/L bacteriological agar) supplemented with ampicillin (100 μg/mL) and LB agar plates supplemented with chloramphenicol (34 μg/mL).
15. 37 °C incubator.

2.3.2 Analysis

1. Synthetic primers for colony PCR and sequencing analyses (*see* Table 1).
2. Agarose gel electrophoresis apparatus.
3. Agarose.

Table 1
Primers used for RF cloning and subsequent colony PCR analyses

Task	Genes	Primer name	Primer sequence (5′ ⇒ 3′)
Tandem cloning	Strep-tag	Strep_pET21_F (I)	GTTTAACTTTAAGAAGGAGATATACATATGGCTAGCTGGAGCCACCCGCAG
		Strep_pET21_R (II)	ACGGAGCTCGAATTCGGATCCTTTTTCGAACTGCGGGTGGCTCCA
	Trx	Trx_pET21_F (III)	GGATCCGAATTCGAGCTCCGTCGAAGCGATAAAATTATTCACCTGACTG
		Trx_pET21_R (IV)	TCGAGTGCGGCCGCAAGCTTGGGCCAGGTTAGCGTCGAGG
	YagE	YagETEV_pET21_F (V)	GGCCCAAGCTTGCGGCCGCACTCGA**GAAAACCTGTACTTCCAGGGT**CAAGGAGATCTCATGCCGCAG
		YagECHis_pET21_R (VI)	CGGATCTCAGTGGTGGTGGTGGTGGTGGCAAAGCTTGAGCTGTTGCAG
Cloning at distinct positions	Az	RFAzF	CATCACCATCATCACCACAGCCAGGATCCGCTTCTTAGTCAGCACAGC
		RFAzR	TCGACTTAAGCATTATGCGGCCGCAAGCTTTTAGTCCTCCTCACCCGGGTCC
	ODC	RFODCF	ATTAGTTAAGTATAAGAAGGAGATATACATATGAGCAGCTTTACTAAGGACGAG
		RFODCR	CGCAGCAGCGGTTTCTTTACCAGACTCGAGTTACACATTGATCCTAGCAGAAGCAC
	TFIIEα	RFTFaF_Ol30	CATCACCATCATCACCACAGCCAGGATCCGGCAGACCCAGATGTCCTCAC
		RFTFaR_Ol30	TCGACTTAAGCATTATGCGGCCGCAAGCTTTCACTCAAAGAGGTCCTCAAAC
	TFIIEβ	RFTFbF_Ol30	ATTAGTTAAGTATAAGAAGGAGATATACATATGGATCCAAGCCTGTTGAGAG
		RFTFbR_Ol30	CGCAGCAGCGGTTTCTTTACCAGACTCGAGCTATTTGCTGGAAGTAATGTCAG
Colony PCR	–	T7	ATTAATACGACTCACTATAGGGG
		PetRev	ATGCTAGTTATTGCTCAGCGGT
		pACYCDuetUP1	GGGATCTCGACGCTCTCCC
		DuetDOWN1	CGATTATGCGGCCGTGTACAA

Gene-specific sequences are underlined. Roman numbers in parentheses in the primer column indicate the primers shown in Fig. 2. The Tobacco Etch Virus (TEV) recognition sequence is marked in bold in primer YagETEV_pET21_F (*see* Fig. 3 for details)

4. Ethidium bromide dropper bottle, 0.625 mg/mL stock (Amresco, Solon, OH; Catalog # E406) or equivalent.
5. Tris/acetic acid/EDTA (TAE) buffer, 50× stock solution (2 M Tris–HCl, 1 M acetic acid, 50 mM EDTA, pH 8.0).
6. Microwave oven.
7. Gel-Doc 2000 visualization system (Bio-Rad Laboratories, Hercules, CA; Catalog # 170–8126) or similar.
8. 50 mL round conical centrifuge tubes (BD Falcon, Franklin Lakes, NJ; Catalog # 352077) or equivalent.
9. LB medium and agar plates supplemented with ampicillin (100 μg/mL) and LB agar plates supplemented with chloramphenicol (34 μg/mL).
10. 37 °C microbiological incubator (Function line, Haraeus/ Thermo Fisher Scientific; Catalog # 50042308) or similar.
11. 37 °C shaker-incubator.
12. 0.2 mL PCR tubes.
13. Ready-mix PCR master kit, 2× stock (Amplicon, Skoulunde, Denmark; Catalog # 180301) or equivalent.
14. Temperature-controlled microcentrifuge.
15. Sterile toothpicks.
16. Glycerol (J.T. Baker, Phillipsburg, NJ, USA; Catalog # 2136–01) or equivalent.

3 Methods

3.1 Restriction-Free (RF) Cloning

The RF cloning is a two-stage process. The initial stage is PCR amplification of the gene or DNA element of interest. The source of target DNA for amplification can be a plasmid DNA, PCR product, chromosomal DNA, cDNA library, or any other available source (*see* **Note 3**). The purified PCR product(s) is (are) used as a set of mega-primers for incorporation into the destination vector. In the protocol described below we primarily focus on simultaneous assembly of multiple DNA fragments; however, it can be obviously used for insertion of a single DNA fragments into a circular plasmid. Two examples are described for simultaneous multicomponents assembly [9]. In the first, we describe simultaneous RF cloning of two genes at distinct positions within the expression vector pACYCDuet-1 (Figs. 1 and 2), whereas in the second example (Fig. 3) the focus is on tandem multicomponents assembly into the expression vector pET21a.

1. Set up amplification reaction of the target gene(s) in 0.2 mL tubes, in a final volume of 50 μL, as follows (*see* **Note 4**):

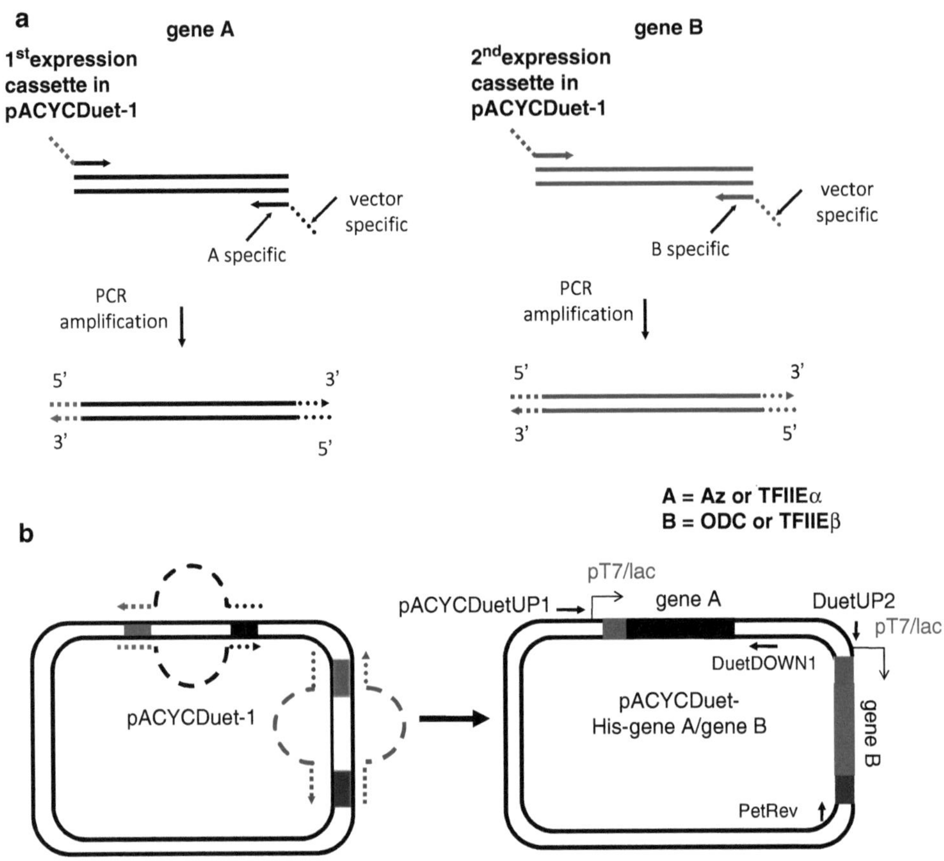

Fig. 1 Schematic representation of the strategy for simultaneous RF cloning. (**a**) Gene A and gene B are shown in *black* and *red lines*, respectively. Primers are indicated by *arrows* containing gene-specific sequences (marked in the same color as the gene) and vector-specific sequences indicated by *dotted lines*. The vector-specific sequences flanking gene A are shown as *red squares* and *purple circles*, whereas for gene B, the sequences are shown as *blue squares* and *green circles*. Gene A represents in our study Antizyme (Az) or transcription factor TFIIEα (TFIIEα) and gene B represents Ornithine Decarboxylase (ODC) or transcription factor TFIIEβ (TFIIEβ). (**b**) Schematic representation of integration of gene A and gene B into the two expression cassettes in the expression vector pACYDuet-1. Color coding of the genes and flanking regions are as in (**a**). *Arrows* indicate primers used in colony PCR screening (*see* Fig. 2 and Table 1) (reproduced from ref. 9 with permission from Elsevier) (color figure online)

Component	Stock solution	Volume/50 μL reaction	Final concentration
Target DNA	10 ng/μL	2 μL	0.4 ng/μL
Forward primer	25 μM	1 μL	0.5 μM
Reverse primer	25 μM	1 μL	0.5 μM
dNTPs mix	10 mM each	1 μL	200 μM each
Phusion HF Buffer	5×	10 μL	1×
Phusion DNA Polymerase	2 U/μL	0.8 μL	0.032 U/μL
Ultrapure water		34.2 μL	

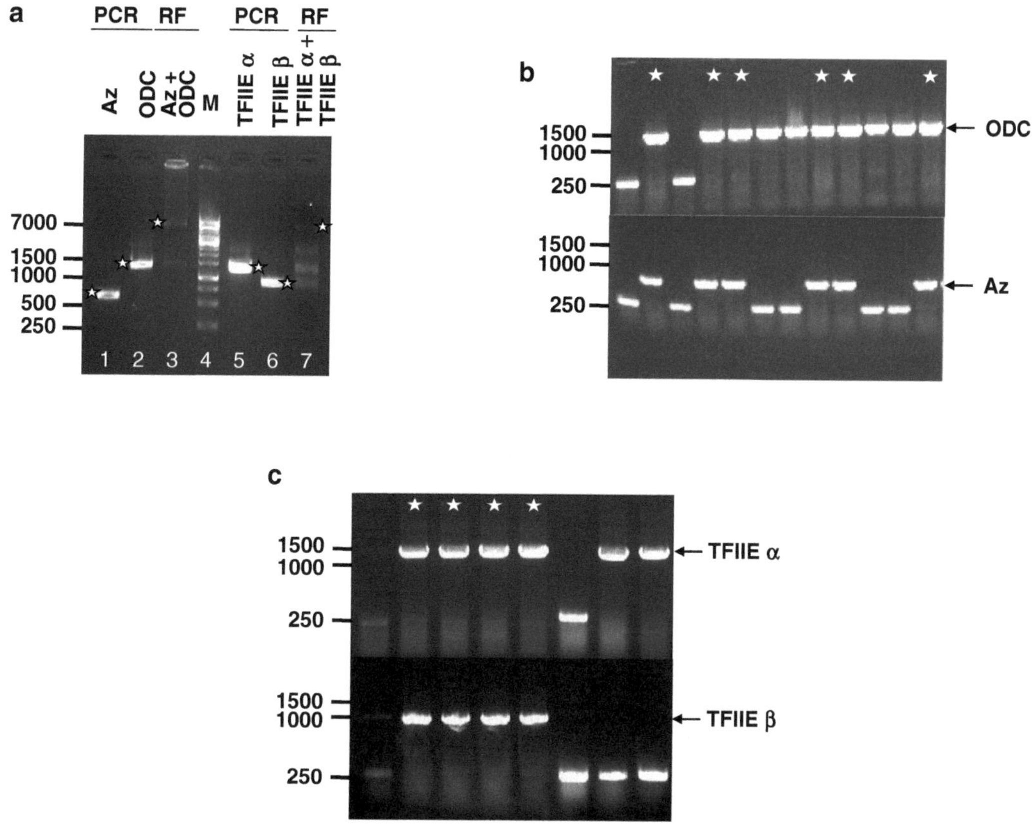

Fig. 2 Analysis of simultaneous RF reactions. (**a**) Samples from PCRs of Az (lane 1), ODC (lane 2), TFIIEα (lane 5), and TFIIEβ (lane 6) and samples following the RF amplification of Az and ODC (lane 3) and of TFIIEα and TFIIEβ (lane 7) were analyzed on a 1 % agarose gel. *Stars* indicate position of positive PCR and RF products. (**b**, **c**) Colony PCR analysis for Az and ODC integration (**b**) and for TFIIEα and TFIIEβ integration (**c**). Twelve and eight colonies were analyzed using pACYCDuetUP1 and DuetDOWN1 primers for integration of Az or TFIIEα, respectively, and DuetUP2 and PetRev primers for integration of ODC or TFIIEβ, respectively. The expected sizes of PCR products for Az, ODC, TFIIEα, and TFIIEβ are ~640, ~1,450, ~1,380, and ~940 bp, respectively. *Arrows* indicate the expected positions of Az, ODC, TFIIEα, and TFIIEβ. PCR products were analyzed on a 1 % agarose gel. *Stars* indicate colonies in which simultaneous integration was observed for Az and ODC and for TFIIEα and TFIIEβ. DNA molecular weight marker (in base-pairs) is shown on the *left* (reproduced from ref. 9 with permission from Elsevier)

2. Mix the 0.2 mL tubes gently and spin briefly in a microcentrifuge.
3. Perform amplification reaction using the following parameters (*see* **Note 4**):

Step	Temperature (°C)	Time (min:s)	Cycles
Denaturation	95	1:00	1
Denaturation Annealing Elongation	95 60 72	0:30 1:00 1:30	25
Elongation Cooling	72 10	6:00 ∞	1

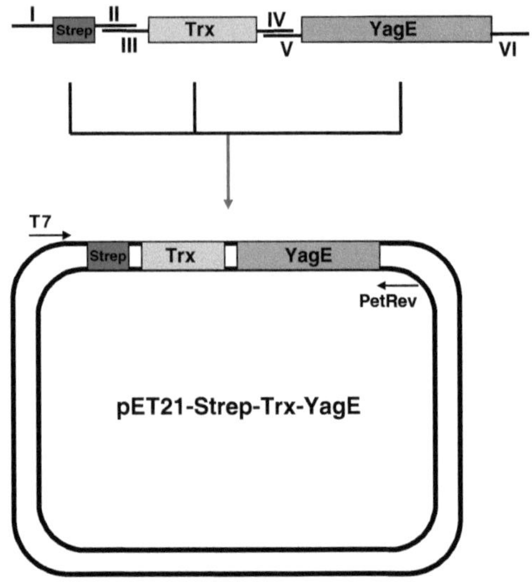

Fig. 3 Tandem multicomponents assembly of pET21-Strep-Trx-YagE. Schematic representation of three components, strep-tag, Thioredoxin (Trx), and YagE gene (located in the *E. coli K12* genome that is part of prophage *CP4-6* [16]). The flanking regions of each component are indicated by *Roman numerals* (*see* Table 1 for sequence information). One set of forward and reverse primers (marked by I and II) was designed to amplify the strep-tag sequence and flanking 27 and 21 nucleotides, respectively. The second set of primers (marked by III and IV) was designed to amplify the Trx gene and an additional 24 and 21 flanking nucleotides, respectively. The flanking nucleotides in the forward primer (III) overlap by 21 nucleotides with the flanking nucleotides of the strep-tag reverse primer (primers II). The third set of primers (primers V and IV) was designed to amplify the YagE gene, with additional 21 nucleotides of TEV recognition sequence 5′ to YagE (marked in bold in Table 1), and 25 and 27 flanking nucleotides, respectively. The flanking nucleotides in the forward primer (primer V) overlap by 21 nucleotides with the flanking nucleotides of the Trx reverse primer (primer IV). The resulting assembled plasmid was designated as pET21-Strep-Trx-YagE. *Black arrows* indicate primer positions used for colony PCR screening (*see* Table 1 for sequence information). The different components are not drawn to scale

4. Following completion of the amplification reaction analyze aliquots (2–3 μL) from each reaction on a 1 % agarose gel (*see* **Note 5**).
5. Purify the PCR product(s) (*see* **Note 6**) and measure DNA concentration using a NanoDrop spectrophotometer.
6. Assemble the PCR products (mega-primers) and the destination vector in 0.2 mL tubes, in a final volume of 50 μL, as follows (*see* **Note 7**):

Component	Stock solution	Volume/50 μL reaction	Final concentration
Destination plasmid	10 ng/μL	1 μL	0.2 ng/μL
PCR product I	*X*ng/μL	*X*μL	2 ng/μL
PCR product II	*X*ng/μL	*X*μL	2 ng/μL
Forward primer	1 μM	1 μL	20 nM
Reverse primer	1 μM	1 μL	20 nM
dNTPs mix	10 mM each	1 μL	200 μM each
Phusion HF Buffer	5×	10 μL	1×
Phusion DNA Polymerase	2 U/μL	0.8 μL	0.032 U/μL
Ultrapure water		Complete to 50 μL	

7. Mix the 0.2 mL tubes gently and spin briefly in a microcentrifuge.
8. Perform the assembly reaction using the following parameters (*see* **Note 8**):

Step	Temperature (°C)	Time (min:s)	Cycles
Denaturation	95	0:30	1
Denaturation Annealing Elongation	95 60 72	0:30 1:00 5:00	30
Elongation Cooling	72 10	7:00 ∞	1

9. Following completion of the amplification reaction analyze aliquots (10 μL) from each reaction on a 1 % agarose gel (*see* **Note 9**).
10. Transfer 10 μL from the RF reactions into a new 1.5 mL tube. Add 1 μL DpnI, spin briefly, and incubate for 1–2 h at 37 °C.
11. Spin briefly the DpnI-treated reaction using a microcentrifuge.
12. Transform the DpnI-treated reaction into competent DH5α *E. coli* cells. Add 1–10 μL of reaction into 90–100 μL competent cells (*see* **Note 10**). Mix gently. Incubate on ice for 30–60 min. Heat the tube at 42 °C for 1 min and return to ice. Add 1 mL of LB and incubate with shaking for 1 h at 37 °C.
13. Plate transformed mix on LB agar plates with the appropriate antibiotic. In the examples presented in this manuscript, for cloning of Antizyme (Az) and Ornithine Decarboxylase (ODC) or TFIIEα and TFIIEβ into the two expression cassettes

of the expression vector pACYCDuet-1 (Figs. 1 and 2), LB-chloramphenicol (34 μg/mL) plates were used. For tandem assembly of the Strep-Trx-YagE into pET21a (Fig. 3) we used LB-ampicillin (100 μg/mL) plates.

14. Incubate plates overnight in a 37 °C incubator.

3.2 Analysis of Clones

On the following day, select 6–10 single colonies from the transformation plate, and perform colony PCR to identify positive colonies as described below (*see* **Note 11**). If no colonies are observed on the plate consult **Note 12** for details on how to proceed.

1. Set up PCRs in 0.2 mL tube, in a final volume of 20 μL, as follows (*see* **Note 13**):

Component	Stock solution	Volume/20 μL reaction	Final concentration
Forward primer	25 μM	1 μL	500 μM
Reverse primer	25 μM	1 μL	500 μM
Master mix solution	2×	10 μL	1×
Ultrapure water		8 μL	

2. Using a sterile toothpick select single colonies and transfer the bacteria into the labeled 0.2 mL PCR tubes. Swirl well.
3. Using the same toothpick maintain each clone by inoculating liquid or plate containing the appropriate antibiotic (*see* **Note 11**).
4. Mix gently the 0.2 mL PCR tubes and spin briefly.
5. Perform colony PCR using the following parameters (*see* **Note 13**):

Step	Temperature (°C)	Time (min:s)	Cycles
Denaturation	95	1:00	1
Denaturation Annealing Elongation	95 60 72	0:30 1:00 1:30	25
Elongation Cooling	72 10	6:00 ∞	1

6. Analyze PCR products on 1 % agarose gel to verify that the DNA product obtained is of the correct size. Select 2–3 individual positive colonies (*see* **Note 14**).

7. Inoculate positive clones into 10 mL LB medium supplemented with antibiotic in 50 mL conical tubes. Concentration of the antibiotic depends on the type of the antibiotic used.
8. Grow bacteria with shaking at 37 °C, for 16–20 h.
9. Harvest cells and extract DNA using a plasmid DNA purification kit (*see* **Note 15**).
10. Measure DNA concentration using a NanoDrop spectrophotometer.
11. Send the 2–3 positive selected clones for DNA sequencing.
12. Transform plasmids from each positive clone to competent BL21(DE3) *E. coli* (*see* **Note 16**) cells and proceed with protein expression experiments as described previously [13, 14].

4 Notes

1. Primers (forward and reverse) for the RF reaction include at the 5′-end a vector-specific sequence, complementary to the site of integration into the recipient vector, and at the 3′-end a sequence complementary to the gene of interest used for amplification of the gene [9, 11] (*see* Table 1 for primer sequences). The total primer length is 50–60 nt. The length of the overlapping sequence to the destination vector can range from 20 to 40 nt with the recommended length of 30 nt. The length of the gene-specific sequence is variable and should be designed to satisfy the melting temperature (Tm) requirements. The Tm for the gene-specific sequence should be between 60 and 70 °C, where A or T, and G or C, each contributes to the Tm 2 and 4 °C, respectively. Synthetic primers were ordered from Integrated DNA Technologies (IDT; Leuven, Belgium) or Sigma-Genosys (Rehovot, Israel). Primers up to 60 nt are ordered with only basic desalting purification. Longer primers are purified either by HPLC or by SDS-PAGE.
2. We found that preparation of competent cells according to the procedure previously described [15] is best suited for the high efficiency of the RF reaction. In this procedure, cells are grown at 18 °C prior to harvesting and preparation of the competent cells. 90–100 μL of competent cells are used for transformation of 1–10 μL of DpnI-treated RF reaction.
3. On a routine basis we use plasmid DNA obtained from various sources as the template for amplification of the gene of interest. However, we have also used other sources for amplification such as chromosomal DNA from various microorganisms and cDNA libraries. When using DNA sources other than plasmid, optimization of the PCR amplification might be essential (*see* **Note 4**).

4. The components and the reaction conditions for the PCR amplification can be modified if needed. Addition of dimethyl sulfoxide (DMSO), 5–10 % *v*/*v*, might be essential for successful reaction if the target DNA has high GC content. Performing a gradient PCR or changing the annealing temperature might be essential, in certain cases, to obtain successful amplification. If large DNA fragments are amplified increasing the elongation time might be essential.
5. Analyzing the PCR amplification reactions on agarose gel is essential step to ensure that the correct size PCR product was obtained as the major band. If no PCR product was obtained, reexamine the primer design. In addition, make sure that source DNA used is correct and if needed alter the PCR conditions (*see* **Note 4**).
6. We highly recommend purifying the PCR product(s) directly from the PCR mix and not from the agarose gel. We have noticed that purifying the PCR product following extraction from the agarose gel resulted in loss of much of the DNA and reduced yield of gene integration into the destination vector in subsequent steps. Removal of the PCR product from an agarose gel is an optional step when multiple bands are obtained in the initial PCR amplification step.
7. In certain cases altering the components for the integration step might be essential. If addition of dimethyl sulfoxide (DMSO) was essential for PCR amplification in the first stage (*see* **Note 4**), it should be used in this stage, as well. On a routine basis, we used 20 ng from the destination vector and 100 ng from each PCR product (mega-primers). However, increasing the amount of the mega-primers (up to 250–300 ng per reaction) might improve, in certain cases, the efficiency of integration. If more than two PCR products (as specified in the protocol) are used for simultaneous integration into the destination vector use the same DNA concentration for the additional mega-primers.
8. If the total size of the newly synthesized vector (DNA insert(s) plus destination vector) is large (>15 kb) increasing the elongation time might be essential. The recommended time for the elongation step for the Phusion DNA polymerase is 15–30 s/kb. In addition, optimization of the integration reaction using gradient PCR might be essential, primarily when multiple mega-primers with various length and complexity are used simultaneously.
9. Analyzing the RF reaction on agarose gel is an optional, but highly recommended, step. The analysis should give a strong indication on how successful the assembly reaction worked. If high molecular band, corresponding to the newly synthe-

sized plasmid, is observed (*see* Fig. 2a), it is highly likely that the reaction was successful. We recommend on proceeding to DpnI treatment followed by transformation even if no upper band is observed. We have observed in many cases that the reaction worked well even if no clear upper band was detected following the agarose gel electrophoresis analysis.

10. The amount of the DpnI-treated RF reactions used for transformation varies between 1 and 10 μL. If we observed an intense high molecular DNA band corresponding to the newly synthesized plasmid following the agarose gel analysis (*see* Fig. 2a) only 1–2 μL are sufficient to obtain ample but distinct colonies for subsequent analysis. However, if no upper DNA band is observed, it is recommended to use the entire 10 μL DpnI-treated reaction for transformation.
11. Before selecting clones for DNA sequencing it is advisable to perform colony PCR to ensure that the selected clones harbor the target genes. The colony PCRs are carried out using forward and reverse primers, derived from sequences flanking the cloning sites in the destination vector. Alternatively, a combination of a specific primer from the insert and a primer from the destination vector is used. It is advisable to perform a negative control PCR where the destination vector without the target gene is used. Before performing the PCR, it is essential to maintain the selected clones by striking on a selective plate or into a liquid media. For multiple DNA fragments integration at distinct positions, two sets of primers are used for amplification of the same colony (Fig. 2). Alternatively, only the two most distal primers (in the case described in Figs. 1 and 2, pACYCDuetup1 and PetRev) are used for identification of clones in which the two DNA fragments are present.
12. The absence of colonies following transformation step is a clear indication for a failure either at the DNA assembly stage or in the transformation procedure. In order to identify the source of failure follow the following guidelines: Remove 1–2 μL from the RF reaction and perform PCR and analysis as described in Subheading 3.2, using flanking primers (T7 and PetRev, in the case of cloning into pET-derived vectors). Analyze PCR on agarose gel.
    - If a PCR product at the expected size is observed, the RF reaction was successful. In this case, the transformation stage has failed. Reexamine the efficiency of the competent cells (*see* **Note 2**) and make sure that the antibiotic used for selection was correct. An additional possibility for failure is that the gene product cloned is toxic to the bacterial cells and its expression, even at very low levels, is detrimental to cell growth. In this case, addition of 1 % glucose to

the selection plates may alleviate the problem. However, re-cloning of your gene into a tighter expression system may be required to obtain transformants following the cloning stage.

- If no PCR product or PCR product corresponding to the empty vector was obtained it is likely that the RF integration reaction has failed. In this case reexamine the primers designed for the RF cloning. If the primer design is correct, it is likely that the reaction conditions should be changed (*see* **Notes 7** and **8**). Repeat the RF reaction with freshly prepared components. Make sure to re-dilute the destination plasmid from a verified stock solution.
- When several DNA fragments are used for simultaneous integration, it is possible that not all new clones will harbor all the different DNAs (Fig. 2b, c). If only partial integration is observed the reaction conditions should be changed. We have observed that changing the annealing temperature in the RF reaction (*see* Subheading 3.1) can have a positive effect on multiple DNA fragments assembly.

13. When multiple colony PCRs are performed in parallel it is recommended not to pipette each of the reaction components separately into each PCR tube. In this case a solution mix including all the reaction components can be assembled for the number of reactions needed. Mix the solution and transfer 20 μL aliquots into the 0.2 mL PCR tubes and proceed with inoculation of the bacterial colonies. If needed, PCR components and or reaction conditions should be optimized to adjust the target gene amplified (*see* **Note 4**).
14. If primers flanking the target gene are used for colony PCR analysis, the PCR products observed from positive clones should run at a higher molecular weight compared to that of the parental plasmid.
15. It is recommended to prepare a glycerol stock from the selected clones (final glycerol concentration 20–25 % *v*/*v*). The stock should be stored at −80 °C until used. Avoid freezing–thawing cycles when using −80 °C glycerol stock.
16. Transformation to *E. coli* BL21(DE3) cells (or its derivatives) is needed for expression of pET-derived or other T7-controlled vectors. However, other strains should be considered, depending on the expression vector used.

Acknowledgments

We thank Prof. J.L. Sussman, Prof. I. Silman, Prof. G. Schreiber, and Prof. Yigal Burstein for their continuous support throughout the study. The ISPC is supported by the Divadol Foundation.

References

1. Benoit RM, Wilhelm RN, Scherer-Becker D et al (2006) An improved method for fast, robust, and seamless integration of DNA fragments into multiple plasmids. Protein Expr Purif 45:66–71
2. Berrow NS, Alderton D, Sainsbury S et al (2007) A versatile ligation-independent cloning method suitable for high-throughput expression screening applications. Nucleic Acids Res 35:e45
3. Aslanidis C, de Jong PJ (1990) Ligation-independent cloning of PCR products (LIC-PCR). Nucleic Acids Res 18:6069–6074
4. Gileadi O, Burgess-Brown NA, Colebrook SM et al (2008) High throughput production of recombinant human proteins for crystallography. Methods Mol Biol 426:221–246
5. Neilan BA, Tillett D (2002) Enzyme-free cloning of PCR products and fusion protein expression. Methods Mol Biol 192:125–132
6. Li MZ, Elledge SJ (2007) Harnessing homologous recombination in vitro to generate recombinant DNA via SLIC. Nat Methods 4:251–256
7. Li MZ, Elledge SJ (2012) SLIC: a method for sequence- and ligation-independent cloning. Methods Mol Biol 852:51–59
8. Gibson DG, Young L, Chuang RY et al (2009) Enzymatic assembly of DNA molecules up to several hundred kilobases. Nat Methods 6:343–345
9. Unger T, Jacobovitch Y, Dantes A et al (2010) Applications of the Restriction Free (RF) cloning procedure for molecular manipulations and protein expression. J Struct Biol 172:34–44
10. Erijman A, Dantes A, Bernheim R (2011) Transfer-PCR (TPCR): a highway for DNA cloning and protein engineering. J Struct Biol 175:171–177
11. van den Ent F, Lowe J (2006) RF cloning: a restriction-free method for inserting target genes into plasmids. J Biochem Biophys Methods 67:67–74
12. Bond SR, Naus CC (2012) RF-Cloning.org: an online tool for the design of restriction-free cloning projects. Nucleic Acids Res 40:W209–W213
13. Peleg Y, Unger T (2008) Application of high-throughput methodologies to the expression of recombinant proteins in *E. coli*. Methods Mol Biol 426:197–208
14. Peleg Y, Unger T (2012) Resolving bottlenecks for recombinant protein expression in *E. coli*. Methods Mol Biol 800:173–186
15. Inoue H, Nojima H, Okayama H (1990) High efficiency transformation of *Escherichia coli* with plasmids. Gene 196:23–28
16. Manicka S, Peleg Y, Unger T et al (2008) Crystal structure of YagE, a putative DHDPS-like protein from *Escherichia coli* K12. Proteins 71:2102–2108

Chapter 7

A Single-Tube Assembly of DNA Using the Transfer-PCR (TPCR) Platform

Ariel Erijman, Julia M. Shifman, and Yoav Peleg

Abstract

DNA cloning is a basic methodology employed for multiple applications in all life-science disciplines. In order to facilitate DNA cloning we developed Transfer-PCR (TPCR), a novel approach that integrates in a single tube, PCR amplification of the target DNA from an origin vector and its subsequent integration into the destination vector. TPCR can be applied for incorporation of DNA fragments into any desired position within a circular plasmid without the need for purification of the intermediate PCR product and without the use of any commercial kit. TPCR reaction is most efficient within a narrow range of primer concentrations. Adaptation of the TPCR should facilitate, simplify, and significantly reduce time and costs for DNA assembly, as well as protein engineering studies. In the current publication we describe a detailed protocol for implementation of the TPCR method for DNA assembly.

Key words DNA cloning, Transfer-PCR (TPCR), Restriction-Free (RF) cloning, Ligation-Independent Cloning (LIC)

1 Introduction

DNA cloning is a basic methodology employed in all life-science disciplines and is essential for many biological and biochemical applications. The traditional Ligation-Dependent Cloning (LDC) approach, which includes digestion of the destination vector and the insert by compatible restriction enzymes and their subsequent assembly by ligation, is still widely used. However, with a rising number of functional and structural studies involving proteins, there is a need for development of efficient and robust techniques, which can be adapted for high-throughput platforms. In recent years, a variety of new alternative methodologies for DNA cloning were reported based on Ligation-Independent Cloning (LIC) principles. These methodologies are gradually replacing the traditional LDC approach [1]. Several of the LIC procedures are based on recombination between the insert and the destination vector such as, for example, the Gateway-system (Invitrogen, Carlsbad, CA) (Chapter 14) that

Svein Valla and Rahmi Lale (eds.), *DNA Cloning and Assembly Methods*, Methods in Molecular Biology, vol. 1116, DOI 10.1007/978-1-62703-764-8_7, © Springer Science+Business Media New York 2014

utilizes site-specific recombination. Alternative recombination-based technologies such as the In-Fusion™ system (Chapter 15) [2] (Clontech, Mountain View, CA) and the Sequence and Ligation-Independent Cloning (SLIC) (Chapter 2) [3] rely on homologous recombination. Other LIC procedures use complementary single-strand overhangs to combine the vector and the insert [4–6]. A different LIC approach for DNA cloning is based on whole plasmid amplification of the insert and the plasmid [7–11] and designated as Restriction-Free (RF) cloning [10, 11]. In RF cloning, the gene of interest is PCR-amplified using two primers, each containing a target-specific sequence and an extension that overlaps the insertion sites in the destination vector. Following purification, the double-stranded PCR product is used as a set of megaprimers for the second reaction. In this step, each of the DNA strands anneals to the destination vector at a predesigned position and is extended in a linear-amplification reaction. The two new DNA strands form a double-stranded nicked plasmid. The parental methylated DNA is then removed by DpnI treatment and the newly synthesized plasmid, containing the DNA insert, is introduced into *Escherichia coli* cells where the nicked DNA is sealed by endogenous enzymatic activity. Recently, we have expanded the applications of the RF cloning for diverse molecular manipulations including simultaneous cloning at distinct positions and multicomponent assembly [10].

In order to further facilitate and simplify DNA cloning we developed the Transfer-PCR (TPCR) platform as an attractive and efficient alternative to the currently available procedures for recombinant DNA cloning [12]. Cloning by TPCR combines, in a single tube, PCR amplification of the gene of interest from a donor vector, generation of an intermediate PCR product, and its subsequent integration into the recipient vector (Fig. 1). As in the previously described RF cloning, DpnI is used to remove traces of the donor vector. TPCR allows precise and seamless integration of the DNA insert into any destination vector, at any position and without any additional unnecessary sequences. We have developed the TPCR platform for both DNA cloning and multiple-site-targeted mutagenesis [12]. In the current protocol we describe a detailed procedure for implementation of the TPCR for DNA assembly.

2 Materials

2.1 Bacterial Strains

For DNA cloning and plasmid preparation procedures, *E. coli* DH5α (Agilent Technology, Stratagene division, Santa Clara, CA) was used. For protein expression *E. coli* BL21(DE3) (Novagen/EMD Millipore Chemicals, Darmstadt, Germany) was employed.

2.2 Vectors

For the experiments described in the methods section the expression vector pET28-TevH [13] was used as a destination vector (*see* **Note 1**).

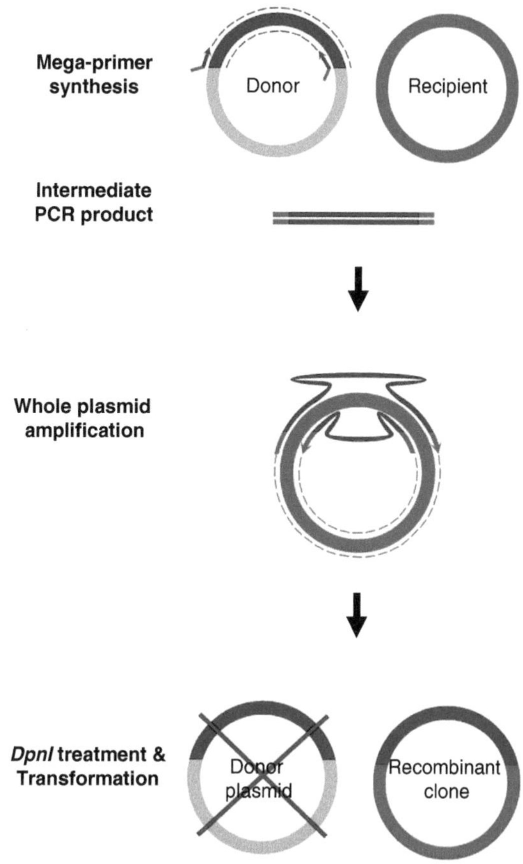

Fig. 1 Schematic presentation of the Transfer-PCR (TPCR) process for DNA cloning. The target gene for cloning is marked in *blue*. The donor and the recipient plasmids are marked in *green* and *brown*, respectively. Primers are indicated by *arrows* that contain a gene-specific sequence at the 3′ end (*solid blue line*) and a vector-specific sequence corresponding to the integration sites on the recipient vector, at the 5′ end (*solid brown line*). *Dashed blue line* indicates formation of an amplification product. X over the donor plasmid indicates elimination of the *E. coli*-derived plasmid by DpnI treatment. Stages of the TPCR process are described on the *left* (reproduced from [12] with permission from Elsevier) (color figure online)

The donor vector for the IFNα8 encoding gene was pPIC9K-IFNα8 [12].

2.3 TPCR Cloning Reaction and Analysis

2.3.1 TPCR Cloning Reaction

1. Phusion high fidelity thermo-stable DNA polymerase, 5× Phusion HF buffer, and dimethyl sulfoxide (DMSO) (New England Biolabs, Ipswich, MA; Catalog # M0530S).
2. dNTP set (Fermentas/Thermo Fisher Scientific, Glen-Burnie, MD; Catalog # R0181). Prepare 10 mM dNTPs stock solution in sterile double-distilled water.

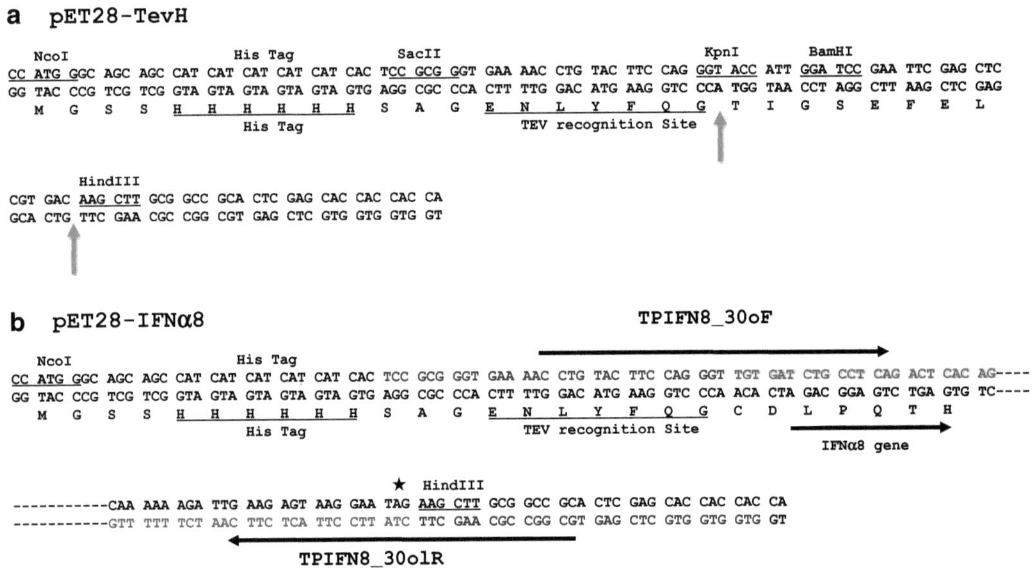

Fig. 2 Cloning of the IFNα8 gene into the recipient vector pET28-TevH by the Transfer-PCR (TPCR) reaction. (**a**) Sequence of the expression cassette in the pET28-TevH. The cassette includes N-terminal 6xHis-tag, TEV protease recognition site (sequences are *underlined* and *marked*), and multiple cloning sites (some of the restriction enzyme sites are indicated and *underlined*). *Arrows* indicate the planned integration sites of the IFNα8 gene into the expression cassette. (**b**) Sequence of the expression cassette following integration of the IFNα8 gene. Only the N- and C-terminal sequences of IFNα8 are shown. Primer names and directions are shown. Forward (TPIFN8_30olF) and reverse (TPIFN8_30olR) primers include recipient vector sequences (marked in *blue color*) and IFNα8-specific sequences (marked in *red color*). *Asterisk* indicates the translation stop codon (TAG) of the IFNα8 gene (color figure online)

3. Synthetic primers (forward and reverse) for TPCR cloning. Both primers are diluted with TE buffer (10 mM Tris–HCl, 1 mM EDTA, pH 8) or sterile double-distilled water, to 1 μM stock solution (*see* Fig. 2, **Note 2**, and Subheading 3.1 for primer design).
4. Plasmid DNA purification kit (QIAprep spin mini prep kit, Qiagen, Hilden, Germany; Catalog # 27106) or equivalent.
5. Donor and recipient plasmids. Both plasmids are diluted with TE buffer (10 mM Tris–HCl, 1 mM EDTA, pH 8) or sterile double-distilled water, to 10 ng/μL stock solution.
6. Spectrophotometer (Nano-Drop/Thermo Fisher Scientific, Wilmington, DE; Catalog # ND-1000) or equivalent.
7. Thermocycler machine (SensoQuest, Goettingen, Germany; Labcycler, Triple Block 3x21).
8. DpnI (New England Biolabs, Ipswich, MA; Catalog # R0176S) or equivalent.

Table 1
Primers used for TPCR cloning of IFNα8 gene into pET28-TevH and subsequent analyses

Gene/task	Recipient plasmid	Primer name (length)	Primer sequence (5′ ⇒ 3′)
IFNα8	pET28-TevH	TPIFN8_30olF (53 bases)	TCCGCGGGTGAAAACCTGT ACTTCCAGGGTTGTGATC TGCCTCAGACTCACAG
		TPIFN8_30olR (57 bases)	GTGGTGGTGCTCGAGTGCG GCCGCAAGCTTCTATTCC TTACTCTTCAATCTTTTTG
Colony PCR	–	T7 (23 bases)	ATTAATACGACTCACTATAGGGG
		PetRev (22 bases)	ATGCTAGTTATTGCTCAGCGGT

IFNα8 sequences are underlined (marked in red in Fig. 2b). The T7 and PetRev primers are used for colony PCR (*see* Subheading 3.3) and sequencing analyses

9. High efficiency *E. coli* DH5α competent cells for heat shock transformation (*see* **Note 3**).
10. Sterile spreaders.
11. Lysogeny Broth (LB) agar plates supplemented with Kanamycin (30 μg/mL).
12. 37 °C incubator.

2.3.2 Analysis

1. Synthetic primers for colony PCR and sequencing analyses (*see* Table 1).
2. Agarose gel electrophoresis apparatus.
3. Agarose I (Amresco, Solon, OH; Catalog # 0710) or equivalent.
4. Ethidium bromide dropper bottle, 0.625 mg/mL stock (Amresco, Solon, OH; Catalog # E406) or equivalent.
5. Tris/acetic-acid/EDTA (TAE) buffer, 50× stock (Bio-Rad Laboratories, Hercules, CA; Catalog # 161–0743) or equivalent.
6. Microwave oven.
7. Gel-Doc 2000 visualization system (Bio-Rad Laboratories, Hercules, CA; Catalog # 170–8126) or similar.
8. 50 mL round conical centrifuge tubes (BD Falcon, Franklin Lakes, NJ; Catalog # 352077) or equivalent.
9. Lysogeny Broth (LB) liquid medium supplemented with Kanamycin (30 μg/mL).

10. 37 °C microbiological incubator (Function line, Haraeus/ Thermo Fisher Scientific; Catalog # 50042308) or similar.
11. 37 °C shaker-incubator (Multitron, Infors HT, Basel, Switzerland) or similar.
12. 0.2 mL PCR tubes.
13. Ready-mix PCR master kit, 2× stock (Amplicon, Skoulunde, Denemark; Catalog # 180301) or equivalent.
14. Temperature-controlled microcentrifuge (Eppendorf, Hamburg, Germany; Catalog # 5417R) or equivalent.
15. Sterile toothpicks.
16. Glycerol (J.T. Baker, Phillipsburg, NJ, USA; Catalog # 2136–01) or equivalent.

3 Methods

3.1 Primer Design

Primer design is a critical step for success of the TPCR reaction. Primers include at the 5′-end a vector-specific sequence, complementary to the site of integration into the recipient vector, and at the 3′-end a sequence complementary to the gene of interest used for amplification of the gene [12]. The total primer length should be 50–60 bases. The length of the overlapping sequence to the destination vector can range from 20 to 40 bases with the recommended length of 30 bases (*see* Fig. 2). The length of the gene-specific sequence is variable and should be designed to satisfy the melting temperature (Tm) requirements. The Tm for the gene-specific sequence should be between 60 and 70 °C, where A or T, and G or C, each contributes to the Tm 2 and 4 °C, respectively.

3.2 Cloning by Transfer-PCR (TPCR)

For the TPCR reaction we use a two-stage amplification process. In the first stage, 13 cycles are performed to amplify the target DNA. In the second stage, additional 20 longer amplification cycles are used for incorporation of the PCR product into the destination vector (*see* **Note 4**). The most critical parameter for efficient assembly of DNA fragments using the TPCR reaction is the final primer concentration which should be in the range of 10–20 nM for both the forward and the reverse primers [12]. This primer concentration is much lower than that being used for a routine PCR amplification, which is in the range of 0.4–0.8 μM for each primer. The example described in the protocol for cloning by TPCR is transferring the gene encoding IFNα8 from the donor plasmid pPIC9K-IFNα8 into the expression vector pET28-TevH (Figs. 2 and 3).

1. Set up TPCR reactions in 0.2 mL tubes, in a final volume of 50 μL, as follows (*see* **Note 5**):

Component	Stock solution	Volume/50 μL reaction	Final concentration
Donor plasmid	10 ng/μL	1 μL	0.2 ng/μL
Recipient plasmid	10 ng/μL	1 μL	0.2 ng/μL
Forward primer	1 μM	1 μL	20 nM
Reverse primer	1 μM	1 μL	20 nM
dNTPs mix	10 mM each	1 μL	200 μM each
Phusion HF buffer	5×	10 μL	1×
Phusion DNA polymerase	2 U/μL	0.8 μL	0.032 U/μL
Deionized, sterile water		34.2 μL	

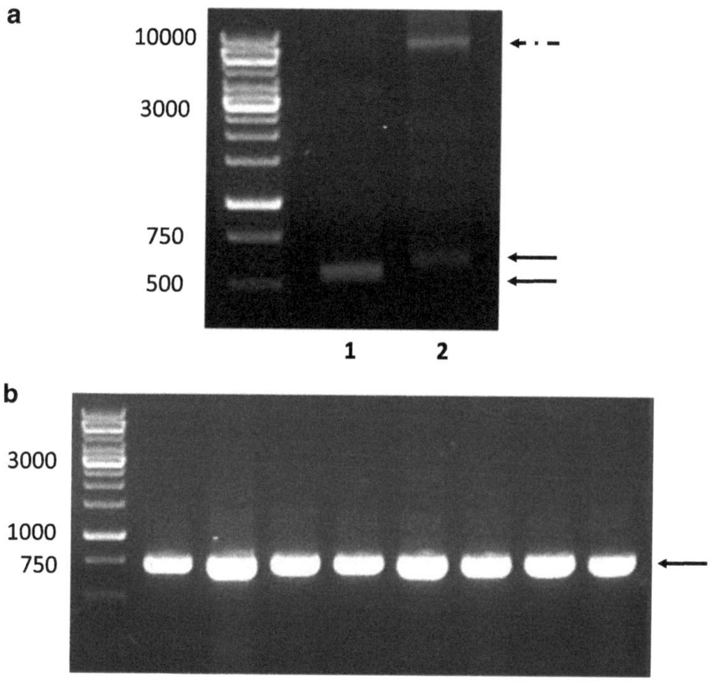

Fig. 3 Analysis of the TPCR reactions. (**a**) Agarose gel analysis of two TPCR reactions. 10 μL of the reaction was analyzed on 1 % agarose gel. Lane 1, TPCR reaction for integration of IFNα8 into pET28-TevH (*see* Subheading 3.2); lane 2, analysis of additional, non-specified, TPCR reaction. *Broken arrow* indicates the position of the newly synthesized plasmid. *Solid arrows* indicate the position of the intermediate PCR products. (**b**) Colony PCR analysis of random colonies, using the T7 and PetRev primers (*see* Subheading 3.3 and Table 1), following the TPCR reaction for generation of the expression vector pET28-IFNα8. A *solid black arrow* indicates the position of the PCR products following the colony PCR analysis. The actual size of the PCR products obtained is in agreement with the expected size, 759 base pairs. Molecular weight marker, in base pairs, is shown on the *left*

2. Mix the 0.2 mL tubes gently and spin briefly in a microcentrifuge.
3. Perform TPCR reaction using the following parameters (*see* **Note 6**):

Step	Amplification stage	Temperature (°C)	Time (min:s)	Cycles
Denaturation		95	1:00	1
Denaturation	I	95	0:30	13
Annealing		60	1:00	
Elongation		72	1:30	
Denaturation	II	95	0:30	20
Annealing		67	1:00	
Elongation		72	4:00	
Elongation		72	7:00	1
Cooling		10	∞	

4. Following completion of the TPCR reaction analyze aliquots (10 μL each) from the TPCR reaction mixture on a 1 % agarose gel for the appearance of the newly synthesized plasmid (*see* Fig. 3 and **Note 7**).
5. Transfer 10 μL from the TPCR reaction into a new 1.5 mL tube. Add 1 μL DpnI, spin briefly, and incubate for 2 h at 37 °C.
6. Spin briefly the DpnI-treated reaction using a microcentrifuge.
7. Transform the DpnI-treated reaction into competent DH5α *E. coli* cells. Add 1–10 μL of reaction into 90–100 μL competent cells (*see* **Notes 3** and **8**). Mix gently. Incubate on ice for 45–60 min. Heat the tube at 42 °C for 1 min and return to ice. Add 1 mL of LB and incubate with shaking for 1 h at 37 °C.
8. Plate transformed mix on LB plates with the appropriate antibiotic. For cloning of IFNα8 into pET28-TevH (Figs. 2 and 3) use LB-Kanamycin (30 μg/mL) plates.
9. Incubate plates overnight in a 37 °C incubator.

3.3 Analysis of Clones

On the following day, select 6–8 single colonies from the transformation plate, and perform colony PCR to identify positive colonies as described below (*see* **Note 9**). If no colonies are observed on the plate consult **Note 10** for details on how to proceed.

1. Set up PCR reactions in 0.2 mL tube, in a final volume of 20 μL, as follows (*see* **Note 11**):

Component	Stock solution	Volume/20 μL reaction	Final concentration
Forward primer	25 μM	1 μL	500 μM
Reverse primer	25 μM	1 μL	500 μM
Master mix solution	2×	10 μL	1×
Deionized, sterile water		8 μL	

2. Using a sterile toothpick select single colonies and transfer the bacteria into the labeled 0.2 mL PCR tubes. Swirl well.
3. Using the same toothpick maintain each clone by inoculating liquid or plate containing the appropriate antibiotic (*see* **Note 9**).
4. Mix gently the 0.2 mL PCR tubes and spin briefly.
5. Perform colony PCR reaction using the following parameters:

Step	Temperature (°C)	Time (min:s)	Cycles
Denaturation	95	1:00	1
Denaturation Annealing Elongation	95 60 72	0:30 1:00 1:30	25
Elongation Cooling	72 10	6:00 ∞	1

6. Analyze PCR reaction products on 1 % agarose gel to verify that the DNA product obtained is of the correct size. Select 2–3 individual positive colonies (*see* **Note 12**).
7. Inoculate positive clones into 10 mL LB medium supplemented with antibiotic in 50 mL conical tubes. Concentration of the antibiotic depends on the type of the antibiotic used (e.g., 30 μg/mL of Kanamycin for cloning of IFNα8 into pET28-TevH).
8. Grow bacteria with shaking at 37 °C, for 16–20 h.
9. Harvest cells and extract DNA using a plasmid DNA purification kit (*see* **Note 13**).
10. Measure DNA concentration using a Nano-Drop spectrophotometer.
11. Send the 2–3 positive selected clones for DNA sequencing (*see* **Note 14**).
12. Transform plasmids from each positive clone to competent BL21(DE3) *E. coli* (*see* **Note 15**) cells and proceed with protein expression as described previously [13, 14].

4 Notes

1. The expression vector pET28-TevH is derived from the vector pET28a (Novagen, EMD Millipore, Billerica, MA). pET28-TevH was constructed by insertion of the *Nco*I–*Bam*HI cassette, described in Fig. 2a, into the corresponding sites in pET28a [13].
2. Synthetic primers are ordered from Integrated DNA Technologies (IDT; Leuven, Belgium) or Sigma-Genosys (Rehovot, Israel). Primers up to 60 bases are ordered with only basic desalting purification. Longer primers are purified either by HPLC or by SDS-PAGE.
3. We found that preparation of competent cells according to the procedure previously described [15] is best suited for the high efficiency of the TPCR reaction. In this procedure, cells are grown at 18 °C prior to harvesting and preparation of the competent cells. 90–100 μL of competent cells are used for transformation of 1–10 μL of DpnI-treated TPCR reaction.
4. On a routine basis we perform TPCR cloning following the procedure described in the method section. We do not optimize primer concentration or other parameters for each individual cloning reaction. Changes are made only if we encounter difficulties in the cloning reaction.
5. The components of the TPCR reaction can be modified if needed. Addition of DMSO, 5–10 % *v*/*v*, might be essential for successful reaction if either the gene or the destination vector has high GC content.
6. If large destination vectors are used (>10 kb) increasing the elongation time in the second amplification stage is recommended. The recommended extension time for Phusion polymerase is 15–30 s per 1 kb plasmid DNA.
7. Analyzing the TPCR reaction by agarose gel electrophoresis is an optional step that we highly recommend. This analysis should give an indication on how well the TPCR reaction worked. If a high molecular band, corresponding to the newly synthesized plasmid, is observed (Fig 3a, lane 2)—it is highly likely that the reaction was successfully completed. If no clear upper band is observed (Fig. 3a, lane 1) there is still a chance that the reaction succeeded as was observed by us in several cases. Hence, we recommend proceeding to the transformation stage even if no clear band is observed on the gel.
8. The amount of the DpnI-treated TPCR reaction used for transformation varies between 1 and 10 μL. If we observed an intense high molecular DNA band corresponding to the newly synthesized plasmid following the agarose gel analysis (*see* Subheading 3.2, **Note** 7, and Fig. 3a, lane 2) only 1–2 μL are

sufficient to obtain ample but distinct colonies for subsequent analysis. However, if no upper DNA band is observed (*see* Subheading 3.2, **Note 7**, and Fig. 3a, lane 1), it is recommended to use the entire 10 μL reaction for transformation.

9. Colony PCR is an optional step, although highly recommended. Before selecting clones for DNA sequencing it is advisable to perform colony PCR to ensure that the selected clones harbor the target gene. The colony PCR reactions are carried out using forward and reverse primers, derived from sequences flanking the cloning sites in the vector (*see* Table 1). Alternatively, a combination of a specific primer from the insert and a primer from the expression vector is used. It is advisable to perform a negative control PCR reaction where the destination vector without the target gene is used instead of the clonal DNA. Before performing the PCR reaction, it is essential to maintain the selected clones by striking on a selective plate or into a liquid media.
10. The absence of colonies following transformation step is a clear indication for a failure either at the cloning stage or in the transformation procedure. In order to identify the source of failure follow the following guidelines: Remove 1–2 μL from the TPCR reaction and perform PCR reaction and analysis as described in Subheading 3.3, using flanking primers (T7 and PetRev, in the case of cloning into pET-derived vectors). Analyze PCR reaction on agarose gel.
    - If a PCR product at the expected size is observed, the TPCR reaction was successful. In this case, the transformation stage has failed. Reexamine the efficiency of the competent cells (*see* **Note 3**) and make sure that the antibiotic used for selection was correct. An additional possibility for failure is that the gene product cloned is toxic to the bacterial cells and its expression, even at very low levels, is detrimental to cell growth. In this case, addition of 1 % glucose to the selection plates may alleviate the problem. However, re-cloning of your gene into a tighter expression system may be required to obtain transformants following the cloning stage.
    - If no PCR product was obtained it is likely the TPCR reaction has failed. In this case reexamine the primers designed for the TPCR cloning (*see* Subheading 3.1, **Note 2**, and Fig. 2). To determine if the first stage of the TPCR reaction (megaprimer synthesis) worked, set up PCR reaction as described in Subheading 3.2 but without the recipient plasmid. Analyze 2–3 μL from the reaction on 1 % agarose gel. If following the agarose gel analysis no band at the expected size of the target gene is observed it is likely that the reaction conditions should be changed (e.g., add DMSO 5–10 % (*v*/*v*)). However, if a PCR product corresponding to the size of the target gene is obtained without

the recipient plasmid, the first stage of the reaction was successful. Hence, repeat the TPCR reaction with freshly prepared components. Make sure to re-dilute the donor and the recipient plasmids from a verified stock solution.

11. When multiple colony PCR reactions are performed in parallel it is recommended not to pipette each of the reaction components separately into each PCR tube. In this case a solution mix including all the reaction components can be assembled for the number of reactions needed. Mix the solution and transfer 20 μL aliquots into the 0.2 mL PCR tubes and proceed with inoculation of the bacterial colonies as described in Subheading 3.3.
12. PCR products originating from positive clones should run at a higher molecular weight compared to that of the parental plasmid (Fig. 3b). The expected size of the PCR products should be calculated in advance.
13. It is recommended to prepare a glycerol stock from the selected clones (final glycerol concentration 20–25 % *v/v*). The stock should be stored at −80 °C until used. Avoid freezing–thawing cycles when using −80 °C glycerol stock.
14. DNA sequencing can be performed using either vector-specific primers, such as T7 and PetRev (*see* Table 1), or gene-specific primers (*see* **Note 2**).
15. Transformation to *E. coli* BL21(DE3) cells is needed for expression of pET-derivatives, such as pET28-IFNα8. However, other strains should be considered, depending on the expression vector used.

Acknowledgments

We thank Prof. J. Sussman, Prof. I. Silman, Prof. G. Schreiber, and Prof. Yigal Burstein for their continuous support throughout the study. The ISPC is supported by the Divadol Foundation. This research was supported by the ISF grant 1372/10 (J.M.S.), Deutsche Forschungsgemeinschaft grant EI 423/2-1 (J.M.S.), and the Abisch Frenkel foundation (J.M.S).

References

1. Graslund S, Nordlund P, Weigelt J et al (2008) Protein production and purification. Nat Methods 5:135–146
2. Benoit RM, Wilhelm RN, Scherer-Becker D et al (2006) An improved method for fast, robust, and seamless integration of DNA fragments into multiple plasmids. Protein Expr Purif 45:66–71
3. Li MZ, Elledge SJ (2007) Harnessing homologous recombination in vitro to generate recombinant DNA via SLIC. Nat Methods 4:251–256

4. Aslanidis C, de Jong PJ (1990) Ligation-independent cloning of PCR products (LIC-PCR). Nucleic Acids Res 18:6069–6074
5. Gileadi O, Burgess-Brown NA, Colebrook SM et al (2008) High throughput production of recombinant human proteins for crystallography. Methods Mol Biol 426:221–246
6. Neilan BA, Tillett D (2002) Enzyme-free cloning of PCR products and fusion protein expression. Methods Mol Biol 192:125–132
7. Chen GJ, Qiu N, Karrer C et al (2000) Restriction site-free insertion of PCR products directionally into vectors. BioTechniques 28:498–500, 504–495
8. Geiser M, Cebe R, Drewello D et al (2001) Integration of PCR fragments at any specific site within cloning vectors without the use of restriction enzymes and DNA ligase. BioTechniques 31:88–90, 92
9. Miyazaki K, Takenouchi M (2002) Creating random mutagenesis libraries using megaprimer PCR of whole plasmid. BioTechniques 33:1033–1034, 1036–1038
10. Unger T, Jacobovitch Y, Dantes A et al (2010) Applications of the Restriction Free (RF) cloning procedure for molecular manipulations and protein expression. J Struct Biol 172:34–44
11. van den Ent F, Lowe J (2006) RF cloning: a restriction-free method for inserting target genes into plasmids. J Biochem Biophys Methods 67:67–74
12. Erijman A, Dantes A, Bernheim R et al (2011) Transfer-PCR (TPCR): a highway for DNA cloning and protein engineering. J Struct Biol 175:171–177
13. Peleg Y, Unger T (2008) Application of high-throughput methodologies to the expression of recombinant proteins in *E. coli*. Methods Mol Biol 426:197–208
14. Peleg Y, Unger T (2012) Resolving bottlenecks for recombinant protein expression in *E. coli*. Methods Mol Biol 800:173–186
15. Inoue H, Nojima H, Okayama H (1990) High efficiency transformation of *Escherichia coli* with plasmids. Gene 96:23–28

Chapter 8

Circular Polymerase Extension Cloning

Jiayuan Quan and Jingdong Tian

Abstract

High-throughput genomics, proteomics, and the emerging field of synthetic biology demand ever more convenient, economical, and efficient technologies to assemble and clone genes, gene libraries, and synthetic pathways. Here, we describe an extremely simple, efficient, and cost-effective cloning method, circular polymerase extension cloning (CPEC), for complex, combinatorial, or multi-fragment assembly as well as routine cloning. This method uses a single polymerase to assemble and clone multiple inserts with any vector in a one-step reaction in vitro. No restriction digestion, ligation, or single-stranded homologous recombination is required.

Key words Molecular cloning, Genetic assembly, Library cloning, DNA polymerase

1 Introduction

Molecular cloning is a foundational technology for molecular biology and biotechnology. Pioneered by the restriction digestion and ligation-based methods [1–3], new cloning technologies have continuously been invented and evolved to suit various requirements and applications. Circular polymerase extension cloning (CPEC) is a simple, efficient, and economical circular DNA assembly and cloning method developed to meet the ever-increasing demand from high-throughput genomics, proteomics, and synthetic biology. Compared with existing cloning strategies, either sequence-dependent or -independent, CPEC offers significant benefits by combining simplicity, efficiency, versatility, and cost-effectiveness in one method [4]. In addition to routine single-gene cloning, CPEC is ideal for a wide variety of other applications, including complex gene library cloning, high-throughput expression cloning, and multi-way assembly of genetic pathways [5].

CPEC is a single-tube, one-step reaction that normally takes 5–10 min to complete for everyday laboratory cloning. The method is directional, sequence-independent, and ligase-free.

Svein Valla and Rahmi Lale (eds.), *DNA Cloning and Assembly Methods*, Methods in Molecular Biology, vol. 1116, DOI 10.1007/978-1-62703-764-8_8, © Springer Science+Business Media New York 2014

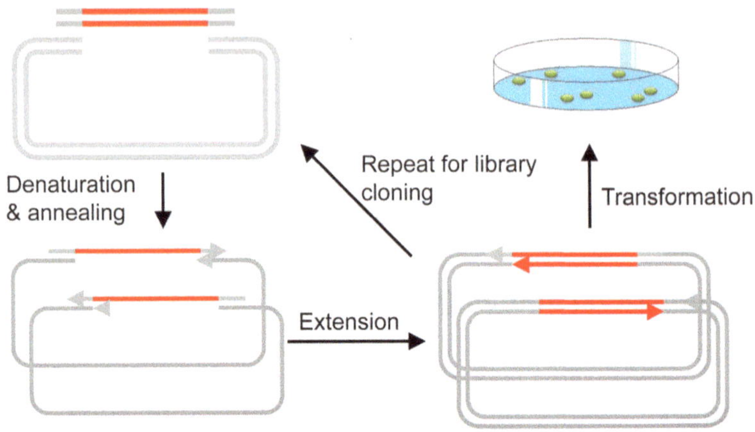

Fig. 1 A schematic diagram of the proposed CPEC mechanism for cloning a single insert. The vector and the insert share overlapping regions at the ends. After denaturation and annealing **(Step 1)**, the hybridized insert and vector extend using each other as a template until they complete a full circle and reach their own 5′-ends **(Step 2)**. The final completely assembled plasmid has two nicks, one on each strand, at the positions marked by an *arrow head*. They can be used for transformation **(Step 3)** with or without further purification. For library cloning, the cycle may be repeated in order to increase the yield of complete plasmids

It uses the polymerase extension mechanism [6] to join overlapping DNA fragments into a double-stranded circular form, such as a plasmid. In a typical CPEC reaction, linear double-stranded insert(s) and vector are first heat-denatured; the resulting single strands then anneal with their overlapping ends and extend using each other as a template to form double-stranded circular plasmids. In CPEC, all overlapping regions between insert(s) and the vector are unique and carefully designed to have very similar and high melting temperatures (*Tm*), which eliminates vector reannealing and concatenation of inserts and makes CPEC very efficient and accurate. The low concentrations of fragments in the reaction favor plasmid circularization and effectively prevent plasmid concatenation. After the CPEC reaction, the perfectly formed double-stranded circular plasmids, with one nick in each strand, can be directly transformed into competent host cells (Fig. 1).

Complex library cloning and multi-way pathway assembly require high cloning efficiency and accuracy. Although other relevant cloning methods only allow the overlapping fragments to anneal or recombine once, CPEC allows multiple annealing–extension cycles that not only increase the chance of hybridization but also permanently join the fragments through polymerase extension, thereby maximizing the cloning efficiency. Whereas the other relevant cloning methods perform the critical annealing/incubation step under ambient temperature, which tends to cause nonspecific hybridization and leads to compromised cloning efficiency and accuracy, CPEC designs the overlapping ends to have very similar *Tm* (±2 °C) and performs the annealing step at high, stringent temperatures (typically in the range of 55–65 °C) to ensure highest accuracy in multi-way assembly and complex library cloning. Unlike PCR, CPEC does not amplify sequences and therefore does not propagate errors with an increased number of thermal cycles.

The combinatorial library cloning strategy using CPEC is illustrated in Fig. 2. In this example, two libraries are cloned simultaneously into a single vector for expression or functional screens to identify the best combinatorial sequences. It is anticipated that such screens will be performed more and more frequently in synthetic biology applications to construct and identify the optimal macromolecular complexes or gene networks. So far, CPEC is the only in vitro method that works well in our hands for combinatorial library cloning [5]. The successful development of this cloning strategy will highly accelerate the process of protein expression screening using large quantity of library gene variants and hence the development of synthetic biology applications as such screens will be performed more and more frequently to construct and identify the optimal macromolecular complexes or gene networks.

2 Materials

2.1 Reagents

1. Linearized cloning vector: Commercial or custom-designed vectors can be used.
2. Cloning insert or insert library: Cloning insert can be prepared by PCR or restriction digestion from a particular plasmid or DNA template. Insert library is often assembled using oligo libraries synthesized with a DNA synthesizer in house or from commercial providers such as IDT.
3. dNTP mix (dATP, dCTP, dGTP, and dTTP) (e.g., Bioline, cat. no. BIO-39043).
4. Oligonucleotide primers: Custom DNA primer synthesis is available from commercial suppliers such as IDT. Prepare stock solutions of primers (e.g., 100 μM) using sterile DNase/ RNase-free water. Prepare aliquots of 10× working solution (e.g., 10 μM) and store at −20 °C to prevent contamination of stock and repeat freeze–thaw cycles.
5. Phusion High-Fidelity DNA polymerase with 5× Phusion HF buffer (Finnzymes, cat. no. F-530).
6. Nuclease-free water (Sigma-Aldrich, cat. no. W4502).
7. Taq DNA polymerase with ThermoPol buffer (New England Biolabs, cat. no. M0267).

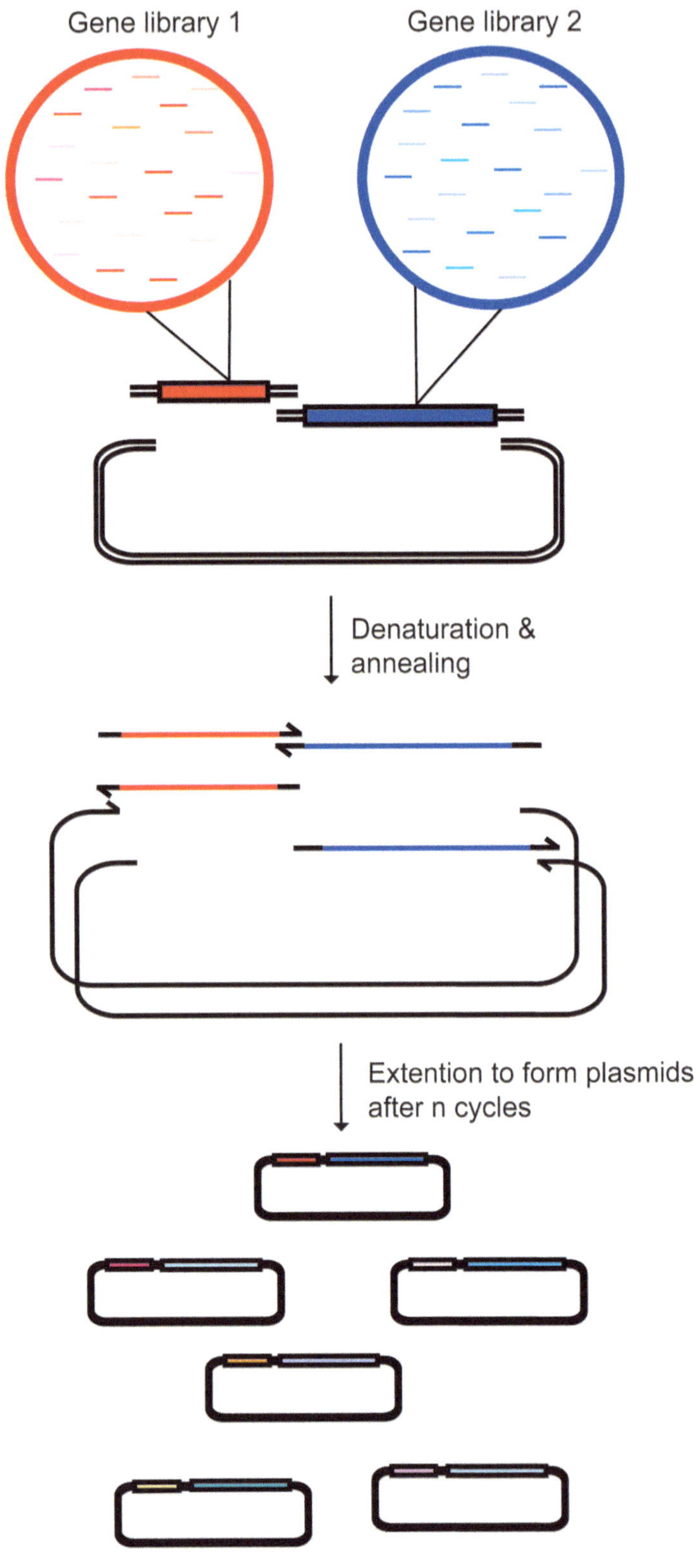

Fig. 2 Schematic diagram of CPEC of combinatorial gene libraries. Two gene libraries are cloned in frame into a vector. The vector and the inserts share overlapping regions at the ends. In each CPEC cycle, after denaturation and annealing, the hybridized inserts and vector extend using each other as a template until they complete a full circle. The assembled plasmid library can be directly used for transformation into competent cells

8. 1× Tris Acetate-EDTA buffer: Mix 900 mL DNase/RNase-free water with 100 mL 10× TAE (Sigma-Aldrich, cat. no. T9650).
9. Ethidium bromide solution (Sigma-Aldrich, cat. no. E1510).
10. 1–2 % (wt/vol) agarose gel with ethidium bromide: Weigh 1.5–3 g agarose (Denville Scientific Inc., cat. no. CA3510-8) and mix with 150 mL 1× Tris Acetate-EDTA buffer in a flask. Heat in a microwave for 1 min and shake the flask. Repeat two more times or until agarose is completely dissolved. Add ethidium bromide to the gel solution to a final concentration of 0.5 μg/mL. Let the solution cool to about 60 °C and then pour it into a casting tray containing a comb and allow it to solidify at room temperature.
11. 6× gel loading dye, blue (New England Biolabs, cat. no. B7021S).
12. 1 kb DNA ladder (New England Biolabs, cat. no. N3232S; Bio-Rad, cat. no. 170–8204) (*see* **Note 1**).
13. 100 bp DNA ladder (New England Biolabs, cat. no. N3231S) (*see* **Note 1**).
14. LB agar (Sigma, cat. no. L3027-1KG).
15. GC5 competent cells (Genesee Scientific, cat. no. 42–653).
16. S.O.C. medium (Cellgro, cat. no. 46-003-CR).
17. Qiaprep Spin MiniprepKit (Qiagen, cat. no. 27106).
18. E.Z.N.A. gel extraction kit (Omega, cat. no. D2501).
19. ExoSAP-IT for PCR clean-up (Affymetrix, cat. no. 78200).

2.2 Equipment

1. Thermal Cycler.
2. Centrifuge.
3. Electrophoresis apparatus.
4. UV transilluminator.
5. FluoChem multi-Imaging system (Alpha Innotech).
6. NanoDrop spectrophotometer (ND-1000, Thermo Scientific).
7. 37 °C shaker.
8. 37 °C cabinet incubator.
9. Water bath.
10. Pipettes.
11. Pipette tips.
12. Microcentrifuge tubes (1.7 mL; Axygen Scientific, cat. no. MCT-175-C).
13. PCR tubes (Denville Scientific, cat. no. 18064).
14. Petri dish (15 × 150 mm; BD, cat. no. 351058).
15. Culture tubes (17 × 100 mm; VWR, cat. no. 60818–703).

16. Cell spreaders.
17. Microplate reader and software (GENios-Basic, Tecan Group).
18. Microwave.

3 Methods

3.1 Preparation of Insert or Insert Library(s)

1. Design PCR primers for the insert or insert library so that they hybridize with the ends of linear vector. The hybridizing parts can either be already incorporated in the insert DNA or be added in this PCR step as overhangs in the primers. The homolog overhangs can range from 0 to 40 bp in order to make the annealing temperature of the hybridizing sequence between vector and insert to be over 55 °C.
2. Set up the PCR on ice as described below. Note that the amount of library DNA that is needed as template is generally larger than that required for standard PCR (*see* **Note 2**).

Initial concentration	Volume per 50 μL reaction	Final amount in 50 μL reaction
Phusion HF buffer (5×)	10 μL	1×
dNTP mix (40 mM)	1 μL	0.8 mM
Phusion High-Fidelity DNA polymerase (2 U/μL)	0.5 μL	1 U
Forward primer (10 μM)	2.5 μL	0.5 μM
Reverse primer (10 μM)	2.5 μL	0.5 μM
DNA template (variable)	Variable	Variable (*see* **Note 2**)
Nuclease-free water	Up to 50 μL	–

3. Run the PCR under the following conditions. For the Phusion enzyme, the annealing temperature should be 3 °C higher than the lower *Tm* of the two primers; *Tm* should be calculated using the part of primer that hybridizes with the insert and using the nearest-neighbor method [7].

Cycle number	Denature	Anneal	Extend
1	98 °C, 30 s		
2–31	98 °C, 10 s	(*Tm*+3)°C, 30 s	72 °C, 15 s/kb
32			72 °C, 5 min

4. Combine 50 μL of the finished PCR with 10 μL of 6× DNA loading dye and run the product on a 1–1.5 % agarose gel. The time required for electrophoresis is dependent on the length of the expected product.

5. Visualize the resolved product with a UV transilluminator or a gel imaging system. A prominent and unique band at the desired position of the length of the insert should be seen.
6. Cut the PCR product band from the gel with a clean sharp razor and purify the insert DNA using a commercial kit (e.g., E.Z.N.A. gel extraction kit). Measure the concentration of the purified insert DNA with the NanoDrop spectrophotometer.

3.2 Preparation of Linear Vector

1. Design PCR primers for the vector so that they hybridize with the ends of the insert DNA. The hybridizing parts can either be already incorporated in the vector or be added in this PCR step as overhangs in the primers. The homolog overhangs can range from 0 to 40 bp in order to make the annealing temperature of the hybridizing sequence between vector and insert to be over 55 °C.
2. Set up the PCR on ice as tabulated below:

Initial concentration	Volume per 50 μL reaction	Final amount in 50 μL reaction
Phusion PCR buffer (5×)	10 μL	1×
dNTP mix (40 mM)	1 μL	0.8 mM
Phusion High-Fidelity DNA polymerase (2 U/μL)	0.5 μL	1 U
Forward primer (10 μM)	2.5 μL	0.5 μM
Reverse primer (10 μM)	2.5 μL	0.5 μM
Vector DNA template (variable)	Variable	1 pg–10 ng
Nuclease-free water	Up to 50 μL	

3. Run the PCR with the following conditions. Again, for the Phusion enzyme, the annealing temperature should be 3 °C higher than the lower *Tm* of the two primers; *Tm* should be calculated using the part of primer that hybridizes with the insert and using the nearest-neighbor method [7].

Cycle number	Denature	Anneal	Extend
1	98 °C, 30 s		
2–31	98 °C, 10 s	(*Tm*+3)°C, 30 s	72 °C, 15 s/kb
32			72 °C, 5 min

4. Combine 50 μL of the finished reaction with 10 μL of 6× DNA loading dye and run the product on a 1–1.5 % percent agarose gel. The time for electrophoresis is dependent on the length of the product.

5. Visualize the resolved product with a UV transilluminator or a gel imaging system. A prominent and unique band at the desired position of the length of the linear vector should be seen.
6. Cut the PCR product band from the gel with a clean sharp razor and purify DNA using a commercial kit (e.g., E.Z.N.A. gel extraction kit). Measure the concentration of the purified vector DNA with the NanoDrop spectrophotometer.

3.3 CPEC

3.3.1 CPEC of a Single Insert

1. Set up the cloning reaction on ice as below. The amount of insert required will be dependent on its size and should be calculated to maintain an insert:vector molar ratio between 1:1 and 2:1.

Initial concentration	Volume per 20 μL reaction	Final amount per 20 μL reaction (*see* Note 3)
Phusion HF buffer (5×)	4 μL	1×
dNTP mix (40 mM)	0.4 μL	0.8 mM
Phusion High-Fidelity DNA polymerase (2 U/μL)	0.2 μL	0.4 U
Vector DNA (variable)	Variable	50–100 ng
Insert DNA (variable)	Variable	–
Nuclease-free water	Up to 20 μL	

2. Run the CPEC reaction using the following conditions (*see* **Note 4**) (Fig. 3):

Cycle number	Denature	Anneal	Extend
1	98 °C, 30 s		
2	98 °C, 10 s	($Tm+3$)°C, 30 s (*see* **Note 5**)	72 °C, 15 s/kb (*see* **Note 6**)
3			72 °C, 5 min

3.3.2 CPEC of a Single Library

1. Set up the cloning reaction on ice as below. The amount of insert required will be dependent on its size and should be calculated to maintain an insert:vector molar ratio between 1:1 and 2:1.

Initial concentration	Volume per 20 μL reaction	Final amount per 20 μL reaction (*see* Note 3)
Phusion HF buffer (5×)	4 μL	1×
dNTP mix (40 mM)	0.4 μL	0.8 mM
Phusion High-Fidelity DNA polymerase (2 U/μL)	0.2 μL	0.4 U
Vector DNA (variable)	Variable	50–100 ng
Insert library DNA (variable)	Variable	–
Nuclease-free water	Up to 20 μL	

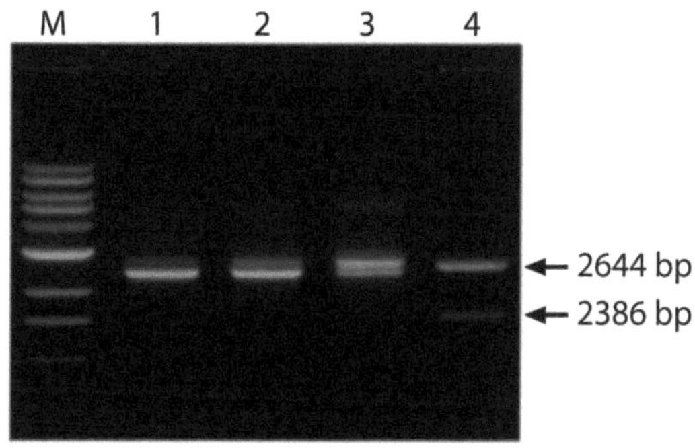

Fig. 3 CPEC of a 258 bp gene. The image shows gel electrophoresis analysis of the CPEC reaction product after 1, 2, and 5 cycles (lanes 1–3) together with a full-length plasmid (lane 4) and 1 kb DNA ladder (lane M, New England Biolabs). The assembled full-length plasmid is 2,644 bp shown as the *upper band* of the two prominent bands close to each other. The *lower band* of the two is the empty vector of 2,386 bp that has not been incorporated within reaction cycles

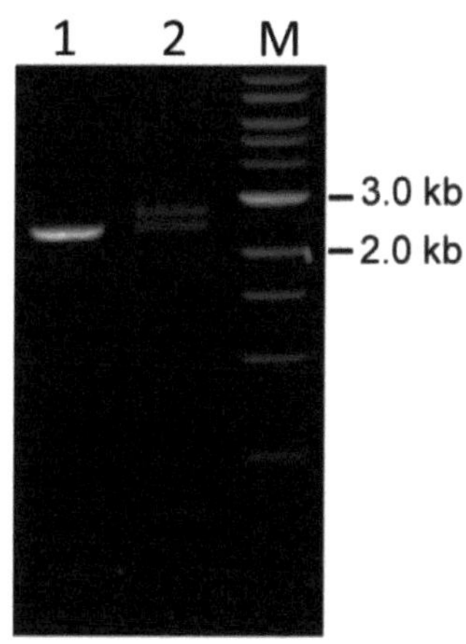

Fig. 4 Agarose gel examination of the CPEC product for a 258 bp gene library after 5 thermal cycles (lane 2) together with linear vector (lane 1) and 1 kb DNA ladder (lane M, New England Biolabs), respectively

2. Run the CPEC reaction using the following conditions (*see* **Note 7**) (Fig. 4):

Cycle number	Denature	Anneal	Extend
1	98 °C, 30 s		
2–6	98 °C, 10 s	(*Tm*+3)°C, 30 s (*see* **Note 5**)	72 °C, 15 s/kb (*see* **Note 6**)
7			72 °C, 5 min

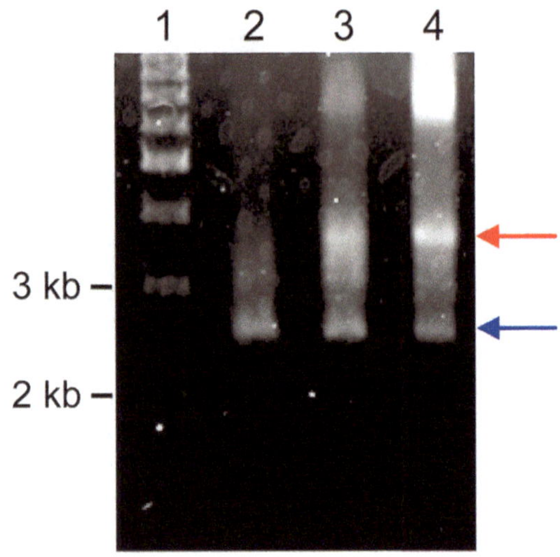

Fig. 5 CPEC of combinatorial gene libraries. Lanes 2, 3, and 4 show the CPEC reaction products after 5, 10, and 20 thermal cycles. The *upper arrow* indicates the full-length cloning product of a 306 bp gene library and a 741 bp gene library as well as the vector (2.5 kb); the *lower arrow* indicates the remaining empty vector. Lane 1 is the 1-kb ladder from Bio-Rad

3.3.3 CPEC of Combinatorial Gene Libraries

1. Set up the cloning reaction on ice as below. The amount of insert required will be dependent on its size and should be calculated to maintain an insert:vector molar ratio between 1:1 and 2:1 for each insert library.

Initial concentration	Volume per 20 μL reaction	Final amount per 20 μL reaction (*see* Note 3)
Phusion HF buffer (5×)	4 μL	1×
dNTP mix (40 mM)	0.4 μL	0.8 mM
Phusion High-Fidelity DNA polymerase (2 U/μL)	0.2 μL	0.4 U
Vector DNA (variable)	Variable	50–100 ng
Insert library DNA (variable)	Variable	–
Nuclease-free water	Up to 20 μL	

2. Run the CPEC reaction using the following conditions (*see* **Note 8**) (Fig. 5):

Cycle number	Denature	Slow ramp anneal	Anneal	Extend
1	98 °C, 30 s			
2–31	98 °C, 10 s	70 to (*Tm*+3) °C (0.1 °C/s)	(*Tm*+3) °C, 2 min (*see* **Note 5**)	72 °C, 15 s/kb (*see* **Note 6**)
32				72 °C, 5 min

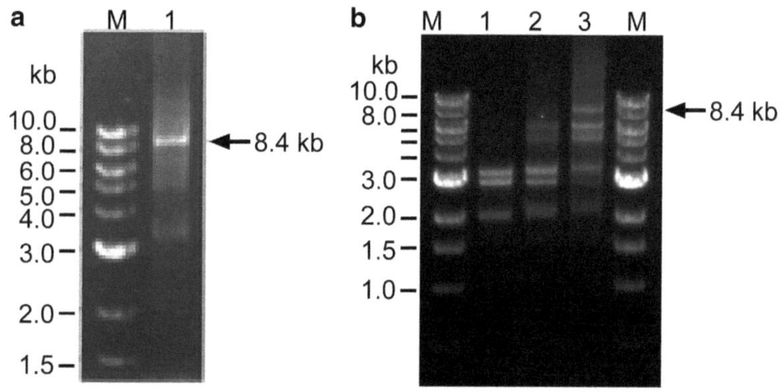

Fig. 6 (**a**) Gel electrophoresis analysis of the final assembly product after a 20-cycle CPEC for four fragments of 3,280, 2,959, 2,040, and 171 bp, respectively. The final full-length plasmid is 8,360 bp as shown in lane 1. (**b**) Gel electrophoresis analysis of the multi-way CPEC reaction. 20 μL is taken out of the reaction after 2, 5, and 10 cycles and separated on a 0.8 % agarose gel (lanes 1, 2, and 3). The starting lengths of the four fragments are 3,280, 2,959, 2,040, and 171 bp, respectively. Discrete bands representing extension products joining neighboring pieces to form longer and longer intermediates are clearly visible. The 171-bp band is not visible from the gel image. Lane M in both figures is 1 kb DNA ladder (New England Biolabs)

3.3.4 CPEC Assembly of Multicomponents

1. Set up the cloning reaction on ice as below. The amount of insert required will be dependent on its size and should be calculated to maintain an insert:vector molar ratio between 1:1 and 2:1 for each insert library.

Initial concentration	Volume per 20 μL reaction	Final amount per 20 μL reaction (*see* Note 3)
Phusion HF buffer (5×)	4 μL	1×
dNTP mix (40 mM)	0.4 μL	0.8 mM
Phusion High-Fidelity DNA polymerase (2 U/μL)	0.2 μL	0.4 U
Vector DNA (variable)	Variable	50–100 ng
Insert library DNA (variable)	Variable	–
Nuclease-free water	Up to 20 μL	

2. Run the CPEC reaction using the following conditions (*see* **Note 9**) (Fig. 6):

Cycle number	Denature	Anneal	Extend
1	98 °C, 30 s		
2–21	98 °C, 10 s	(*Tm*+3) °C, 2 min (*see* **Note 5**)	72 °C, 15 s/kb (*see* **Note 6**)
22			72 °C, 5 min

3. Combine 10 μL of the finished cloning reaction with 2 μL of 6× DNA loading dye and run the product on a 1–1.5 % agarose gel to assess whether the CPEC reaction is successful. The time for electrophoresis is dependent on the length of the expected product. The rest of the cloning product can be stored at −20 °C for up to 1 year.
4. Visualize the resolved products with a UV transilluminator or a gel imaging system (*see* **Note 10**).

3.4 Transformation of the Cloning Product

1. Thaw the required number of tubes containing high-efficiency competent cells of your choice on ice. We recommend following the manufacturer's recommended procedures for transformation; **Steps 2–9** are based on the supplier's protocol for transformation of GC5 competent cells.
2. Add 1–5 μL of the cloning reaction to 50 μL of GC5 competent *Escherichia coli* cells in a microcentrifuge tube and gently tap the tube to ensure an even distribution of DNA in the solution.
3. Incubate the tubes on ice for 30 min.
4. Heat shock the cells for 45 s in a 42 °C water bath and then leave the tubes on ice for 2 min (*see* **Note 7**).
5. Add 450 μL of room temperature S.O.C. medium to each transformation reaction.
6. Shake the tubes in a shaker incubator at 225 rpm at 37 °C for 1 h.
7. Spread 50–100 μL of transformation reaction evenly on each 15-cm LB-agar plate containing appropriate antibiotics (depending on the vector used) and incubate at 37 °C overnight (16–18 h).
8. The next day, pick a desired number (varies according to the application) of isolated colonies from each plate and culture each colony in 3–10 mL of LB broth with antibiotics for 16–18 h (overnight).
9. Take 700 μL of the overnight bacterial culture and mix it well with 300 μL of 50 % (vol/vol) sterile glycerol in a cryovial to prepare 15 % (vol/vol) glycerol stocks and store at −80 °C.

3.5 Colony PCR and/or Sequencing

1. Colony PCR is performed to determine whether the picked colonies have insert(s) of the correct size. Colonies picked directly from the plate or from overnight cultures can be used as the template. Set up a PCR on ice as follows:

Components	Final volume per 30 μL reaction	Final amount per 30 μL reaction
ThermoPol reaction buffer (10×)	3 μL	1×
dNTP mix (40 mM)	0.6 μL	0.8 mM
Taq DNA polymerase (5 U/μL)	0.3 μL	1.5 U

(continued)

(continued)

Components	Final volume per 30 μL reaction	Final amount per 30 μL reaction
Bacteria colony or overnight culture	1 μL if using colony culture	–
Vector forward primer (10 μM)	0.6 μL	0.2 μM
Vector reverse primer (10 μM)	0.6 μL	0.2 μM
Nuclease-free water	Up to 30 μL	

2. Run the PCR using the following conditions. The *Tm* will depend on the vector–primer pair.

Cycle number	Denature	Anneal	Extend
1	95 °C, 5 min		
2–26	95 °C, 30 s	*Tm* °C, 30 s	68 °C, 1 min/kb
27			68 °C, 5 min

3. Combine 10 μL of the finished reaction with 2 μL of 6× DNA loading dye and run the product on a 1–1.5 % (wt/vol) agarose gel. The time for electrophoresis is dependent on the length of the expected product.
4. Visualize the resolved product with a UV transilluminator and determine the presence of inserts and their size. Inserts of correct sizes can be further used for sequencing or restriction digestion if needed.
5. Prepare sequencing samples by first purifying the colony PCR with ExoSAP-IT. Set up a reaction on ice as following the manufacturer's recommended procedures:

Components	Final volume per 7 μL reaction
Colony PCR product	5 μL
ExoSAP-IT	2 μL

6. Incubate the mixture at 37 °C for 15 min to degrade remaining primers and nucleotides.
7. Incubate at 80 °C for 15 min to inactivate ExoSAP-IT. Purified colony PCRs are now ready for direct sequencing, either in house or by commercial services.

4 Notes

1. Both 1 kb and 100 bp DNA ladder are diluted to 250 μg/mL with distilled water and 6× gel loading dye (supplied with the DNA ladder). For a 5 mm wide lane, 2 μL of the mixture

should be loaded onto the agarose gel. The amount of the mixture should be scaled up or down, depending on the width of the agarose gel.

2. For plasmid DNA, the amount in a 50-μL reaction is 1 pg–10 ng. For assembled library DNA, the amount in a 50-μL reaction is 1–50 ng depending on the length of the insert. Usually for library DNA, testing a series of template concentrations is recommended.
3. The 20-μL reaction volume has been tested for cloning of at least two complex gene libraries. For cloning simpler inserts, such as a single fragment or one library, lower volumes can be used. The volume can also be scaled up to 50 μL, especially if the researcher wants to analyze partial cloning product on an agarose gel.
4. For single-gene CPEC, Cycle 2 can be repeated 1–9 more times to increase the amount of cloning product. For gene insert of longer than 1 kb, at least 5 cycles are recommended.
5. *Tm* is calculated based on each pair of the hybridizing parts of the insert and vector and use the lowest *Tm* of all *Tm*s.
6. Extension time is calculated based on the full length of the cloning product. For example, if a 1 kb insert is to be cloned into a 4 kb vector, then extension time should be calculated based on 5 kb and hence 75 s for Phusion DNA polymerase.
7. For single-library CPEC, Cycles 2–6 can be extended to up to 20 cycles to increase the amount of cloning product. For gene insert of longer than 1 kb, at least 10 cycles are recommended.
8. Due to the complexity of the combinatorial library cloning, at least 30 cycles should be used.
9. For multicomponent CPEC, slow ramp annealing is not necessary. However, if the cloning product is less than expected, slow ramp annealing as in combinatorial library CPEC should be attempted.
10. This step is to confirm the success of the cloning process. A prominent band should be present on the agarose gel representing the total length of the vector plus inserts. Sometimes a high-molecular-weight smear and/or additional bands representing excessive or unincorporated vector or inserts may also be visible but should not affect the subsequent transformation step.

Acknowledgement

This work was supported by NIH grant R01HG005862 and the National Basic Research Program of China (2011CBA00800, 2012CB721100) to J.T.

References

1. Smith HO, Wilcox KW (1970) A restriction enzyme from *Hemophilus influenzae*. I. Purification and general properties. J Mol Biol 51:379–391
2. Danna K, Nathans D (1971) Specific cleavage of simian virus 40 DNA by restriction endonuclease of *Hemophilus influenzae*. Proc Natl Acad Sci U S A 68:2913–2917
3. Cohen SN et al (1973) Construction of biologically functional bacterial plasmids *in vitro*. Proc Natl Acad Sci U S A 70: 3240–3244
4. McArthur GH 4th, Fong SS (2010) Toward engineering synthetic microbial metabolism. J Biomed Biotechnol 2010:459760
5. Quan J, Tian J (2009) Circular polymerase extension cloning of complex gene libraries and pathways. PLoS One 4:e6441
6. Horton RM et al (1989) Engineering hybrid genes without the use of restriction enzymes: gene splicing by overlap extension. Gene 77:61–68
7. Breslauer KJ et al (1986) Predicting DNA duplex stability from the base sequence. Proc Natl Acad Sci U S A 83:3746–3750

Chapter 9

Golden Gate Cloning

Carola Engler and Sylvestre Marillonnet

Abstract

DNA assembly methods are essential tools for biological research and biotechnology. Therefore various methods have been developed to clone DNA fragments of interest. Conventional methods usually require several cloning steps to generate a construct of interest. At each step, a single DNA fragment is transferred from a donor plasmid or PCR product to a recipient vector. In the past few years, a number of methods have been developed to facilitate and speed up this process. One of these methods, Golden Gate cloning, allows assembling up to nine fragments at a time in a recipient plasmid. Cloning is performed by pipetting in a single tube all plasmid donors, the recipient vector, a type IIS restriction enzyme and ligase, and incubating the mix in a thermal cycler. Despite the simplicity of the cloning procedure, the majority of clones obtained after transformation contain the expected construct. Using Golden Gate cloning however requires the use of carefully designed donor and recipient plasmids. We provide here a protocol describing how to design these plasmids and also describe the conditions necessary to perform the assembly reaction.

Key words DNA assembly, DNA shuffling, Combinatorial, Hierarchical, Type IIS restriction enzymes, Seamless cloning, Modular cloning, Synthetic biology

1 Introduction

In the past few years several methods have been developed to allow assembly of multiple DNA fragments in a single cloning step [1–7]. Most of these methods are based on recombination between homologous sequences present at the ends of the DNA fragments to assemble. These methods have the advantage of allowing seamless assembly of any sequence of choice irrespective of the presence of restriction enzyme recognition sites. A limitation is however the need for overlapping sequences of at least 15 nucleotides at the ends of the fragments. Assembly of non-overlapping DNA fragments therefore requires adding terminal extensions or using bridging oligonucleotides [7]. This is a limiting factor for combinatorial assembly of multiple DNA fragments of interest since a correspondingly large number of bridging oligonucleotides will be required.

Svein Valla and Rahmi Lale (eds.), *DNA Cloning and Assembly Methods*, Methods in Molecular Biology, vol. 1116, DOI 10.1007/978-1-62703-764-8_9, © Springer Science+Business Media New York 2014

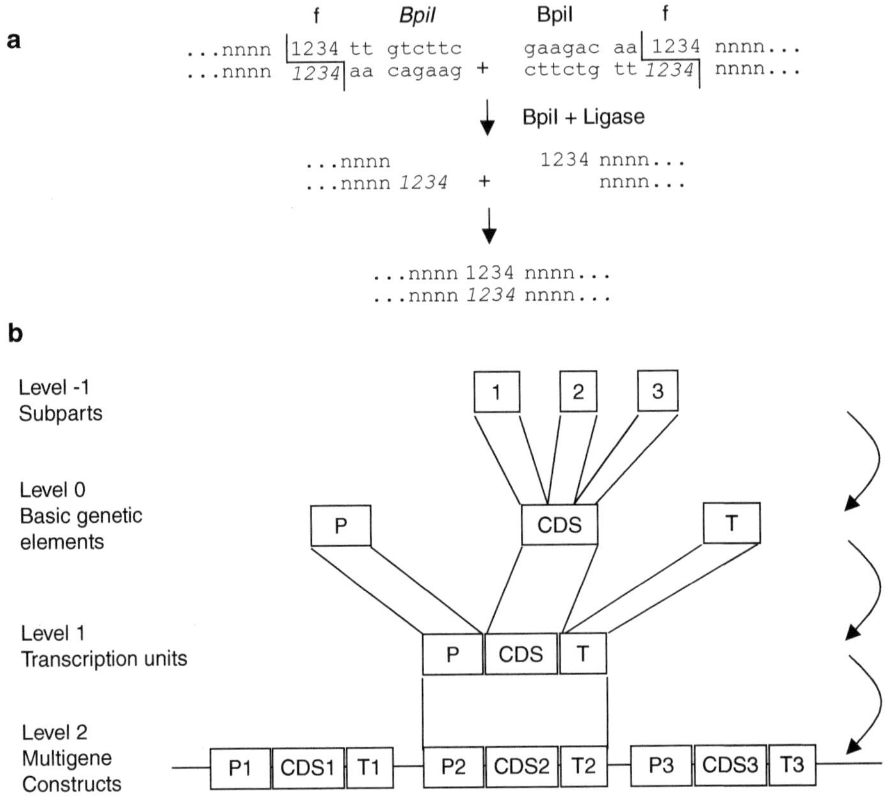

Fig. 1 Golden Gate cloning principle and use for the MoClo system. (**a**) Principle of Golden Gate cloning. Digestion of two DNA fragments containing the same four nucleotide sequence (f, standing for fusion site) flanked by a type IIS restriction site such as BpiI leads to generation of complementary overhangs. Ligation of these two fragments leads to a sequence lacking the original type IIS restriction site. The sequence of the cleavage site can be any four nucleotide sequence of choice. Opposite orientations of the BpiI recognition sites are indicated with non-italics and italics. (**b**) Golden Gate cloning is used for all steps of construct assembly with the MoClo system. Each cloning step is performed using a similar assembly reaction, except that different type IIS enzymes must be used for successive levels of assembly. This is because each cloning step results in a construct that lacks restriction sites for the type IIS enzyme used. Cloning from level −1 to level 0 can be used for gene or promoter shuffling, to make various gene fusions, and to remove internal type IIS restriction sites from native sequences for cloning of level 0 modules. *P* promoter, *CDS* coding sequence, *T* terminator

We have previously developed a method that allows assembly of up to nine DNA fragments in a single cloning step, but that does not require homology at the ends of the fragments to assemble, except for three or four nucleotides of sequence that consist of a restriction enzyme cleavage site [4, 8]. The principle of this method, called Golden Gate cloning, is based on the ability of type IIS enzymes to cleave outside of their recognition sites, allowing two DNA fragments flanked by compatible restriction sites to be digested and ligated seamlessly [1, 9] (Fig. 1a). Since the ligated product of interest no longer contains the original type IIS recognition site, it will not be subject to redigestion in a restriction–ligation reaction. However, all other products that reconstitute the original site will be redigested, allowing their components to be made available for further ligation. This leads to increasing formation of the desired product with increasing time of incubation, and results in a high cloning efficiency. Since the sequence of the overhangs at the ends of the digested fragments can be chosen to be any four nucleotide sequence of choice, multiple compatible DNA fragments can be assembled in a defined linear order in a single restriction–ligation step.

Golden Gate cloning can be used for gene shuffling and for assembly of any construct of interest. To facilitate construct engineering, a modular cloning system (MoClo) that uses Golden Gate cloning for all assembly steps was developed [10]. The base of the MoClo system consists of libraries of standard level 0 modules (for details, please refer to the Fig. 1 legend) that each contains a defined genetic element such as a promoter, a 5′ untranslated region, and a coding sequence or a terminator (Fig. 1b). Transcription units are assembled from level 0 modules using a first assembly reaction and multigene constructs assembled from transcription units using a second cloning step. The process can be repeated to generate constructs containing larger numbers of transcription units. To be suitable for Golden Gate cloning for all assembly steps, level 0 modules internal sequences should not contain restriction sites for any of the type IIS enzymes used for MoClo. Level 0 modules lacking type IIS restriction sites can be cloned from native sequences using Golden Gate cloning.

We provide here a protocol for Golden Gate cloning, using as example the construction of MoClo level 0 modules. The protocol requires amplifying several fragments by PCR using primers designed to mutate internal restriction sites, cloning the amplified fragments and sequencing them, and finally assembling the cloned fragments in a compatible destination vector. The protocol provided here can also be used for gene shuffling or for combinatorial assembly of various sequences of interest [11].

2 Materials

2.1 PCR

1. Novagen KOD Hot Start DNA polymerase (Merck KGaA, Darmstadt), supplied with 10× buffer, 25 mM $MgSO_4$ and 2 mM dNTPs, or any other high fidelity DNA polymerase.
2. Custom-made primers can be ordered from many commercial vendors.
3. NucleoSpin® Extract II kit (Macherey Nagel, Düren), for purification of PCR products.

2.2 Cloning

1. Restriction endonuclease SmaI (10 U/μL) (NEB, New England Biolabs, Inc., Ipswich, MA, USA), supplied with 10× NEBuffer 4 (200 mM Tris-Acetate pH 7.5, 100 mM magnesium acetate, 500 mM potassium acetate, 10 mM dithiothreitol).

2. Restriction endonuclease BsaI (10 U/μL) (NEB), supplied with 10× NEBuffer 4.
3. Restriction endonuclease BpiI (10 U/μL) (Fermentas GmbH, St. Leon-Rot), supplied with 10× Buffer G (100 mM Tris–HCl pH 7.5, 100 mM $MgCl_2$, 500 mM NaCl, 1 mg/mL BSA).
4. T4 DNA Ligase 3 U/μL or T4 DNA Ligase (HC) 20 U/μL (Promega, Mannheim), both supplied with 10× ligation buffer (300 mM Tris–HCl pH 7.8, 100 mM $MgCl_2$, 100 mM DTT, 10 mM ATP).
5. For measuring of DNA concentration, we use the NanoDrop ND2000 (Peqlab, Erlangen).
6. Lysogeny Broth (LB) Medium: 1 % bacto-tryptone, 0.5 % yeast extract, 1 % NaCl in deionized water, adjusted to pH 7.0 with 5 N NaOH. For plates, 1.5 % agar is added.
7. Antibiotics carbenicillin (used instead of ampicillin) and kanamycin: filter-sterilized stocks of 50 mg/mL in H_2O (stored in aliquots at −20 °C) are diluted 1:1,000 (final concentration: 50 μg/mL) in an appropriate amount of medium after the medium has been autoclaved and cooled down. For spectinomycin, a stock of 40 mg/mL is made and is used at a final concentration of 100 μg/mL (dilution 1:400).
8. 5-Bromo-4-chloro-3-indolyl-β-D-galactopyranoside (X-gal): stock solution of 20 mg/mL in dimethylformamide (DMF). For preparation of plates, the stock is diluted 1:500 (final concentration: 40 μg/mL) in an appropriate amount of LB agar after autoclaving/melting and cooling down.

2.3 Screening of Colonies

1. NucleoSpin® Plasmid Quick Pure (Macherey Nagel, Düren) for preparation of miniprep DNA.
2. Restriction endonucleases (NEB or Fermentas), all supplied with 10× buffer and if necessary also with 100× BSA (dilute 1:10 and store in aliquots at −20 °C).
3. DNA ladder: GeneRuler™ 1 kb DNA Ladder Plus (Fermentas) is used as marker for gel electrophoresis.
4. 50× TAE buffer: 242.0 g Tris, 57.1 mL acetic acid, and 100 mL 0.5 M EDTA, pH 8.0, in 1 L of deionized water.
5. For preparation of gels for electrophoresis, agarose (0.7–1.5 %) in 1× TAE is melted in a microwave oven and one drop of a 0.025 % ethidium bromide solution (Carl Roth GmbH, Karlsruhe) is added per 100 mL of melted agarose.
6. Running buffer for agarose gels is 1× TAE.
7. Gels are checked visually using a Syngene GelVue transilluminator (VWR, Darmstadt), and pictures are taken by using a Quantity One® gel analysis software (Biorad).
8. DNA maps of plasmids are made by using the Vector NTI software (Invitrogen).

2.4 Sequencing

1. DNA constructs that need to be sequenced are sent to an external contractor. Sequence data are analyzed using the DNASTAR's Lasergene software.
2. Primers M13RP (CAGGAAACAGCTATGACC) and/or M13FP (TGTAAAACGACGGCCAGT) are used for sequencing of inserts cloned in pUC19-derived vectors.

3 Methods

Level 0 modules of the MoClo system must conform to a standard structure. All modules are flanked by two BsaI restriction sites in inverted orientation (Figs. 2 and 3) and should not contain restriction sites for the enzymes BsaI, BpiI, and Esp3I in internal

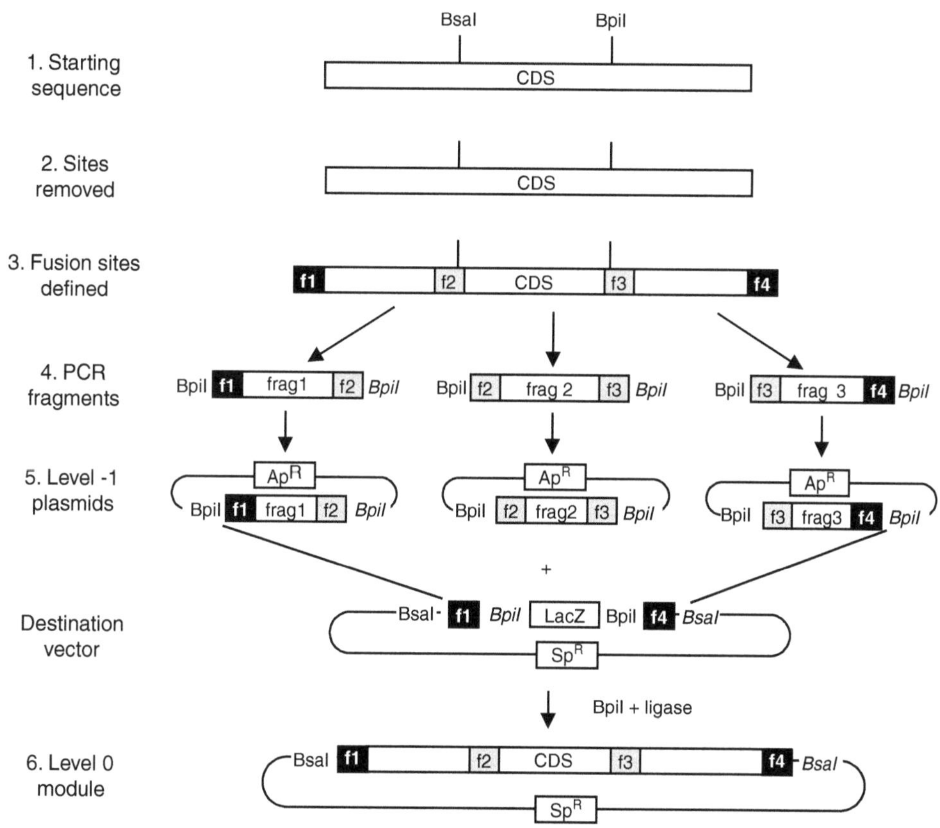

Fig. 2 Overview of intermediate steps and sequences for construction of level 0 modules. An example of a starting native sequence containing two type IIS restriction sites (for BpiI and BsaI) is depicted in [1]. The two restriction sites are mutated by introducing two single nucleotide substitutions. Fusion sites (shown in *grey*, f2 and f3) are then defined overlapping or near the mutated sites. Two fusion sites of standard sequence (shown in *black*, f1 and f4) are added at the beginning and the end of the sequence. These two sites must be compatible with two fusion sites in the destination vector. **Steps 2** and **3** are performed in silico to design the sequence that needs to be assembled. All following steps describe the physical construction of the module. Ap^R, Sp^R: ampicillin and spectinomycin resistance markers. Opposite orientations of the BpiI or BsaI recognition sites are indicated with *non-italics* and *italics*

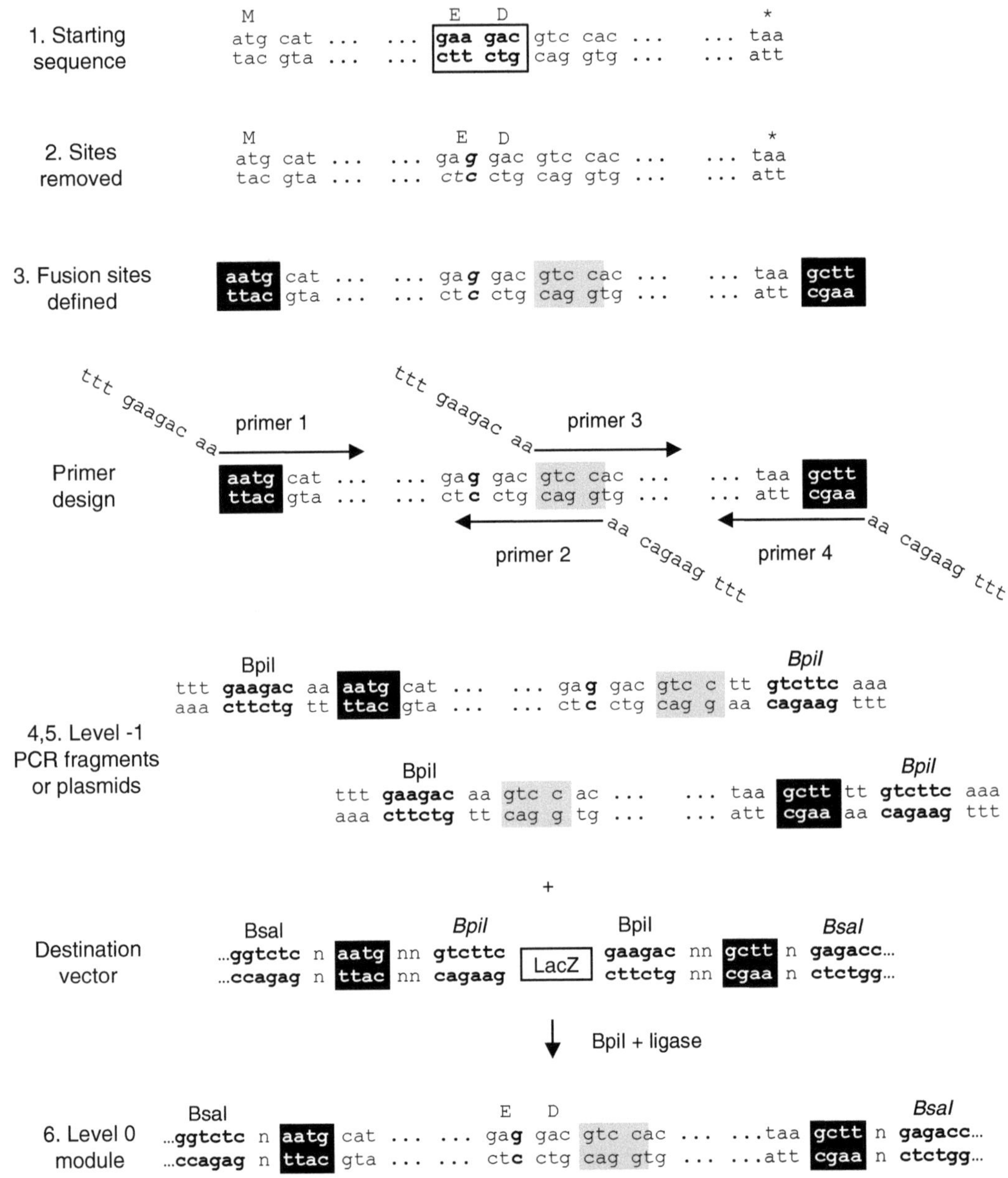

Fig. 3 Depiction of sequence details of the DNA fragments and vectors shown in Fig. 2. For simplicity, removal of a single BpiI site is shown. In **step 3**, the fusion site gtcc does not overlap with the introduced mutation (nucleotide in *italics*), but could be selected on this sequence as well

sequences. Generation of such modules using Golden Gate cloning requires the following steps.

3.1 Domestication of the Target Sequence

The first step consists of modifying the sequence of interest to remove any unwanted restriction site (a process called domestication [12]). This step is performed in silico using a program such as Vector NTI, or any other program of choice. A single nucleotide substitution is sufficient to remove any type IIS restriction site (an example is shown in **step 2** in Fig. 3). For removal of sites within coding sequences, it is always possible to make a silent mutation that will not change the protein sequence. However, for noncoding sequences such as promoters where regulatory sequence motifs may not necessarily be known, it may be necessary to later check that the introduced mutation has not affected the genetic function encoded by the sequence of interest.

3.2 Selection of Fusion Sites and Primer Design

Fusion sites are selected at each point where the sequence needs to be split in several sub-fragments. This includes locations where restriction sites need to be removed, and any position where several variant sequences may need to be recombined. Finally, fusion sites may also be defined at several locations to split a large sequence into several smaller ones. This may be useful for cloning of large level 0 modules, since it is sometimes more efficient to sequence and screen several small fragments at −1 level than sequencing a larger fragment at level 0.

Restriction sites are eliminated by amplification with primers containing a mismatch in the sequence that binds the target sequence (Figs. 2 and 3). $n+1$ fragments are usually amplified by PCR to remove n restriction sites (*see* **Note 1**). To be able to reconstitute the entire sequence from the amplified fragments, all primers have an extension that contains a BpiI restriction site. The sequence at the cleavage site must correspond exactly to a sequence from the target sequence (the fusion site) to avoid introducing unwanted mutations to the sequence of interest. Cleavage of the amplified fragments using BpiI will release a fragment containing only target sequence, flanked on each side by four nucleotide DNA overhangs derived from the fusion site. The fusion site may be designed to overlap the mutated site but may also be selected near the mutated site (as shown in Fig. 3).

Two fusion sites are also needed at the beginning and the end of the sequence of interest. These two fusion sites must be compatible with two fusion sites in the destination vector. These sites have a standard sequence (AATG and GCTT for MoClo level 0 modules for coding sequences, Fig. 3), and are therefore not necessarily found in native sequences. For example, AATG overlaps with the ATG start codon but an A that is not necessarily native must be added upstream of the ATG; GCTT is usually not part of the sequence of interest and is added after the stop codon (fusion sites of standard sequence are shown in black in Figs. 2 and 3).

Fusion sites must be carefully selected to all have a different sequence to avoid assembly of the amplified fragments in the wrong order. It is important to check that all fusion sites also do not match the complementary sequence of the other fusion sites, since this would sometimes lead to ligation of two inappropriate fragments, one in the inverse orientation. For example, choice of the sequence

ATTC will preclude the choice of the sequence GAAT for any of the other sites. Another requirement is to avoid the 16 palindromic sequences (for four nucleotide fusion sites), since palindromic DNA ends are compatible with themselves in the other orientation. For enzymes that cleave on a four nucleotide sequence, 240 possible sequences are therefore available.

3.3 PCR Amplification of the Modules

1. A PCR mix is set up following the manufacturer's instructions. For example, using KOD polymerase (*see* **Note 2**), the following conditions are used: 1 μL plasmid DNA (5–20 ng/μL), 5 μL of 10× buffer, 3 μL of 25 mM $MgSO_4$, 5 μL of 2 mM dNTPs, 1.5 μL each of 10 μM sense and antisense primers, and 1 μL of KOD Hot Start DNA polymerase (10 U/μL, final concentration 0.02 U/μL) in a total reaction volume of 50 μL.
2. PCR is performed using the following cycling conditions: (1) incubation at 95 °C for 2 min for polymerase activation, (2) denaturation at 95 °C for 20 s, (3) annealing at 58 °C for 10 s the temperature for the annealing step can be adjusted for specific primers, but the temperature of 58 °C usually works well for most primers, (4) extension at 70 °C, the duration depends on the length of the expected fragment (from 10 s/kb for fragments smaller than 500 bp up to 25 s/kb for fragments larger than 3 kb, see manufacturer's instructions); **steps 2–4** are repeated 34 times and are followed by a final extension step at 70 °C for 20 s–2 min (depending on fragment length). The reaction is then kept at 12 °C until taken out of the thermal cycler.
3. Of the PCR product obtained, 2 μL is analyzed by gel electrophoresis to make sure that a product of the correct size has been amplified.
4. The amplified fragment is purified from remaining primers, potential primer dimers, and remaining polymerase enzyme by using the NucleoSpin® Extract II kit following the kit protocol. DNA is eluted from the column with 30–50 μL of elution buffer (5 mM Tris–HCl, pH 8.5). In case several bands were amplified rather than only the expected fragment, the same kit can also be used to cut and extract the appropriate DNA fragment from an agarose gel.

3.4 Blunt-End Cloning of the Modules

Cloning of the modules before assembly is optional since PCR fragments can be assembled directly in a level 0 cloning vector (*see* **Note 3**). However, as mentioned above, cloning of level −1 fragments may be preferable to facilitate cloning of large level 0 modules. Cloning fragments before assembly may also be useful if one wants to assemble several sequence fragment variants combinatorially.

Cloning of the level –1 fragments can be performed using commercial kits such as the pGEM-T (Promega), pJET (Fermentas), and the TOPO® TA (Invitrogen) kits. PCR products can also be cloned efficiently using blunt-end cloning with a protocol that uses restriction–ligation [13, 14]. The vector chosen for cloning of level –1 fragments should fulfill two requirements: (1) it should preferably not contain any restriction site for the type IIS enzyme that will be used for the following assembly step, i.e., BpiI in the present example (*see* **Note 4**), and (2) the vector backbone should have a different selection marker from the destination vector used for the next step of assembly (level 0 modules have a spectinomycin resistance marker; therefore, level –1 cloning vectors can have an ampicillin resistance marker). Since many cloning vectors have a BsaI restriction site in the ampicillin resistance gene (for example, pGEM-T or pJET or pUC19), we have made a modified pUC19 vector that lacks this site (*see* **Note 5**).

1. Add 0.5 μL of vector (50 ng), 1 μL of PCR product (50–100 ng), 2 μL of 10× ligation buffer, 1 μL of SmaI enzyme (10 U), 1 μL of T4 DNA ligase (3 U), and 14.5 μL of water (total volume of 20 μL) into a tube. The reaction mix is incubated for 1–2 h at room temperature or in a 25 °C incubator (*see* **Note 6**).
2. The entire ligation mix is transformed into DH10B chemically competent cells and plated on LB plates containing X-gal and the appropriate antibiotic.
3. White colonies (or sometimes pale blue when small inserts are cloned) are picked and inoculated in 5 mL of LB medium containing the appropriate antibiotic.
4. Plasmid DNA is extracted using the NucleoSpin® Plasmid Quick Pure kit following the manufacturer's instructions.
5. Plasmid DNA can be checked by restriction enzyme digestion using BpiI, followed by analysis of the digested DNA by agarose gel electrophoresis.
6. DNA from two minipreps is sent for sequencing using primers M13RP and/or M13FP.
7. When a correct sequence has been identified, DNA concentration of the plasmid prep is measured using the NanoDrop ND2000.

3.5 Construction of the Destination Vector

A destination vector compatible with the entry modules needs to be made. The vector should contain two BpiI sites compatible with the two fusion sites present at the beginning of the first level –1 fragment and the end of the last fragment (*see* Figs. 2 and 3). The vector backbone should not contain any additional BpiI restriction site and should have an antibiotic resistance gene different from the one used for cloning of level −1 modules. Additionally, the vector may contain a lacZα fragment to allow blue-white selection of the resulting clones. For MoClo vectors, two restriction sites for a second type IIS enzyme (BsaI) are placed flanking the two fusion sites to allow the level 0 modules to be further subcloned (for assembly of transcription units, for example). If the assembled modules do not need to be subcloned, restriction sites for this second enzyme may be omitted.

3.6 Golden Gate Assembly

Once entry constructs and the recipient vector are made and sequenced, assembling the fragments only requires pipetting all components into a reaction mix and incubating the mix in a thermal cycler.

1. A restriction–ligation is set up by pipetting into a tube 40 fmol (approximately 100 ng, *see* **Note 7**) of each level −1 module (or PCR fragment) and of the vector, 2 μL 10× ligation buffer, 10 U (1 μL) of BpiI, and either 3 U (1 μL) of ligase for assembly of 2–4 modules or 20 U (1 μL) HC ligase for assembly of more than 4 modules (final volume of 20 μL).
2. The restriction–ligation mix is incubated in a thermal cycler. For assembly of 2–4 level −1 modules, incubation for 60–120 min at 37 °C is sufficient. If more modules are ligated together, the incubation time is increased to 6 h, or cycling is used as following: 2 min at 37 °C followed by 3 min at 16 °C, both repeated 30–50 times (*see* **Note 8**).
3. Restriction–ligation is followed by a digestion step (5 min at 37 °C for BpiI or 50 °C if BsaI is used for cloning) and then by heat inactivation for 5 min at 80 °C. The final incubation step at 80 °C is very important and is needed to inactivate the ligase at the end of the restriction–ligation. Omitting this step would lead to religation of some of the insert and plasmid backbone fragments when the reaction vessel is taken out of the thermal cycler and would lead to a higher proportion of colonies containing incorrect constructs.

3.7 Transformation of the Constructs into Competent Cells

The entire ligation is transformed into chemically competent DH10B cells (*see* **Note 9**).

1. Thaw frozen chemically competent cells (100 μL per tube) on ice.
2. Add the entire ligation to the cells, and incubate on ice for 30 min.
3. Incubate 90 s at 42 °C in a water bath.
4. Let the cells recover on ice for 5 min.
5. Add 1 mL of LB medium to the cells, and incubate the tube at 37 °C in a shaker-incubator (150 rpm) for 45 min to 1 h.
6. After incubation, plate 25–100 μL of the transformation on LB agar plates containing antibiotic and X-gal.

7. Incubate the plates overnight at 37 °C. Many white and very few blue colonies should be obtained.
8. Pick a few white colonies from the plate (*see* **Note 10**).
9. Check by restriction digest using BsaI (for MoClo level 0 modules) or any other suitable enzyme.

Level 0 modules that have been made from sequenced level −1 constructs do not need to be sequenced again. Modules that have been directly assembled from PCR products of course need to be sequenced. Level 0 modules are then ready for further assembly using BsaI and ligase and MoClo vectors (described in ref. 10).

4 Notes

1. Two restriction sites located next to each other can be mutated using a single long primer. Also, the fusion site may be selected between the two sites than need to be mutated, and the two mutations introduced by a single mismatch in each of the two primers required at this fusion site. Therefore, the number of PCR fragments that needs to be amplified to remove *n* restriction sites may sometimes be lower than $n+1$.
2. KOD polymerase is a useful enzyme as it has proofreading activity and does not add any nucleotide at the end of the amplified fragments (unlike Taq polymerase). Fragments amplified with KOD polymerase therefore have blunt ends, which is a prerequisite for blunt-end cloning using SmaI.
3. Direct assembly of PCR products in a destination vector is possible [15]. It is however recommended to purify the PCR products using a column to remove DNA polymerase and primer dimers. Indeed some of the primer dimers are flanked by two fusion sites (these are part of the primers) and can therefore be cloned, resulting in incorrect constructs. The final constructs may also contain PCR-induced mutations and therefore need to be sequenced.
4. The presence of a BpiI site in the vector backbone of level −1 modules would not prevent assembling them using Golden Gate cloning, as the final construct will not contain this vector backbone. However, the presence of a BpiI site in all level −1 vector backbones would lead to continuous digestion and religation at this site, which would unnecessarily consume some ATP from the ligation mix.
5. A simple strategy, enzymatic inverse PCR [16], can be used to eliminate the internal BsaI site in pUC19. The entire plasmid is amplified with two primers designed to introduce a mutation in the BsaI site: primers bsarem1 (ttt ggtctc a ggtt ctcgcggtatcattgcagc) and bsarem2 (ttt ggtctc a aacc acgctcaccggctccag).

The primers are themselves flanked by two BsaI restriction sites that form two compatible overhangs after BsaI enzyme digestion. After amplification of the entire plasmid with both primers, the PCR is purified with a column to remove remaining polymerase and nucleotides. The amplified fragment is subjected to restriction–ligation with BsaI and ligase and is then transformed in *E. coli*.

6. SmaI cannot be used for cloning of fragments containing an SmaI restriction site. However, another restriction enzyme that produces blunt ends, such as EcoRV, could be used as well (a cloning vector containing a unique EcoRV site in the polylinker is however required). Alternatively, cloning of such products using a commercial kit might be simpler.
7. In practice, if all module plasmids and the vector have approximately the same size (4–5 kb), simply adding 100 ng of DNA of each module set and of the vector will work relatively well. However, when plasmids or PCR fragments with widely different sizes are used, calculating an equimolar amount should provide a higher cloning efficiency. The following relation (from the NEB catalog) can be used: 1 μg of a 1,000 bp DNA fragment corresponds to 1.52 pmol. Therefore, the volume of DNA to pipet (in μL) to have 40 fmol is given by the equation: 40 (fmol) × size (bp) of the DNA fragment/(concentration (ng/μL) × 1,520).
8. For cloning of level 0 modules, which usually does not require assembling too many fragments, and where only a few positive clones are needed (for screening by sequencing), any program should work efficiently, including with continuous incubation at 37 °C. Optimization of the assembly conditions is more important for construction of libraries by DNA shuffling where cloning efficiency needs to be higher.
9. Any other *E. coli* strain can also be used. If higher transformation efficiency is required, for DNA shuffling for example, the restriction–ligation mix can be transformed in electrocompetent *E. coli* cells. In this case, DNA from the restriction–ligation mix should first be ethanol-precipitated and resuspended in 10 μL of water.
10. Two colonies should be sufficient for constructs that are assembled from sequenced level −1 modules, since most colonies should contain the correct fragment, and clones with a correct restriction pattern do not need to be sequenced.

Acknowledgment

The authors would like to thank Dr. Stefan Werner for critical reading of this manuscript.

References

1. Lebedenko EN, Birikh KR, Plutalov OV et al (1991) Method of artificial DNA splicing by directed ligation (SDL). Nucleic Acids Res 19: 6757–6761
2. Li MZ, Elledge SJ (2007) Harnessing homologous recombination in vitro to generate recombinant DNA via SLIC. Nat Methods 4:251–256
3. Gibson DG, Benders GA, Axelrod KC et al (2008) One-step assembly in yeast of 25 overlapping DNA fragments to form a complete synthetic Mycoplasma genitalium genome. Proc Natl Acad Sci U S A 105:20404–20409
4. Engler C, Kandzia R, Marillonnet S (2008) A one pot, one step, precision cloning method with high throughput capability. PLoS One 3:e3647
5. Shao Z, Zhao H, Zhao H (2009) DNA assembler, an in vivo genetic method for rapid construction of biochemical pathways. Nucleic Acids Res 37:e16
6. Gibson DG, Glass JI, Lartigue C et al (2010) Creation of a bacterial cell controlled by a chemically synthesized genome. Science 329: 52–56
7. Tsvetanova B, Peng L, Liang X et al (2011) Genetic assembly tools for synthetic biology. Methods Enzymol 498:327–348
8. Engler C, Gruetzner R, Kandzia R et al (2009) Golden gate shuffling: a one-pot DNA shuffling method based on type IIs restriction enzymes. PLoS One 4:e5553
9. Szybalski W, Kim SC, Hasan N et al (1991) Class-IIS restriction enzymes—a review. Gene 100:13–26
10. Weber E, Engler C, Gruetzner R et al (2011) A modular cloning system for standardized assembly of multigene constructs. PLoS One 6:e16765
11. Weber E, Gruetzner R, Werner S et al (2011) Assembly of designer TAL effectors by Golden Gate cloning. PLoS One 6:e19722
12. Sarrion-Perdigones A, Falconi EE, Zandalinas SI et al (2011) GoldenBraid: an iterative cloning system for standardized assembly of reusable genetic modules. PLoS One 6:e21622
13. Bolchi A, Ottonello S, Petrucco S (2005) A general one-step method for the cloning of PCR products. Biotechnol Appl Biochem 42: 205–209
14. Liu ZG, Schwartz LM (1992) An efficient method for blunt-end ligation of PCR products. Biotechniques 12:28–30
15. Kotera I, Nagai T (2008) A high-throughput and single-tube recombination of crude PCR products using a DNA polymerase inhibitor and type IIS restriction enzyme. J Biotechnol 137:1–7
16. Stemmer WP, Morris SK (1992) Enzymatic inverse PCR: a restriction site independent, single-fragment method for high-efficiency, site-directed mutagenesis. Biotechniques 13: 214–220

Chapter 10

Design and Construction of Multigenic Constructs for Plant Biotechnology Using the GoldenBraid Cloning Strategy

Alejandro Sarrion-Perdigones, Jorge Palaci, Antonio Granell, and Diego Orzaez

Abstract

GoldenBraid (GB) is an iterative and standardized DNA assembling system specially designed for Multigene Engineering in Plant Synthetic Biology. GB is based on restriction–ligation reactions using type IIS restriction enzymes. GB comprises a collection of standard DNA pieces named "GB parts" and a set of destination plasmids (pDGBs) that incorporate the multipartite assembly of standardized DNA parts. GB reactions are extremely efficient: two transcriptional units (TUs) can be assembled from several basic GBparts in one T-DNA less than 24 h. Moreover, larger assemblies comprising 4–5 TUs are routinely built in less than 2 working weeks. Here we provide a detailed view of the GB methodology. As a practical example, a Bimolecular Fluorescence Complementation construct comprising four TUs in a 12 kb DNA fragment is presented.

Key words Synthetic biology, DNA assembly, Type IIS enzymes, Plant biotechnology, Multigene constructs, Multigene engineering

1 Introduction

Synthetic Biology aims to engineer living systems with functions not found in nature, thus incorporating new features to existing living organisms [1, 2]. Complex protein production, metabolic engineering, or genetic networks often need the integration of several transgenes into the biological model, what requires more than single transcriptional units (TUs) building but the construction of complex multigenic structures. Custom DNA synthesis [3] as well as other strategies involving homologous and site-specific recombination [4–6] have been reported as successful methodologies for multigene engineering. However, neither they produce reusable units nor facilitate the set up of combinatorial strategies. An alternative approach is the modular construction of genetic devices using standardized DNA parts. Modular design facilitates combinatorial

Svein Valla and Rahmi Lale (eds.), *DNA Cloning and Assembly Methods*, Methods in Molecular Biology, vol. 1116, DOI 10.1007/978-1-62703-764-8_10, © Springer Science+Business Media New York 2014

engineering, as standard DNA parts can be easily exchanged improving the possibilities of the building process.

GoldenBraid (GB) [7, 8] is a cloning system that follows a modular construction strategy. GB allows the binary combination of multipartite assemblies using an extremely simple set of rules. GB makes use of the multipartite Golden Gate cloning method (Chapter 9) [9, 10] to generate a multipartite assembly of standardized basic DNA parts, which are then incorporated to a double loop cloning design that allows binary assembly of multipartite constructs. In this way, GB technology enables the standardization of Golden Gate for its use in Synthetic Biology. This is achieved with a small toolbox consisting of a maximum of only eight destination plasmids and a limited number of simple assembly rules. This new cloning system facilitates the reusability of DNA parts and assembled devices to efficiently build complex constructs.

In this chapter we describe in depth the methodology of the GB cloning system, giving the details on the reactions to be performed and providing special hints to facilitate the assembly of multigene constructs. As an example, we show here the design and assembly of four genes/TUs in one binary destination vector and its use to perform a Bimolecular Fluorescence Complementation (BiFC) analysis [11]. In order to validate the BiFC GB constructs, the interaction between the *Arabidopsis thaliana* transcriptional factors FRUTITFULL (FUL) and SUPRESSOR OF OVER EXPRESSION OF CO (SOC1) will be tested [12, 13]. The final BiFC GB constructs comprise four transcriptional units: in addition to the usual BiFC units (N and C terminal portions of the Yellow fluorescent Protein), we have incorporated a "monitoring/silencing suppressor module" to improve the performance of transient expression assays in *Nicotiana benthamiana*. This special module contains the reporter *Renilla* [13] used as internal reporter to normalize the efficiency of the transient transformation, and the Tomato Bushy Stunt Virus silencing suppressor p19 [14] used to inhibit the effects of gene silencing.

2 Materials

2.1 GoldenBraid Destination Vectors (pDGBs) and Cloning Methodology

Previously described assembly systems allow standardization but the resulting units cannot be easily reused in subsequent assembly reactions. A solution to this limitation, described as GB [7], was to insert a loop (braid) in the cloning design, so that the expression plasmids from first level become entry plasmids for second level assemblies and vice versa. To this end, two types of destination plasmids were designed, namely level α and level Ω. The key point in GB design is that, while all plasmids contain two restriction/recognition sites for two different type IIS enzymes, level α and level Ω plasmids are designed to have their sites in inverted orientations. Plasmids also carry different resistance markers for efficient counter selection.

Although in the most basic setup only four pDGBs are needed to establish the double loop, we build eight different pDGBs in order to make possible the assembly of TUs in reverse orientation. Numbers and letters serve to identify each destination plasmid according to the flanking overhangs left by BsaI and BsmBI digestion, respectively:

1. Level α plasmids are used as destination plasmids in BsaI GB-reactions. These are pDGB_A12C, pDGB_C12B, pDGB_A21C, and pDGB_C21B. In most cases only pDGB_A12C or pDGB_C12B will be used unless there is any interest in assembling TUs in reverse orientation (in this case the choice would be pDGB_A21C or pDGB_C21B). This group of plasmids contains kanamycin as resistance marker and is both used for the multipartite assembly of GBparts and for the binary combination of TUs.
2. Level Ω plasmids are used as destination plasmids in BsmBI-GB reactions. These are pDGB_1AB3, pDGB_3AB2, pDGB 1BA3, and pDGB 3BA2. Regularly, only pDGB_1AB3 or pDGB_3AB2 will be selected by users unless they need to assemble TUs in reverse orientation. This group of plasmids incorporates spectinomycin as bacterial resistance.

The mechanism of GB is shown in Fig. 1. Standard parts are normally assembled in level α plasmids, for example into pDGB_A12C. The resulting composite parts can be combined with other structures assembled in the complementary pDGB_C12B using any of the level Ω plasmids as destination vectors for the assembly. In a second assembly round, composite parts assembled using level Ω plasmids can be combined together inside a level α plasmids provided that they share a common sticky end. As it can be observed, GB works as an endless iteration of binary assemblies keeping the ability of incorporating more units, with the only theoretical limit of the capacity of the destination vector backbone.

2.2 GoldenBraid Parts

"GBparts" are functional DNA fragments flanked by fixed 4 nt overhangs which are generated by cleavage with the type IIS enzyme BsaI. As described in Golden Gate assembly (Chapter 9) [10], the overhangs determine the relative position of a GBpart in a multipartite assembly. In the most general case, we considered three functional standard categories when building TUs: promoters (which include the transcription origin and 5′ UTR), coding regions (CDS), and terminators (which include 3′ UTR and polyA site and transcriptional stop signal). Each category corresponds to a relative position in the assembly (although users can define

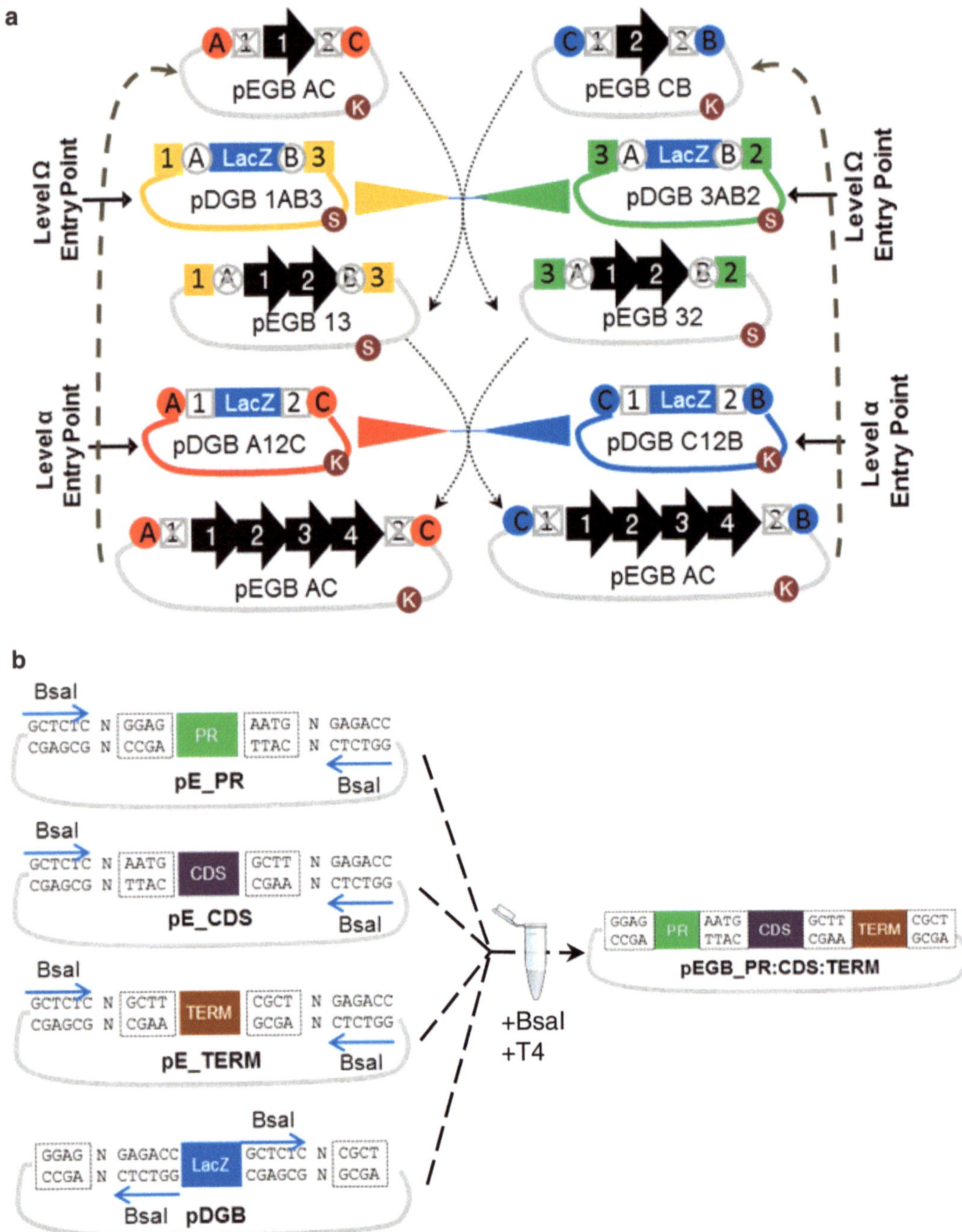

Fig. 1 The mechanism of GoldenBraid. (**a**) TUs (e.g., 1, 2) assembled in complementary level α plasmids can be used as entry vectors for a level Ω binary assembly, provided that they share a common BsmBI sticky end (*encircled* C). Similarly, constructs assembled using paired level Ω plasmids can be used as entry vectors for a subsequent level α assembly, as they share a BsaI sticky end (*squared* 3). Level α and level Ω can alternate indefinitely creating increasingly complex structures, as depicted by the *arrows* closing the double loop. *Encircled* K and S represent kanamycin resistance gene and spectinomycin resistance gene, respectively. (**b**) Standard parts flanked by fixed BsaI cleavage sites (*boxed*) are multipartite assembled using level α plasmids. Upon assembly, the newly assembled TU(Promoter:CodingSequence:Terminator) remains flanked by BsmBI cleavable sites (not depicted here)

additional intermediate categories). Therefore we arbitrarily assigned flanking 4 nt to each of the three basic categories (*see* Fig. 1b) that have to be used in every GBpart to ensure compatibility between different users.

Basic parts normally come in the form of a circular plasmid, and the insert corresponds to the part itself. Upon BsaI digestion, the part is released from the plasmid leaving 4 nt overhangs, ready to be assembled together with other parts in a BsaI GB-reaction. A large collection of ready-to-assemble standard GBparts has been established and can be consulted at http://www.gbcloning.org.

2.3 Materials for PCR Amplification and TA-Cloning of DNA Pieces

1. Gene-specific Oligonucleotides with GB extensions (*see* Subheading 3.1).
2. Phusion® High-Fidelity DNA Polymerase (ThermoScientific).
3. dNTP mixture (10 mM each dNTP).
4. Milli-Q sterilized water.
5. Biotools DNA Polymerase (Biotools).
6. Agarose electrophoresis gel: 1 % agarose TAE 1 × (40 mM Tris–acetate and 1 mM EDTA).
7. QIAquick PCR Purification Kit (Qiagen).
8. pGEM®-T Easy Vector System (Promega).
9. Thermocycler.
10. 50 % Glycerol for storing the correct assemblies.

2.4 *Escherichia coli* Cell Transformation and Culture

1. House-made competent Cells, One Shot® TOP10 or One Shot® Mach1™ T1R chemically competent *Escherichia coli* kit (Invitrogen).
2. Electroporator and 1 mm electroporation cuvettes.
3. Sterile SOC medium: 2 % tryptone, 0.5 % yeast extract, 10 mM sodium chloride, 2.5 mM potassium chloride, 10 mM magnesium chloride, 10 mM magnesium sulfate, 20 mM glucose.
4. Sterile Lysogeny Broth (LB) medium: 1 % tryptone, 0.5 % yeast extract, 1 % NaCL.
5. Lysogeny Agar (LA) plates: 1 % tryptone, 0.5 % yeast extract, 1 % NaCL, 1.5 % agar. Plates contain the appropriate antibiotics (kanamycin at 50 μg mL^{-1}, ampicillin and spectinomycin at 100 μg mL^{-1}), IPTG (0.5 mM), and X-Gal (40 μg mL^{-1}).
6. A shaker and growing chamber at 37 °C.
7. E.Z.N.A. Plasmid Mini Kit (Omega Bio-tek).

2.5 Materials for Multipartite Assembly Reactions

1. Miniprep-purified GBparts.
2. α-level GB destination vectors.
3. T4 DNA ligase (Promega, Madison, USA), BsaI restriction enzyme (New England Biolabs).

4. Thermocycler.
5. 50 % Glycerol for storing the correct assemblies.

2.6 Materials for Binary Assembly Reactions

1. Miniprep-purified composite parts.
2. α-level and/or Ω-level GB destination vectors.
3. T4 DNA ligase (Promega), BsaI, BsmBI restriction enzyme (New England Biolabs).
4. Thermocycler.
5. 50 % Glycerol for storing the correct assemblies.

2.7 Agrobacterium tumefaciens Cell Transformation and Culture

1. House-made pSOUP *Agrobacterium tumefaciens* strain GV3101 electrocompetent cells.
2. Electroporator and 1 mm electroporation cuvettes.
3. Sterile SOC medium (2 % w/v tryptone, 0.5 % w/v Yeast extract, 10 mM NaCl, 2.5 mM KCl, 10 mM MgCl, 10 mM $MgSO_4$, and 20 mM Glucose).
4. Sterile LB supplemented with 2 mM $MgSO_4$ and 2 mM sucrose.
5. LA plates containing the appropriate antibiotics (tetracycline at 12.5 μg mL^{-1}, gentamicin at 30 μg mL^{-1}, rifampicin and kanamycin at 50 μg mL^{-1}, ampicillin and spectinomycin at 100 μg mL^{-1}).
6. QIAprep Spin Miniprep Kit (Qiagen).
7. Lysozyme (Sigma).

2.8 Plant Transient Transformation

1. Agroinfiltration MES Buffer (10 mM MES pH 5.6, 10 mM $MgCl_2$, 200 μM acetosyringone).
2. Spectrophotometer set at a wavelength of 600 nm and transparent plastic cuvettes.
3. Sterile 1 mL Plastikpak syringes without needle.
4. 30–35 days old *Nicotiana benthamiana* plants (growing conditions: 24 °C day/20 °C night in a 16 h light/8 h dark cycle).

2.9 Bifluorescence Complementation Analysis

1. Confocal microscope, such as TCS SL (Leica).
2. Microscope slides and cover slips.
3. Scalpel.
4. Mounting media for microscopy (Leica Biosystems).

2.10 House-Made DH5α Electro-competent Cells

1. Day 1: Streak out frozen glycerol stock of bacterial cells onto an LA plate without antibiotics and grow overnight at 37 °C.
2. Day 2:
   (a) Media Preparation: 2 L of ddH_2O, 1 L of 10 % v/v glycerol, 1.5 L LB. Chill overnight at 4 °C.

(b) Pick a single colony of *E. coli* from the fresh LA plate and inoculate a 15 mL starter culture of LB without antibiotics. Grow overnight at 37 °C.

3. Day 3:

(a) Inoculate 1.5 L of LB media with the 15 mL starter culture and grow for about 3 h in a 37 °C shaker.

(b) Check the OD600 and when it reaches 0.4, chill on ice for 30 min. Chill also the centrifuge bottles.

(c) Distribute the culture in six centrifuge tubes. Harvest the cells by centrifugation at 5,000 rpm for 15 min at 4 °C.

(d) Resuspend each pellet in 250 mL of ice cold water. Shake smoothly. Centrifuge at 5,000 rpm for 15 min at 4 °C.

(e) Decant the supernatant and resuspend each pellet in half the volume so the final volume of the culture is reduced to 750 mL and it can be combined in three centrifuge tubes. Harvest the cells by centrifugation at 5,000 rpm.

(f) Decant the supernatant and resuspend the pellets in 10 mL 10 % glycerol. Transfer the final volume to smaller centrifuge tubes.

(g) Centrifuge the tubes at 5,000 rpm for 15 min at 4 °C and decant the supernatant. Resuspend the final pellets in 2 mL of ice cold 10 % glycerol. Aliquot in ≈50 μL into 1.5 mL tubes and freeze with liquid nitrogen. Store at −80 °C.

2.11 House-Made *Agrobacterium tumefaciens* with pSOUP Electrocompetent Cells

1. Day 1: Streak out frozen glycerol stock of bacterial cells onto an LB plate with rifampicin and tetracycline and grow at 28 °C for 2 days.

2. Day 3:

(a) Media Preparation: 2 L of water, 1 L of 10 % v/v glycerol, 1.5 L LB. Chill at 4 °C.

(b) Pick a single colony from the plate late and inoculate a 5 mL starter culture of LB without rifampicin and tetracycline. Grow for 2 days at 28 °C to saturation.

3. Day 5:

(a) Inoculate 1.5 L of LB media with 1:200 saturated *A. tumefaciens* culture. Grow overnight for about 16 h in a 28 °C shaker. The final OD should be around 1.5.

(b) Distribute the culture in six centrifuge tubes. Harvest the cells by centrifugation at 5,000 rpm for 15 min at 4 °C.

(c) Resuspend each pellet in 250 mL of ice cold 10 % Glycerol ddH_2O. Centrifuge at 5,000 rpm for 15 min at 4 °C.

(d) Decant the supernatant and resuspend each pellet in half the volume so the final volume of the culture is reduced to 750 mL and it can be combined in three centrifuge tubes. Harvest the cells by centrifugation at 5,000 rpm.

(e) Decant the supernatant and resuspend the pellets in 10 mL 10 % glycerol. Transfer the final volume to smaller centrifuge tubes.

(f) Centrifuge the tubes at 5,000 rpm for 15 min at 4 °C and decant the supernatant. Resuspend the final pellets in 2 mL of ice cold 10 % glycerol. Aliquot into 1.5 mL tubes (≈50 μL each) and freeze with liquid nitrogen. Store at −80 °C.

3 Methods

3.1 Domestication of GoldenBraid Basic Parts

A crucial aspect of Synthetic Biology is standardization, which can only be achieved through concerted community effort. In order to ensure that GB users can exchange their parts, user's need to use the same overhangs for the same part categories. We call "domestication" to the adaptation of basic DNA parts to the GB rules. It comprises not only the addition of flanking BsaI sites as indicated earlier, but also the removal of internal BsaI and BsmBI sites. We strongly recommend building BsaI and BsmBI-free parts to facilitate GB reactions, especially for big assemblies (see further in this section for instructions). As GB reactions are extremely efficient, assemblies can still be successful even if the sites are not eliminated. However, this will produce less correct colonies, and for large assemblies involving many pieces your efficiency may be lower. Furthermore, if assembled TUs are going to be reused to build more complex devices, the presence of undomesticated pieces will eventually cause troubles.

As it was described in Subheading 2.2, there are three basic GB categories. In this example, we will build one Terminator-GBPart (pE_T35s; the 35S terminator of the Cauliflower Mosaic Virus), two CDS-GBparts (pE_FUL and pE_SOC1) and two special GBparts that share overhangs with standard promoters but include the split fragments of YFP to enable BiFC experiments (pE_35s:YFN and pE_35s:YFC, consisting on constitutive 35S promoter of CaMV fused to the N-/C-terminal half of YFP). To build these GB parts, follow these steps:

1. Design GB oligonucleotides, incorporating the indicated extensions listed in Table 1, including 20–22 nt from the sequence to be amplified. Note that the overhang preceding a CDS includes the starting ATG codon, and therefore the rest of the gene should be designed in frame with this ATG (*see* **Note 1**).
2. If the part to be domesticated contains no BsaI or BsmBI internal sites, continue this protocol in **step 5**. If any internal BsaI or BsmBI sites have to be removed, continue the protocol in **step 3**.

Table 1
GB Extensions for the three main categories in multigenic assemblies

Category	Forward	Reverse
Promoter (including 35s:YFC and 35s:YFN)	5′ GG**GGTCTC**A*GGAG*-GSP 3′	5′ GG**GGTCTC**A*AATC*-GSP 3′
CDS (including FUL and SOC)	5′ GG**GGTCTC**A*AATG*-GSP 3′	5′ GG**GGTCTC**A*AAGC*-GSP 3′
Terminator (including T35s)	5′ GG**GGTCTC**A*GCTT*-GSP 3′	5′ GG**GGTCTC**A*AGCG*-GSP 3′

BsaI recognition and cutting sites (corresponding to GB overhangs) are marked in **bold** and *italics*, respectively. *GSP* gene-specific primers

3. The process for removing internal type IIS sites is depicted in Fig. 2a where an internal BsmBI site from FUL is eliminated, following a standard overlap extension PCR protocol (OE-PCR). In the case of one internal site, two pairs of primers are required. Design the two external primers (FUL.F1 and FUL.R2 in Fig. 2a) as described above (Subheading 3.1, **step 1**, using extensions shown in Table 1). Design the second (internal) pair of primers (FUL.R1 and FUL.F2) incorporating a nucleotide mismatch so as to mutate the internal Type IIS BsmBI site. Keep in mind that for CDSs, the open reading frame should be maintained (see primers designed for FUL in Table 2). The internal primers must overlap at least 20 nt. Perform the OE-PCR as follows:

   (a) Prepare the first pair of reactions using primer pairs FUL.F1-FUL.R1 and FUL.F2-FUL.R2 (*see* **Note 2**) and a suitable template. Run an electrophoresis gel to verify the success of the PCR (use only 1/5 of the volume) and purify the rest of the reaction using QIAquick PCR Purification Kit (*see* **Note 3**).

   (b) Prepare the overlapping reaction with an equimolar ratio of both PCR fragments as template. Do not add any primers at the beginning. Set your PCR in two different steps:

   - *Step 1*: 10 cycles with an annealing temperature determined by the overlapping region between the two fragments (*see* **Note 4**).

   - *Step 2*: Add 10 μM of the external primers (FUL.F1 and FUL.R2). Set the reaction for 25 cycles with the annealing temperature determined by the external pair of primers. At the end, check the PCR by agarose gel electrophoresis (Fig. 2c, Lane 5). Once it is correct, proceed to **step 6**.

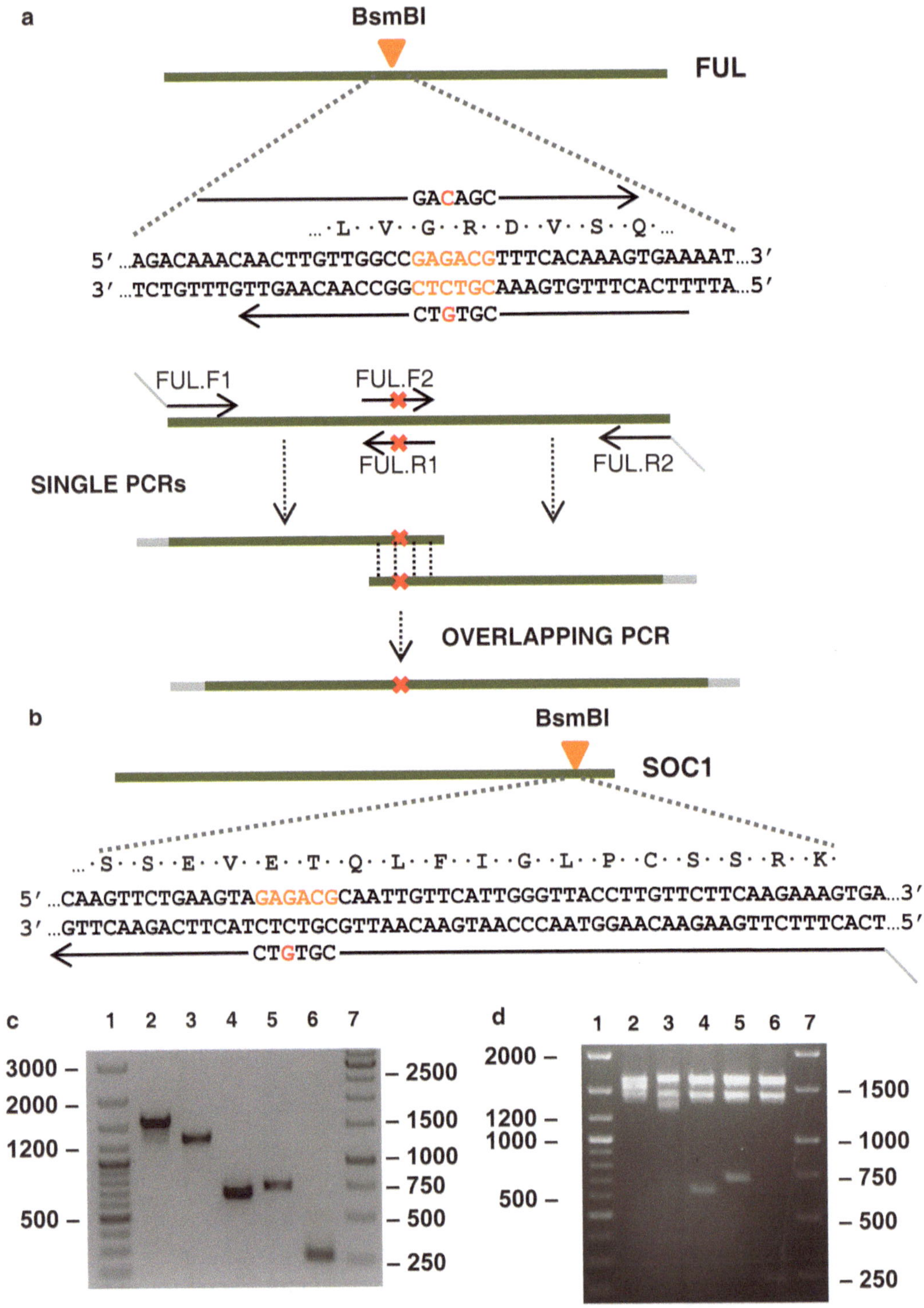

Fig. 2 Domestication of GBparts. (**a**) Overlap extension PCR strategy for FUL domestication through the silent mutation of an internal BsmBI site. Primers FUL.F1 and FUL.R2 are designed to introduce the appropriate BsaI flanking sites. A second pair of primers (FUL.F2 and FUL.R1) are designed to introduce a single nucleotide mismatch (G > C) producing a silent mutation and eliminating the internal BsmBI site. In a first reaction, two fragments sharing 20–25 nt are PCR amplified using primer pairs FUL.F1-FUL.R1 and FUL.F2-FUL.R2; in a

Table 2
GB Oligonucleotides for the amplification of the GBparts used in this chapter

GB part	Primer	Sequence
35s:YFN and 35s:YFC	35s.F1	5′ **GGGGTCTCAGGAG**ACTAGAGCCAAGCTGATCTC 3′
	Linker	5′ **GGGGTCTCACATT**AGCGATCCACCTCCACCAGAT 3′
SOC	SOC.F1	5′ **GGGGTCTCAAATG**GTGAGGGGCAAAACTCA 3′
	SOC.R1	5′ **GGGGTCTCAAAGC**TCACTTTCTTGAAGAACAAGGTAAC CCAATGAACAATTGTGTCTCTACTTCAGAAC 3′
FUL	FUL.F1	5′ **GGGGTCTCAAATG**GGAAGAGGTAGGGTTCA 3′
	FUL.R1	5′ CAAACAACTTGTTGGCCGCGACGTTTCACAAAGTG 3′
	FUL.F2	5′ TTTCACTTTGTGAAACGTCGCGGCCAACAAGTTG 3′
	FUL.R2	5′ **GGGGTCTCAAAGC**TCACTCGTTCGTAGTGGTAGGAC 3′
T35s	T35s.F1	5′ **GGGGTCTCAGCTT**CGGCCATGCTAGAGTCCGCAAA 3′
	T35s.R1	5′ **GGGGTCTCAAGCG**AGGTCACTGGATTTTGGTTT 3′

GB Extensions are marked in **bold**. 35s:YFN and 35s:YFC are amplified using the same pair of oligos as these oligos bind to the 35s and to the linker located after the YFP half

4. If the internal restriction site is close enough to the 5′ or 3′ ends of the GBpart (see example for the GB part SOC1 in Fig. 2b), the situation is solved by making the GB oligo longer, and introducing a mutation in the recognition sequence of the restriction enzyme. Keep in mind that in the case of a CDS, the open reading frame should be maintained. Proceed to **step 5**.
5. PCR the GBpart using a suitable template and specially designed GB primers. Verify the correct amplification by agarose gel electrophoresis. Primers used for the amplifications are listed in Table 2 (Fig. 2c, Lanes 2, 3, 4, and 6).
6. Purify the PCR products using QIAquick PCR Purification Kit (*see* **Note 5**).
7. Add 3′ A overhangs with Taq DNA Polymerase. Set the reaction by mixing 17 μL of the purified PCR, 2 μL 10× reaction buffer, 0.5 μL dNTPs, and 0.5 μL Taq Polymerase.

Fig. 2 (continued) second step, an overlapping PCR using both PRC products as templates and using only FUL.F1 and FUL.R2 primers is performed. The final GB-FUL PCR lacks the internal BsmBI site. (**b**) Removal of a Type IIS site close to the 3′ end of SOC1 by mutating the site on the reverse primer of the GBpart, making it longer than usual. (**c**) Agarose gel electrophoresis of the PCRs for the 5 GBparts described. *Lane 1*: DNA Marker; *Lane 2*: 35s:YFN; *Lane 3*: 35s:YFC; *Lane 4*: SOC1; *Lane 5*: FUL; *Lane 6*: T35s; *Lane 7*: DNA Marker (**d**) BsaI digestion of the 5 GB parts generated. Each of them has three bands, two of them from pGEMT (1,622 and 1,433 pb) and a third one corresponding to the GB part. *Lane 1*: DNA Marker; *Lane 2*: pE35s:YFN; *Lane 3*: pE_35s:YFC; *Lane 4*: pE_SOC1; *Lane 5*: pE_FUL; *Lane 6*: pE_T35s; *Lane 7*: DNA Marker

8. Proceed to ligate the resulting amplicons in pGEMT by adding: 50 ng of pGEMT, 150 ng of the purified PCR, 3u T4 DNA ligase, and 5 μL of the ligase buffer, in a 10 μL reaction (*see* **Note 6**).
9. Incubate the ligation reaction during 1 h at room temperature.
10. Transform 1 μL of the reaction into 50 μL *E. coli* electrocompetent cells, outgrow by adding 500 μL SOC shaking during 1 h in a shaker set in a 37 °C growing chamber. Spread two aliquots (50 and 500 μL) in LB plates containing ampicillin, IPTG, and X-Gal. Incubate overnight in a 37 °C growing chamber.
11. Pick four white colonies and grow them overnight in liquid LB containing ampicillin.
12. Miniprep the cultures and check that the cloned part is correct by restriction digestion analysis. Correct clones from pE_35s:YFC, pE_35s:YFN, pE_FUL, pE_SOC1, and pE_T35s are shown in Fig. 2d (*see* **Note 7**).
13. Sequence the GBpart using M13 forward and reverse universal primers.
14. Store the cells (containing the GBpart) in the form of glycerol stock (15 % glycerol) and DNA miniprep.

3.2 Single TU Assembly in α-Level Plasmids (Multipartite Reaction)

GB constructs are based on restriction–ligation reactions. GBparts have to be combined with the α-level vectors to assemble the functional TU. In this chapter we show two examples of multipartite assembly, which lead to the construction of two TUs, namely 35s:YFN::FUL:T35s and 35s:YFC::SOC1:T35s, from their constitutive GBparts. The remaining two TUs needed to complete a luciferase-monitored BiFC system (namely the "monitoring/silencing suppressor module" comprising constitutively expressed P19 and luciferase TUs) were assembled separately, stored in the GB Collection (you can check the GBparts we have in the collection in www.gbcloning.org), and incorporated to the BiFC system as fully reusable composite parts (*see* Subheading 2.6).

For the Split Fluorescence units (35s:YFN::FUL:T35s and 35s:YFC::SOC1:T35s), three domesticated parts are assembled into the complementary level α vectors as depicted in Fig. 3a, b. The following protocol has to be followed:

1. Prepare the assembly by dispensing in a 10 μL reaction 75 ng of the α-level destination vector, 75 ng of the DNA parts to be assembled, 3u BsaI and 3u T4 DNA ligase.
2. Set the reaction in a thermocycler: 25 cycles × (37 °C 2 min, 16 °C 5 min).
3. Transform 1 μL of the reaction into 50 μL *E. coli* electrocompetent cells, outgrow by adding 500 μL SOC shaking during 1 h in a shaker set in a 37 °C growing chamber. Spread two dilutions (50 and 500 μL) in LB plates containing kanamycin,

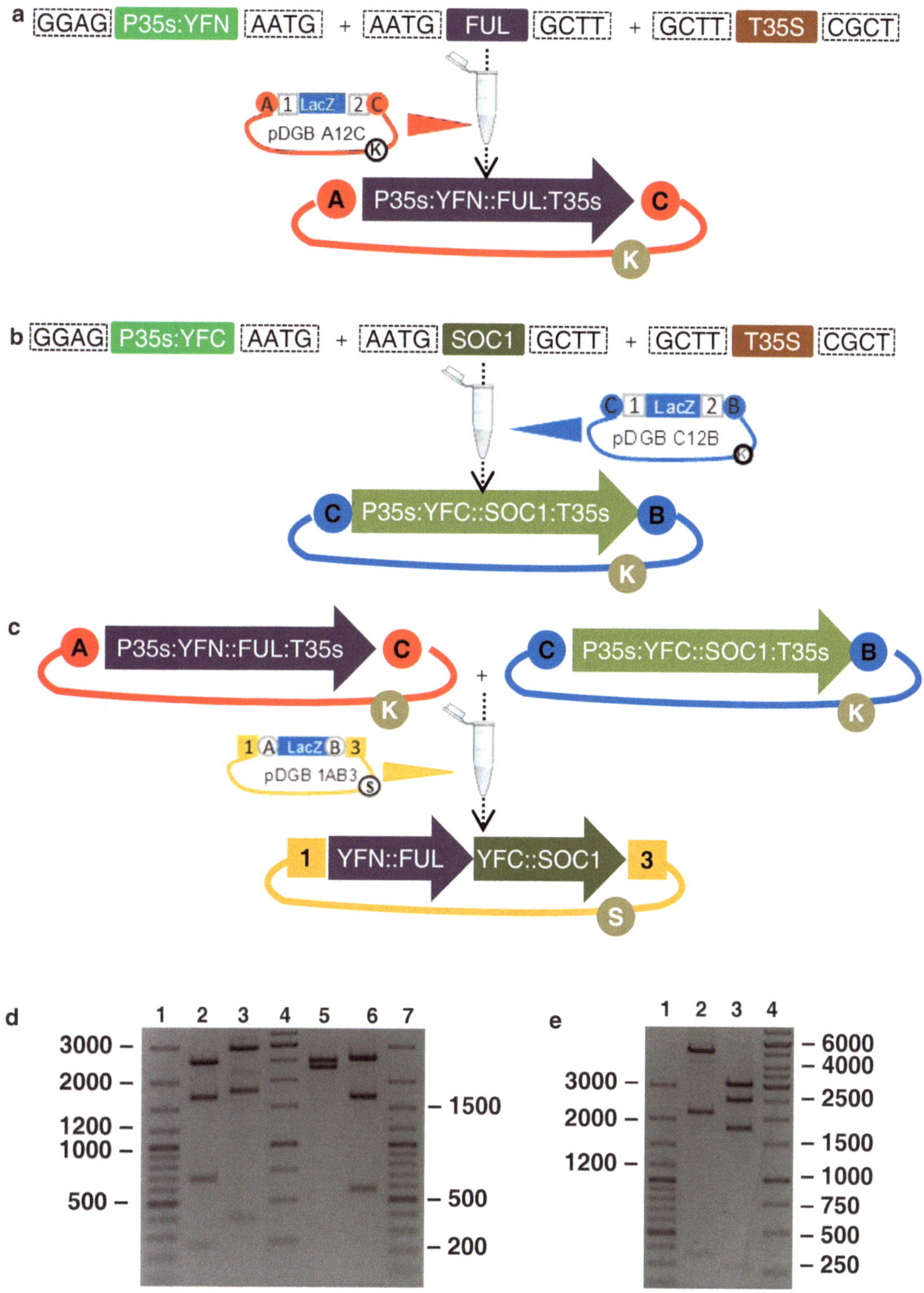

Fig. 3 Multipartite assembly of a transcriptional unit in a level α plasmid and Binary combination of two units in a level Ω plasmid. (**a**) Combination GBparts (pE:35s:YFN, pE_FUL and pE_T35s) to build the TU pEGB_35s:YFN::FUL:T35s in the vector pDGBA12C. (**b**) Combination GBparts (pE:35s:YFC, pE_SOC1 and pE_T35s) to build the TU pEGB_35s:YFC::SOC1:T35s in the vector pDGBC12B. (**c**) Assembly of the complementary TUs pEGB_A- 35s:YFN::FUL:T35s-C and pEGB_C-35s:YFC::SOC1:T35s-B in the destination vector pDGB1AB3. (**d**) Digestion of correct clones of the multipartite assemblies pEGB_A-35s:YFN::FUL:T35s-C (*Lane 2*: BglII; *Lane 23*: PvuI + NcoI) and pEGB_C-35s:YFC::SOC1:T35s-B (*Lane 5*: BglII; *Lane 6*: HindIII) (E) BanI (*Lane 2*) and BglI (*Lane 3*) digestion of a correct clone of the assembly of the two TUs pEGB_1-35s:YFN::FUL:T35s-35s:YFC::SOC1:T35s-3

IPTG, and X-Gal. Incubate overnight in at 37 °C growing chamber.

4. Once colonies are visible, it is possible to distinguish between those carrying intact vectors (blue) and those transformed with your construction (white). Pick four white colonies and grow them overnight in LB containing kanamycin.
5. Isolate DNA by a miniprep method and digest the obtained plasmids. A correct clone from the example constructs A-35s:YFN::FUL:T35s-C and C-35s:YFC:SOC1:T35s are shown in Fig. 3c (*see* **Note 8**).
6. Store the composite units in the form of bacterial glycerol stock (15 % glycerol) and DNA miniprep.

3.3 Multigene Assembly in Ω Level Plasmids (Binary Reaction)

Any composite part GB assembled in α-level plasmids can be combined in Ω-level plasmids. To combine two TUs into a Ω-level plasmid, the right entry and destination plasmids have to be carefully chosen so that the sticky ends are compatible. In an example (*see* Fig. 3b), pEGB_A-35s:YFN::FUL:T35s-C and pEGB_C-35s:YFC::SOC1:T35s-B can be assembled into pDGB 1AB3 as they share a C sticky end.

1. Prepare the assembly reaction by dispensing 75 ng of the Ω-level destination vector, 75 ng of the TUs to be assembled, 3u BsmBI and 3u T4 DNA ligase in a final volume of 10 μL.
2. Set the reaction in a thermocycler: 25 cycles × (37 °C 2 min, 16 °C 5 min).
3. Transform 1 μL of the reaction into 50 μL *E. coli* electrocompetent cells, outgrow by adding 500 μL SOC shaking during 1 h in a shaker set in a 37 °C growing chamber. Spread two dilutions (50 and 500 μL) in LB plates containing spectinomycin, IPTG, and X-Gal. Incubate overnight in a 37 °C growing chamber.
4. Pick four white colonies (blue colonies will contain the intact destination vector) and grow them overnight in LB containing spectinomycin. Obtain DNA minipreps from them and check the assembly by restriction digestion analysis. The digestion of a correct clone from the construct 1- pEGB_1-35s:YFN::FUL: T35s-35s:YFC::SOC1:T35s-3 is shown in Fig. 3e (*see* **Note 5**).
5. Store the DNA construct in the form of bacterial glycerol stock (15 % glycerol) and DNA miniprep.

3.4 Multigene Assembly in α-Level Plasmids (Binary Reaction)

Composite parts GB assembled in Ω-level plasmids can be combined in α-level plasmids as long as entry plasmids are compatible. The previously assembled BiFC TUs (Subheading 3.3) can be combined with a previously assembled, reusable "monitoring/silencing suppressor module" to perform transient expression

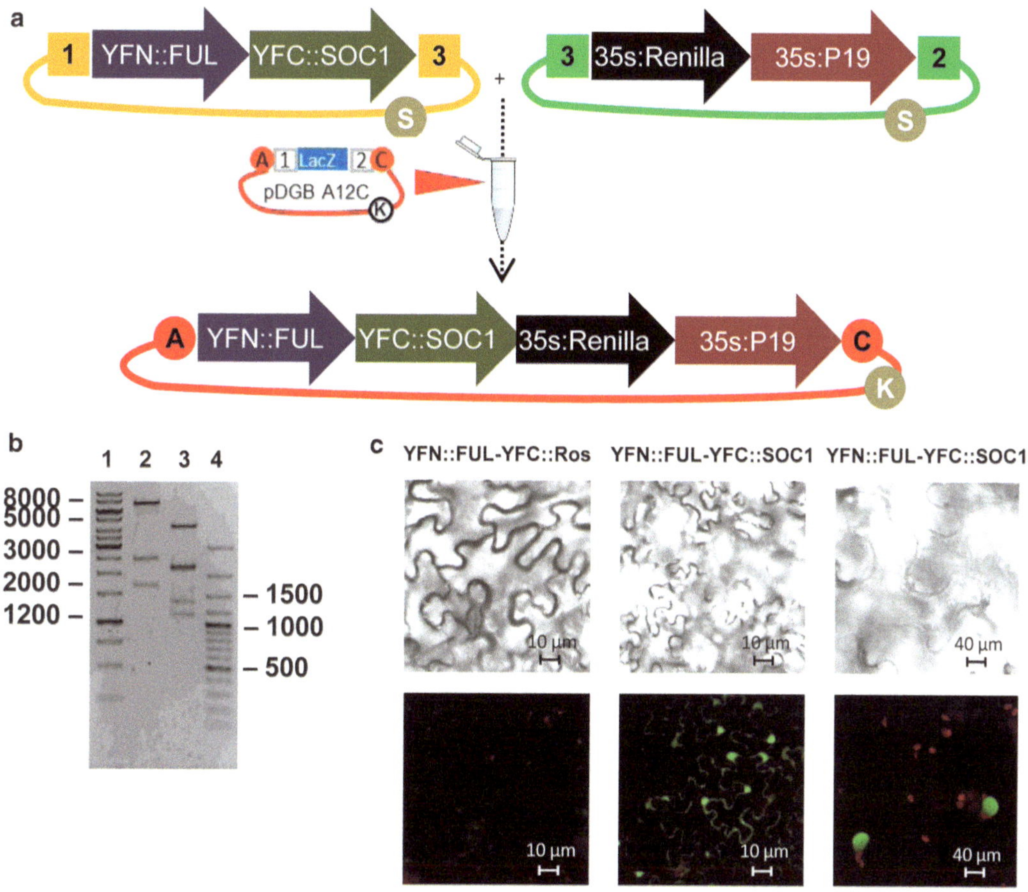

Fig. 4 Multigenic constructs for BiFC. (**a**) Assembly of four TUs in one T-DNA comprising the two units for BiFC and the "monitoring/silencing suppressor module." (**b**) BglII (*Lane 2*) and BanI (*Lane 3*) digestion of the final multigenic construct pEGB_A-35s:YFN::FUL:T35s-35s:YFC::SOC1:T35s-35s:Renilla:TNos-35s:P19:Tnos-C. (**c**) Confocal microscopy expression patterns of the negative (YFN::FUL-YFN:Ros) and the positive (YFN::FUL-YFN:SOC1) BiFC constructs, this latter at two magnifications

assays in *Nicotiana benthamiana*. This special module comprises two TUs, the reporter *Renilla* [14] and the TBSV silencing suppressor p19 [15] (in both cases the expression is directed by the 35S promoter and the Nopaline Synthase terminator). This special module prevents silencing effects and allows monitoring the transformation efficiency using commercial Renilla assay systems. The renilla-p19 module was previously assembled in a similar fashion as it is described in Subheadings 3.2 and 3.3 and was subsequently added to the stored GB collection. The ability to reuse GB devices is one of the strengths of GB system.

To assemble together the YFN-TFC module with the "monitoring/silencing suppressor" module, one, the following steps are followed (depicted in Fig. 4):

1. Prepare the assembly reaction by dispensing 75 ng of the α-destination vector, 75 ng of the TUs to be assembled, 3u BsaI, and 3u T4 Ligase in a final volume of 10 μL.
2. Set your reaction in a thermocycler: 25 cycles × (37° 2 min, 16° 5 min).
3. Transform 1 μL of the reaction into 50 μL *E. coli* electrocompetent cells, outgrow by adding 500 μL SOC shaking during 1 h in a shaker set in a 37 °C growing chamber. Spread two dilutions (50 and 500 μL) in LB plates containing kanamycin, IPTG, and X-Gal. Incubate overnight in a 37 °C growing chamber.
4. Pick four white colonies and grow them overnight in LB containing kanamycin.
5. Check that the construct is correctly assembled by restriction digestion analysis using DNA miniprep from the colonies. The final construct pEGB_A-35s:YFN::FUL:T35s-35s:YFC:: SOC1: T35s-35s:Renilla:TNos-35s:P19:Tnos-C is ready (*see* Fig. 4b) (*see* **Note 5**).
6. Keep a stock of the final construct in the form of bacterial glycerol stock (15 % glycerol) and DNA miniprep.

3.5 Agrobacterium GV3101 Transformation

Once the multigene construct is ready, it has to be transformed into *Agrobacterium tumefaciens*. Some GB destination vectors are based in pGreenII [16], so if that is the case, the *Agrobacterium* strain should carry the helper plasmid pSOUP (*see* **Note 9**).

1. Transform 15 ng of plasmid to 50 μL *A. tumefaciens* electrocompetent cells. Add 500 μL of SOC medium and incubate in a 15 mL tube at 28 °C for 2 h with agitation. Spread two dilutions (50 and 500 μL) in LB plates containing kanamycin (or spectinomycin, according to the resistance of the plasmid), tetracycline, and rifampicin. Incubate for 48 h in a 28 °C growing chamber.
2. Pick four colonies and inoculate 5 mL of LB medium containing the appropriate antibiotics and incubate for 24–36 h.
3. Collect cells by centrifugation and perform a miniprep DNA isolation (*see* **Note 10**).
4. Check correct clones by restriction digestion analysis (*see* **Note 11**) and store the strain in the form of glycerol stock (15 % glycerol).

3.6 Agrobacterium-Mediated Transient Expression Protocol for BiFC Assays in Nicotiana Benthamiana Leaves

For BiFC experiments [11], a transient expression experiment in *Nicotiana benthamiana* leaves can be performed. It is important to include a negative control construct that prevents the reassembly of the fluorescent protein and results only in background fluorescence. The negative construct was also built following the same procedure as the FUL-SOC1 positive construct described in Subheadings 3.3,

3.4 and 3.5 (not shown). The pair of non-interacting proteins was FUL and the *Antirrhinum majus* Rosea1 transcription factors [17]. Agroinfiltration [18], a frequently used technique for transient gene expression, was used for BiFC monitoring. The assay can be performed as follows:

1. Pick the selected clones (positive and negative BiFC) and inoculate a 50 mL culture tube containing 5 mL LB medium supplemented with 2 mM $MgSO_4$, 2 mM sucrose, and the appropriate antibiotics. Grow the cultures at 28 °C with shaking for 48 h.
2. Subculture the clones into a new tube by adding 50 μL of the saturated culture into 5 mL fresh LB medium supplemented with 2 mM $MgSO_4$, 2 mM sucrose, and the appropriate antibiotics. Grow overnight in the same conditions.
3. Pellet the cells (20 min at 3,000 rpm) and resuspend them with Agroinfiltration buffer to an optical density at 600 nm of 0.4.
4. Incubate the cultures at room temperature for 2 h with agitation.
5. *Nicotiana benthamiana* leaves are inoculated by syringe-agroinfiltration in leaves of 30–35 days old plants. After 4–5 days, the tissue can be harvested and the expression of the transgenes analyzed.
6. Cut a 0.5 × 0.5 cm piece of the agroinfiltrated leaves and prepare it using the mounting media on the slide for the Confocal Microscopy TCS SL (Leica) and visualize the positive and negative probes (Fig. 4c).

4 Notes

1. GB oligonucleotides comprise between 30 and 40 nt. There is no need to use highly purified primers if your supplier guarantees you a low error rate.
2. When using Phusion Polymerase, primers greater than 20 nt in length do the annealing for 10–30 s at 3 °C above the melting temperature (T_m) of the lowest primer T_m. However, manufacturer's T_m calculations are not valid for PCRs based on GB oligos as they have an extension that should not be included in T_m calculation. To calculate T_m, you can use the formula $T_m = 4(GC) + 2(AT)$, where GC represents the number of guanine and cytosine, and AT represents the number of adenine and thymine.
3. This purification step is essential for removing primers as the second reaction only uses the external pair of primers.

4. The first annealing temperature for the OE-PCR second step can be calculated using the formula $T_a = 3(GC) + 2(AT)$, where GC represents the number of guanine and cytosine, and AT represents the number of adenine and thymine.
5. It is very important to remove all the Phusion DNA Polymerase as the proofreading activity in Phusion DNA Polymerase is very strong. Any remaining Phusion DNA Polymerase will degrade the A overhangs, thus creating blunt ends again.
6. For optimal ligation efficiency, it is recommended to use fresh PCR products, since 3′A-overhangs will gradually be lost during storage.
7. A BsaI restriction reaction will release the desired GBpart as a single fragment. BsaI is a relatively expensive restriction enzyme so we suggest screening the colonies using EcoRI (pGEM®-T Easy Vector multiple cloning region is flanked by recognition sites EcoRI and NotI, providing a cheap digestion for release of the insert) and to verify BsaI sites only with positive clones.
8. Although each TU assembly will require a different restriction reaction, BglII can be considered as a universal enzyme for pDGBs based on pGreenII as it flanks the GB cassette in both 5′ and 3′ ends.
9. If electrocompetent cells without pSOUP are not available, it is possible to co-transform the final pEGB and pSOUP at the same time but the efficiency will be lower. pGreenII/pSOUP is a binary vector system described by Hellens et al. [16] pSOUP is the helper plasmid that provides the replicase function for the pSa replication origin of pGreen. pGreen will not replicate in *Agrobacterium* if pSOUP is not present.
10. We recommend the use of Qiagen Kit for *Agrobacterium* minipreps. Adding 10 μL of lysozyme together with Solution I and incubating for 10 min at 37 °C will improve the final plasmid yield.
11. To select the restriction enzymes for digesting the *Agrobacterium* minipreps, it is important to consider that both pSOUP and the pEGB will be present. We suggest the use of BglII/EcoRV because it will linearize pSOUP (9.2 kb) and will make at least two cuts in the desired pEGB.

Acknowledgments

We want to thank Dr. Alejandro Ferrando sharing BiFC vectors and Dr. Cristina Ferrandiz for help with protein-protein interaction examples. This work was supported by the Spanish Ministry of Economy and Competitiveness: Grant BIO2010-15384. A. Sarrion-Perdigones is recipient of a Research Personnel in Training (FPI) fellowship.

References

1. Haseloff J, Ajioka J (2009) Synthetic biology, history, challenges and prospects. J R Soc Interface 6(Suppl 4):S389–S391
2. Check E (2005) Synthetic biology, designs on life. Nature 438:417–418
3. Kosuri S, Eroshenko N, LeProust EM et al (2010) Scalable gene synthesis by selective amplification of DNA pools from high-fidelity microchips. Nat Biotechnol 28:1295–1299
4. Ellis T, Adie T, Baldwin GS (2011) DNA assembly for synthetic biology, from parts to pathways and beyond. Integr Biol 3:109–118
5. Gibson DG, Young L, Chuang R-Y et al (2009) Enzymatic assembly of DNA molecules up to several hundred kilobases. Nat Methods 6: 343–345
6. Gibson DG, Glass JI, Lartigue C et al (2010) Creation of a bacterial cell controlled by a chemically synthesized genome. Science 329:52–56
7. Sarrion-Perdigones A, Falconi EE, Zandalinas SI et al (2011) GoldenBraid, an iterative cloning system for standardized assembly of reusable genetic modules. PLoS One 6:e21622
8. Sarrion-Perdigones A, Vilar-Vazquez M et al (2013) GoldenBraid2.0, A comprehensive DNA assembly framework for plant synthetic biology. Plant Physiol 162:1618–1631
9. Engler C, Gruetzner R, Kandzia R (2009) Golden gate shuffling, a one-pot DNA shuffling method based on type IIs restriction enzymes. PLoS One 4:e5553
10. Engler C, Kandzia R, Marillonnet S (2008) A one pot, one step, precision cloning method with high throughput capability. PLoS One 3:e3647
11. Bracha-Drori K, Shichrur K, Katz A et al (2004) Detection of protein-protein interactions in plants using bimolecular fluorescence complementation. Plant J 40:419–427
12. Smaczniak C, Immink RG, Muino JM et al (2012) Characterization of MADS-domain transcription factor complexes in Arabidopsis flower development. Proc Natl Acad Sci U S A 109:1560–1565
13. de Folter S, Immink RG, Kieffer M et al (2005) Comprehensive interaction map of the Arabidopsis MADS Box transcription factors. Plant Cell 17:1424–1433
14. Lorenz WW, McCann RO, Longiaru M et al (1991) Isolation and expression of a cDNA encoding Renilla reniformis luciferase. Proc Natl Acad Sci U S A 88:4438–4442
15. Voinnet O, Pinto YM, Baulcombe DC (1999) Suppression of gene silencing: a general strategy used by diverse DNA and RNA viruses of plants. Proc Natl Acad Sci U S A 96: 14147–14152
16. Hellens RP, Edwards EA, Leyland NR et al (2000) pGreen: a versatile and flexible binary Ti vector for Agrobacterium-mediated plant transformation. Plant Mol Biol 42:819–832
17. Butelli E, Titta L, Giorgio M et al (2008) Enrichment of tomato fruit with health-promoting anthocyanins by expression of select transcription factors. Nat Biotechnol 26: 1301–1308
18. Kapila J, DeRycke R, VanMontagu M et al (1997) An Agrobacterium-mediated transient gene expression system for intact leaves. Plant Sci 122:101–108

Chapter 11

FX Cloning: A Simple and Robust High-Throughput Cloning Method for Protein Expression

Eric R. Geertsma

Abstract

The immense amount of gene sequences available nowadays allows scientist to screen broadly for extraordinary proteins. Reliable cloning tools that allow the parallel processing of many targets are vital for the success of this strategy. The FX cloning procedure detailed here is such a straightforward and efficient tool. It is dedicated to the cloning of open reading frames (ORFs) with the final aim of expressing the corresponding proteins. FX cloning combines attractive features of established high-throughput cloning methods that were thus far not unified in one single method. It facilitates the subcloning of a sequence-verified ORF to a variety of expression vectors, but is sufficiently versatile to accept PCR products as well. Moreover, the common, but undesirable feature of extending target ORFs with long cloning-related sequences is avoided. It leads to the addition of only one amino acid to each side of the protein. As a consequence, only one primer pair or PCR product suffices to generate expression vectors for both N- and C-terminal translational fusions. FX cloning is highly efficient and economical in its use. The method is suited for high-throughput cloning projects and also for everyday cloning of single targets. FX cloning is based on the use of type IIS restriction enzymes and negative selection markers. The full procedure takes place in one pot in less than 3 h and does not require intermediate purification steps nor extensive handling. The method has proven to be very robust and suitable for all common expression systems.

Key words High-throughput cloning, Type IIS restriction enzymes, Subcloning, *ccdB*, *sacB*, Counterselection marker

1 Introduction

The total number of genome sequencing projects currently doubles every 2 years [1]. The vast amounts of genetic information available nowadays provide unique opportunities for scientists in pursuit of rare protcins with exceptional properties, such as structural biologists looking for well-expressed and stable homologs. Identification of such targets requires the screening of many different proteins in distinct expression contexts and thus critically relies on the ability to reliably generate many expression constructs. The parallel cloning of multiple open reading frames (ORFs) places some restrictions on the cloning system. Traditional restriction- and

Svein Valla and Rahmi Lale (eds.), *DNA Cloning and Assembly Methods*, Methods in Molecular Biology, vol. 1116, DOI 10.1007/978-1-62703-764-8_11, © Springer Science+Business Media New York 2014

ligation-based cloning is not well suited for this purpose because the efficiency of the method varies considerably and the frequent occurrence of restriction sites within genes prevents a uniform strategy. Most successful high-throughput cloning methods are either based on DNA recombination or on annealing of complementary single-stranded overhangs [2–5].

Recombination-based methods rely on phage recombinases to insert an ORF into a plasmid. These make use of either specific recombination sequences, like Gateway (Chapter 14) [6], or sequence overlap between vector and insert, like SLiCE (Chapter 16) [7]. Methods based on DNA annealing require long single-stranded overhangs on the insert that are complementary to the overhangs of the vector to form stable duplexes. Generation of these overhangs requires the exonuclease activity of DNA polymerases, as for LIC [8], In-Fusion (Chapter 15) [9], Quick and Clean cloning (Chapter 3) [10], SLIC (Chapter 2) [11], and Gibson assembly (Chapter 1) [12], or alternative approaches, like EFC [13], USER (Chapter 5) [14], and PIPE [15]. Next to recombination- and single-strand annealing-based systems, methods exclusively relying on assembly PCR, such as RF cloning (Chapter 6) [16], CPEC (Chapter 8) [17], and TPCR (Chapter 7) [18] are emerging.

Fragment exchange (FX) cloning was developed as an alternative to these approaches [19]. FX cloning is dedicated to the parallel cloning of multiple ORFs with the final aim of expressing the corresponding proteins. FX cloning combines several attractive features that were thus far not unified in one single cloning method. Like most annealing-based methods, FX cloning is inexpensive and reduces the extension of the ORF with seams resulting from cloning-related sequences. The addition of large, up to nine amino acids long tails at each terminus of the protein, a common feature of cloning systems like Gateway and LIC [6, 8], is thus prevented. As a consequence of the minimal seams, only one primer pair per ORF suffices to generate expression constructs for both N- and C-terminal translational fusions with tags or proteins. Furthermore, FX cloning allows the subcloning of a sequence-verified ORF to different expression vectors, similar to Gateway. This feature assures a correct DNA sequence of the insert in every new expression construct. In contrast, virtually all other cloning approaches accept only PCR products as insert and consequently require resequencing. FX cloning requires only minimal handling and is performed in one pot. Transformation efficiencies are high and only about tenfold less than for intact plasmids. Finally, the method is highly robust and thus even accessible to users lacking prior experience with molecular biology.

FX cloning relies on the use of type IIS restriction enzymes (RE) [20]. These enzymes differ from the commonly used type IIP RE on two important aspects: (1) type IIS RE have a non-palindromic, asymmetric recognition site; and (2) type IIS RE do not have an overlapping recognition and cleavage site but cleave DNA at a defined distance away from their recognition site. The resulting overhang is thus not defined by its sequence but only by its size. For example, for a three nucleotide overhang 64 different combinations can occur. The lack of overlap between the recognition and the cleavage sites results in their physical separation upon digestion (Fig. 1a). FX cloning makes use of the type IIS RE SapI. This enzyme has a comparably long recognition sequence of seven nucleotides and generates a three base long overhang. The long recognition site assures that SapI sites only rarely occur within ORFs. Furthermore, an occasional SapI site within a gene is likely to yield an overhang with a different sequence that only allows reassociation of both fragments.

FX cloning facilitates the initial cloning of a PCR product into a sequencing vector (Fig. 1b). This vector can subsequently be used for subcloning of the sequence-verified ORF to different expression vectors using the identical procedure (Fig. 1c). Initial cloning in a sequencing vector is, however, not essential and PCR products can be cloned straight away in expression vectors as well. Irrespective of the route, the procedure starts with the amplification of the target ORF by PCR using a pair of comparably short primers containing a SapI site. The SapI recognition sites in these primers are oriented so that upon cleavage they are separated from the ORF (Fig. 1d). The resulting overhangs flanking the ORF differ and thus allow directional cloning.

The ORF can be received by an intermediate sequencing vector (pINITIAL) or one or several expression vectors. The pINITIAL vector carries the *ccdB* counterselection marker [21] flanked by SapI sites. These sites are oriented so that the recognition sites remain associated with the pINITIAL backbone and can be used for subcloning. In the expression vectors *ccdB* is flanked by SapI sites in the opposite orientation (Fig. 1e). Here, the recognition sites are removed from the plasmid backbone, which ensures that the seams of cloning-related sequences are in fact only three nucleotides long.

After cleavage, both classes of vectors hold complementary overhangs that hybridize with the overhangs on the insert and are joined by ligation. As the overhangs on the vectors are incompatible with each other, self-ligation is not possible. Subsequent transformation of the ligation mix to a common CcdB-sensitive *E. coli* strain allows only cells with daughter plasmids containing the target ORF to survive. An additional counterselection approach is used only if a pINITIAL derivative holding an ORF supplies the insert for the expression vectors (Fig. 1c). In this case, the cells are plated on sucrose-containing media after transformation. This prevents cells that received the pINITIAL derivative, but not the final expression plasmid from growing as the pINITIAL backbone

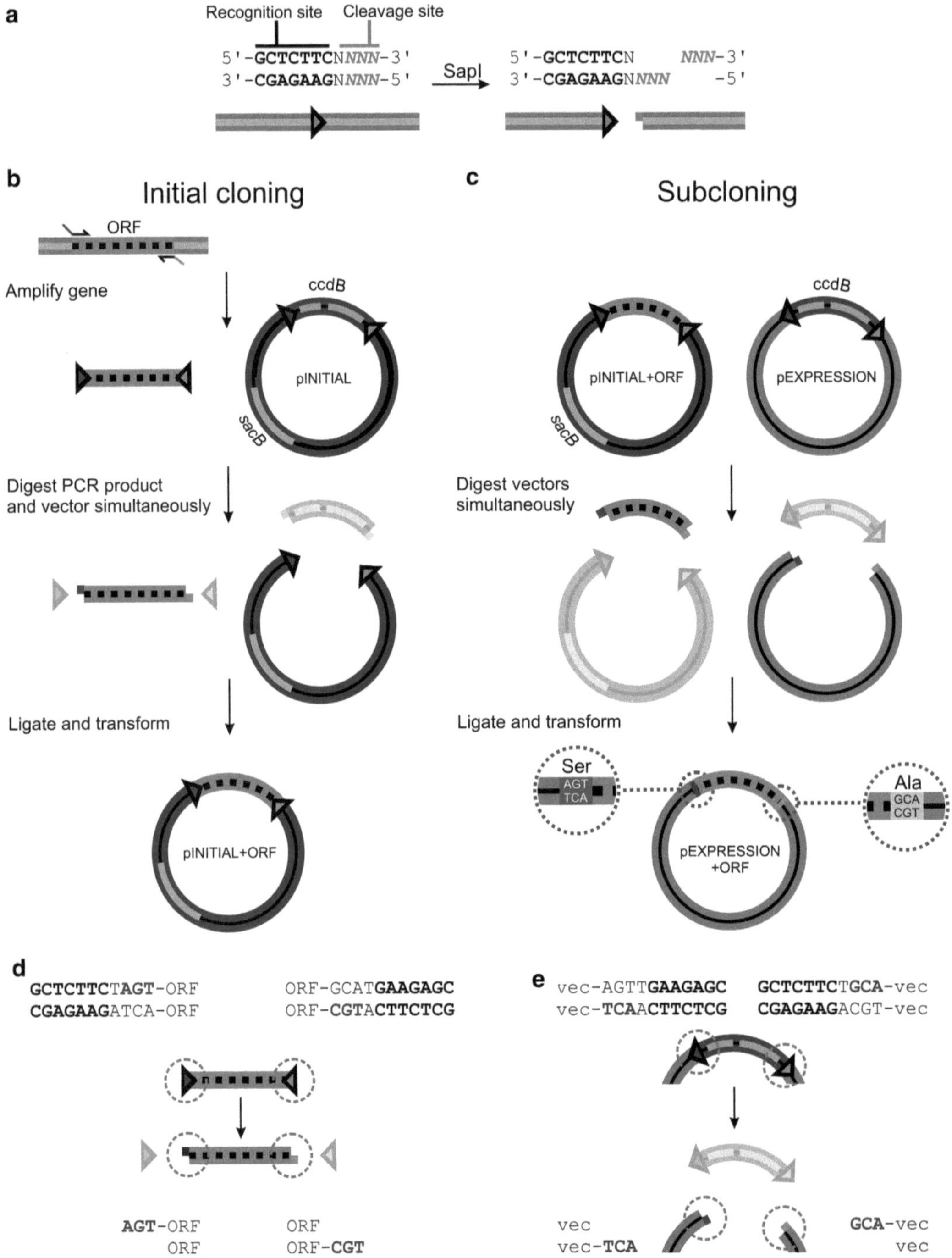

Fig. 1 Schematic overview of the FX cloning method. (**a**) SapI restriction site. The recognition site is shown in *bold* letters; nucleotides constituting the three base-pair single-stranded overhangs are in *italics*. N describes any of the four nucleotides. A schematic view of the cleavage is shown below. *Arrows* indicate the direction of the restriction site. (**b**) Cloning of a PCR product into pINITIAL. The amplified ORF is shown as a *black dotted line*. The direction of the SapI restriction sites is indicated by *arrows*, which are colored corresponding to their respective overhangs generated after cleavage. The genes coding for the counterselection markers *ccdB* and *sacB* on pINITIAL are indicated. (**c**) Subcloning of an ORF into an expression vector (pEXPRESSION). The three nucleotides added to either terminus of the ORF are shown as *insets* (*circle*). The start codon is located upstream of the AGT (Ser) codon. (**d**) Orientation of the SapI cleavage sites in the PCR product and pINITIAL and (**e**) in expression vectors. The single-stranded overhangs generated upon cleavage are shown in *dark* and *light gray*, respectively. Adapted with permission from ref. [19]. Copyright (2011) American Chemical Society

carries the counterselection marker *sacB* which renders *E. coli* sensitive to sucrose [22]. The following contains all relevant information to implement the FX cloning procedure.

2 Materials

2.1 FX Cloning of PCR Products

1. A computer running a Linux-based operating system or a recent Windows version on which the Python programming language (version 2.7) is installed and the scripts `FXprimers.py` and `i_oligo.py` (can be downloaded from http://www.fxcloning.org).
2. Phusion high-fidelity DNA polymerase, dedicated buffer, and dNTPs.
3. Autoclaved ultra-high-purity (UHP) water.
4. Dedicated primers to amplify target ORFs for FX cloning. For ordering, select the lowest synthesis scale offered by the supplier and mere desalting as the purification step. For large sets (>48 primers), order the primers in a 96-well plate. Upon delivery, spin down the primers and dissolve them to a concentration of 100 μM in UHP water. Make a substock at a concentration of 5 μM. The 100 μM stock can be stored for several months at −20 °C. Store the 5 μM stock at 4 °C. Spin down the material before use.
5. DpnI.
6. 50× TAE buffer: 249 g Tris-base, 57.1 mL acetic acid, 100 mL 0.5 M EDTA pH 8.0. Adjust volume to 1 L with water. Store at room temperature. Use a 1× TAE solution for gel electrophoresis.
7. DNA gel extraction kit.
8. FX cloning vectors: available from the author or constructed as described previously [19]. Pro- and eukaryotic expression vectors with different promoters and combinations of tags and gene-fusions have been constructed. A list can be found at http://www.fxcloning.org. All FX cloning vectors should be propagated in a CcdB-resistant *E. coli* strain (such as *E. coli* DB3.1 [6]).
9. SapI and dedicated buffer (*see* **Note 1**).
10. 10 mM ATP pH 7: 10 mM Na_2-ATP, 10 mM $MgSO_4$. Dissolve in 50 mM KPi, pH 7.0, and adjust to a neutral pH (pH 6.5–7.5) with NaOH (*see* **Note 2**). Store in small aliquots at −20 °C.
11. T4 DNA ligase.

Table 1
PCR primer extensions and sequencing primers

Primer name	Sequence	Purpose
ORF forward	`5'atatatGCTCTTCtAGT nnn`	Amplification of target ORF
ORF reverse	`5'tatataGCTCTTCaTGC nnn`	Amplification of target ORF
INITforSQ	`5'ATCTGTTGTTTGTCGGTGAACGC`	Sequencing of 5′ side of insert in pINITIAL
INITrevSQ	`5'TGGCAGTTTATGGCGGGCGT`	Sequencing of 3′ side of insert in pINITIAL

Nucleotides in lowercase can be substituted for other bases if needed for primer optimization. The triplet `nnn` indicates the first ORF-specific codon of the primer. Ideally, it targets the second codon of the ORF (forward primer) or the penultimate codon (reverse primer)

12. LB agar plates: 10 g tryptone, 5 g yeast extract, 10 g NaCl, and 15 g agar. Adjust volume to 1 L with water. Sterilize by autoclaving. Once the medium is cooled to ~60 °C, add the appropriate antibiotics, mix and pour the plates. Store plates at 4 °C.
13. LB medium: 10 g tryptone, 5 g yeast extract, and 10 g NaCl. Adjust volume to 1 L with water. Sterilize by autoclaving. Supplement with the appropriate antibiotics immediately prior to use.
14. Autoclaved 87 % w/v glycerol.
15. Plasmid miniprep kit.
16. Chemical competent cells of an *E. coli* strain sensitive to CcdB (virtually all *E. coli* strains lacking the F plasmid, e.g., *E. coli* MC1061 [23]). Electro-competent cells may be substituted if very high transformation efficiencies are required.
17. Sequencing primers INITforSQ and INITrevSQ (Table 1) to verify sequences cloned into pINITIAL-derivatives or dedicated sequencing primers for sequences cloned into expression vectors.

2.2 Subcloning of Sequence-Verified ORFs

1. 70 % w/v sucrose: 350 g sucrose. Adjust volume to 500 mL with hot water. Mix well and dissolve and sterilize the sucrose by autoclaving. Immediately after autoclaving determine whether the sucrose has completely dissolved and if needed carefully mix the solution until all the material is dissolved.
2. Low salt LB agar supplemented with 7 % w/v sucrose: 10 g tryptone, 5 g yeast extract, 5 g NaCl, and 15 g agar. Adjust volume to 0.9 L with water. Sterilize by autoclaving. Once the medium is cooled to ~60 °C, add 100 mL sterile 70 % w/v sucrose and the appropriate antibiotics, mix and pour the plates. Store plates at 4 °C.

3 Methods

3.1 FX Cloning of PCR Products

1. Design a forward and reverse primer for each gene to be cloned (*see* **Notes 3** and **4**). This is most conveniently done automated. Prepare a plain text file containing a list of all target ORFs including their start and stop codons in the FASTA format (*see* **Note 5**). Run the `FXprimers.py` script and import the resulting `primers.txt` file in spreadsheet software as a tab-delimited file. For each ORF, a forward and reverse primer is indicated that was the result of a short "optimization" subroutine. These primers no longer contain internal stretches of complementary bases that can form strong hairpin structures and interfere with the PCR. The changes in the primer required to remove the hairpins preferably avoid changes in the sequence of the region that anneals with the ORF, but if this cannot be avoided conservative mutations are allowed. The optimized forward and reverse primers do not require any additional inspection and can be ordered right away. In addition, the script performs a quick inspection of each ORF. In the rare situation that an ORF carries a potentially detrimental internal SapI site that requires removal, this is indicated in the column "classification". Alternatively, design the primers "by hand" using the following design rules: the gene-specific part of the primer should be designed so that it: (1) does not contain the start or stop codon of the target gene; and (2) is sufficiently long to anneal in a stable and selective way with the target DNA. The required 5′ extensions of the primers are indicated in Table 1 (*see* **Note 6**). The complete primer should not contain strong secondary structure elements that could interfere with PCR.
2. Amplify the desired ORFs using the Phusion DNA polymerase (*see* **Note 7**). Prepare a 50 μL reaction mixture according to the manufacturer's protocol and add the polymerase just prior to the start of the reaction. Place the sample in a PCR machine preheated to 98 °C and start the reaction. A good starting point for a PCR program is: (a) 30 s at 98 °C; (b) 10 s at 98 °C; (c) 15 s at 61 °C (decrease 0.5 °C per cycle; *see* **Note 8**); (d) *X*s at 72 °C (15–30 s/kb); Repeat (b–d) 14 times; (e) 10 s at 98 °C; (f) 15 s at 53 °C; (g) *X*s at 72 °C; Repeat (e–g) 14 times; (h) 120 s at 72 °C; (i) unlimited at 10 °C.
3. If a plasmid was used as the template, eliminate potential background colonies by exclusively digesting the Dam-methylated plasmid. For this, add 0.5 μL DpnI (5 U) and incubate 30 min at 37 °C. The unmethylated PCR product is not a substrate for DpnI and thus escapes cleavage. Analyze all the material by TAE gel electrophoresis and gel purify the target band using a DNA gel extraction kit (*see* **Notes 9–11**).

Determine the DNA concentration spectrophotometrically. Store the material at −20 °C until use.

4. Prepare 5 mL LB supplemented with the appropriate antibiotics and cultivate *E. coli* DB3.1 cells containing the desired FX cloning vectors overnight at 37 °C (*see* **Note 12**). Isolate the plasmids using a miniprep kit and determine the DNA concentration. Store the material at −20 °C until use.
5. Mix 50 ng of an FX cloning vector with sufficient PCR product to have a vector:insert molar ratio of approximately 1:5 (*see* **Note 13**). Add 1 μL 10× SapI-buffer and adjust the volume to 9 μL with UHP water. Subsequently add 1 μL SapI (2 U) and incubate 1 h at 37 °C (*see* **Note 14**).
6. Heat inactivate SapI for 20 min at 65 °C and allow the sample to cool to room temperature. Add 1.25 μL 10 mM ATP and 1.25 μL T4 DNA ligase (1.25 U) and incubate 1 h at room temperature.
7. Heat inactivate the T4 DNA ligase for 20 min at 65 °C. Transform 5 μL of the ligation mix to 100 μL chemically competent cells of an *E. coli* strain that is CcdB-sensitive (virtually all *E. coli* strains lacking the F plasmid) and use 400 μL medium during the recovery phase. Plate 1 and 10 % aliquots on LB agar supplemented with the appropriate antibiotic (*see* **Note 15**). Incubate the plate overnight at 37 °C.
8. Pick a few single colonies from the plate to inoculate 5 mL LB supplemented with the appropriate antibiotics (*see* **Note 16**). Incubate overnight at 37 °C. Isolate the plasmids with a miniprep kit and determine the DNA concentration. Verify the insert by DNA sequencing. For inserts cloned into a pINITIAL-derivative, primers INITforSQ and INITrevSQ can be used (Table 1). Store the vector at −20 °C until use. For inserts cloned into an expression vector one can proceed as required for the desired expression system.

3.2 Subcloning of Sequence-Verified ORFs from pINITIAL to Expression Vectors

1. Prepare 5 mL LB supplemented with the appropriate antibiotics and cultivate *E. coli* DB3.1 cells containing the desired FX cloning expression vectors overnight at 37 °C. Isolate the plasmids using a miniprep kit and determine the DNA concentration. Store the material at −20 °C until use.
2. Mix 50 ng of an FX cloning expression vector with a sequenced derivative of pINITIAL holding the insert of interest to have an expression vector: sequencing vector molar ratio between 1:3 and 1:5 (*see* **Note 12**). Add 1 μL 10× SapI-buffer and adjust the volume to 9 μL with UHP water. Subsequently add 1 μL SapI (2 U) and incubate 1 h at 37 °C.
3. Heat inactivate SapI for 20 min at 65 °C and allow the sample to cool to room temperature. Add 1.25 μL 10 mM ATP and

1.25 μL T4 DNA ligase (1.25 U) and incubate 1 h at room temperature.

4. Heat inactivate the T4 DNA ligase for 20 min at 65 °C. Transform 5 μL of the ligation mix to 100 μL chemically competent cells of an *E. coli* strain that is CcdB-sensitive (virtually all *E. coli* strains lacking the F plasmid; *see* **Note 17**) and use 400 μL medium during the recovery phase. Plate 1 and 10 % aliquots on low salt LB agar supplemented with 7 % w/v sucrose and the appropriate antibiotic (*see* **Notes 15** and **18**). Incubate the plate overnight at 37 °C.
5. Pick a colony from the plate to inoculate 5 mL LB supplemented with the appropriate antibiotics. Incubate overnight at 37 °C. Isolate the plasmid with a miniprep kit and determine the DNA concentration. The insert does not require additional verification by DNA sequencing. Proceed with the plasmid as required for the desired expression system.

4 Notes

1. FX cloning was established using the enzyme SapI. However, SapI can in principle be replaced with its isoschizomers LguI, PciSI, and BspQI. Note that the latter requires 50 °C for optimal activity.
2. For small volumes, an approximation of the pH is conveniently obtained by pipetting ~1 μL drops on pH strips.
3. ORFs containing an internal SapI site are cloned with approximately tenfold lower efficiency than fragments devoid of SapI sites [19]. Due to the high efficiency of FX cloning, this will still result in many colonies. However, for some applications that require very high transformation efficiencies, such as the preparation of a library of variants of one gene, it will be beneficial to remove an internal SapI beforehand.
4. In case the target ORF will be produced by gene synthesis, internal SapI sites are preferably removed. In addition, the ORF can be ordered already flanked by the 5′ extension indicated in Table 1 (excluding the 5′`tatata` or 5′`atatat` tails). Provided the resulting vector has a different antibiotic resistance marker than the target expression vectors, it can replace the pINITIAL-ORF for the subcloning reaction. In this case the cloning of the ORF in pINITIAL can be skipped.
5. The FASTA format comprises a first line starting with a ">" sign, followed by the name of the sequence, and additional lines describing the sequence. Multiple sequences in FASTA format may be contained in one file.
6. The 5′ extensions of the primers result in the extension of the ORF with triplets coding for an N-terminal Ser and C-terminal

Ala residue. These overhangs were selected because they code for small neutral amino acids and are completely incompatible, thus excluding self-ligation of the vector or concatemer formation of the insert.

7. If genomic DNA is used as a template, delicate handling is required and care should be taken not to compromise the material by repetitive freeze/thawing, vortexing, or vigorous pipetting.
8. Touchdown PCR [24] is recommended as it favors the production of the desired product over products resulting from spurious priming. In addition, as a range of annealing temperatures is used the program does not require much fine-tuning for different targets.
9. Purification of the target PCR product from gel might be omitted if the quality of the PCR was superb. However, regularly small by-products are present and even trace amounts of these are cloned with high efficiency. As a result, more clones need to be analyzed in later steps thus decreasing the throughput. This is not a unique feature of FX cloning but applies to all (high-throughput) cloning procedures.
10. Due to the high efficiency of FX cloning and the use of primers with uniform restriction sites, care should be taken to avoid cross-contamination with PCR products run in adjacent wells. In addition, the gel container should be cleaned and the buffer should be replaced between runs to prevent contamination with PCR products analyzed in previous runs.
11. Should the PCR not yield the desired product in sufficient amounts, consider the use of alternative buffers supplied by the manufacturer, the addition of dimethylsulfoxide (DMSO) or the use of a freshly purified template. More detailed suggestions can be found in the molecular cloning protocol series [25].
12. The choice of FX cloning vector depends on the purpose. To allow subcloning of sequence-verified ORFs, clone the insert first into a pINITIAL-derivative. Preferably the antibiotic marker of the pINITIAL-derivative is different from the expression vectors used later for subcloning as this provides another selection criterium next to sacB counterselection. Derivatives of pINITIAL holding antibiotic markers against kanamycin, chloramphenicol, or tetracycline are available. As most common expression vectors provide resistance against ampicillin or kanamycin, in general pINITIAL-cat (providing chloramphenicol resistance) is recommended. If there is no need for subcloning, the ORFs can be cloned immediately into an FX cloning-compatible expression vector.
13. A facile approximation for a 1:5 molar ratio of vector:insert is calculated using: (amount vector (ng) × size insert (bp) × 5)/size vector (bp) = amount insert (ng) needed.

14. Incubations of small volumes at temperatures above room temperature should be performed in a PCR machine with heated lid or in a waterbath placed inside an incubator. This prevents suboptimal reaction conditions due to excessive condensation at the lid.
15. Use glass beads (5 mm diameter) to conveniently spread the transformed cells over the plate. This has the additional advantage that cross-contamination is prevented. For large target volumes, the spotting of cells is more efficient than plating. Make a serial dilution of the transformed cells using LB in a 96-well plate to obtain columns with undiluted, 10-, 100-, and 1,000-fold diluted cells. Use a multipipet to spot 5 μL aliquots on a square LB agar plate. Spot in duplicate as occasionally neighboring spots fuse. Up to 12 transformations are comfortably spotted on one 120 mm square agar plate.
16. Due to the highly efficient *ccdB* counterselection marker, the fraction of clones without insert will be extremely low. The reason that it is suggested to analyze more than one colony lies in the fact that mutations resulting from the primer synthesis or PCR amplification cannot be excluded.
17. If subsequent protein expression takes place in *E. coli* and if the expression system is sufficiently tight, such as the AraC/P_{BAD}-system [26], the ligation mixture can be transformed immediately to the desired expression strain to avoid plasmid isolations and re-transformation.
18. Supplementing the medium with sucrose counterselects colonies containing an intact pINITIAL-derivative. Counterselection is even more efficient if the pINITIAL-derivative and the expression vector hold different antibiotic markers (*see* **Note 12**).

Acknowledgments

E.R.G. acknowledges a long-term fellowship from the Human Frontier Science Program. Iwan Zimmermann and Margrit Mathys are thanked for critical comments on the manuscript. Mark Schmitz and Carlo Bertozzi are thanked for contributing to the FX cloning website.

References

1. Pagani I, Liolios K, Jansson J et al (2012) The Genomes OnLine Database (GOLD) v. 4: status of genomic and metagenomic projects and their associated metadata. Nucleic Acids Res 40:D571–D579
2. Yang X, Boehm JS, Yang X et al (2011) A public genome-scale lentiviral expression library of human ORFs. Nat Methods 8:659–661
3. Mancia F, Love J (2011) High throughput platforms for structural genomics of integral membrane proteins. Curr Opin Struct Biol 21: 517–522
4. Xiao R, Anderson S, Aramini J et al (2010) The high-throughput protein sample production platform of the Northeast Structural Genomics Consortium. J Struct Biol 172:21–33

5. Gräslund S, Nordlund P, Weigelt J et al (2008) Protein production and purification. Nat Methods 5:135–146
6. Hartley JL, Temple GF, Brasch MA (2000) DNA cloning using *in vitro* site-specific recombination. Genome Res 10:1788–1795
7. Zhang Y, Werling U, Edelmann W (2012) SLiCE: a novel bacterial cell extract-based DNA cloning method. Nucleic Acids Res 40:e55
8. Aslanidis C, de Jong PJ (1990) Ligation-independent cloning of PCR products (LIC-PCR). Nucleic Acids Res 18:6069–6074
9. Berrow NS, Alderton D, Owens RJ (2009) The precise engineering of expression vectors using high-throughput In-Fusion PCR cloning. Methods Mol Biol 498:75–90
10. Thieme F, Engler C, Kandzia R et al (2011) Quick and clean cloning: a ligation-independent cloning strategy for selective cloning of specific PCR products from non-specific mixes. PLoS One 6:e20556
11. Li MZ, Elledge SJ (2007) Harnessing homologous recombination *in vitro* to generate recombinant DNA via SLIC. Nat Methods 4: 251–256
12. Gibson DG, Young L, Chuang R et al (2009) Enzymatic assembly of DNA molecules up to several hundred kilobases. Nat Methods 6: 343–345
13. Tillett D, Neilan BA (1999) Enzyme-free cloning: a rapid method to clone PCR products independent of vector restriction enzyme sites. Nucleic Acids Res 27:e26
14. Geu-Flores F, Nour-Eldin HH, Nielsen MT et al (2007) USERfusion: a rapid and efficient method for simultaneous fusion and cloning of multiple PCR products. Nucleic Acids Res 35:e55
15. Klock HE, Koesema EJ, Knuth MW et al (2008) Combining the polymerase incomplete primer extension method for cloning and mutagenesis with microscreening to accelerate structural genomics efforts. Proteins 71: 982–994
16. van den Ent F, Löwe J (2006) RF cloning: a restriction-free method for inserting target genes into plasmids. J Biochem Biophys Methods 67:67–74
17. Quan J, Tian J (2009) Circular polymerase extension cloning of complex gene libraries and pathways. PLoS One 4:e6441
18. Erijman A, Dantes A, Bernheim R et al (2011) Transfer-PCR (TPCR): a highway for DNA cloning and protein engineering. J Struct Biol 175:171–177
19. Geertsma ER, Dutzler R (2011) A versatile and efficient high-throughput cloning tool for structural biology. Biochemistry 50:3272–3278
20. Szybalski W, Kim SC, Hasan N et al (1991) Class-IIS restriction enzymes–a review. Gene 100:13–26
21. Bernard P, Gabant P, Bahassi EM et al (1994) Positive-selection vectors using the F plasmid *ccdB* killer gene. Gene 148:71–74
22. Recorbet G, Robert C, Givaudan A et al (1993) Conditional suicide system of Escherichia coli released into soil that uses the Bacillus subtilis *sacB* gene. Appl Environ Microbiol 59: 1361–1366
23. Casadaban MJ, Cohen SN (1980) Analysis of gene control signals by DNA fusion and cloning in *Escherichia coli*. J Mol Biol 138:179–207
24. Don RH, Cox PT, Wainwright BJ et al (1991) "Touchdown" PCR to circumvent spurious priming during gene amplification. Nucleic Acids Res 19:4008
25. Sambrook J, Russell DW (2001) Molecular cloning: a laboratory manual. Cold Spring Harbor Laboratory Press, Cold Spring Harbor, NY
26. Guzman LM, Belin D, Carson MJ et al (1995) Tight regulation, modulation, and high-level expression by vectors containing the arabinose P_{BAD} promoter. J Bacteriol 177:4121–4130

Chapter 12

Minimum GC-Rich Sequences for Overlap Extension PCR and Primer Annealing

Mikiko Nakamura, Ayako Suzuki, Hisashi Hoshida, and Rinji Akada

Abstract

PCR is a common method to produce desired DNA fragments from templates. The oligonucleotide primers used for PCR must contain annealing sequences that are usually 20–30 nucleotides long and identical to a part of template DNA. However, primers often contain additional sequences at their 5′ ends, which are restriction enzyme sites, recombination targeting sequences, or overlap sequences for fusion PCR. When these additional sequences are attached to their annealing sequences, the annealing sequences can be shortened. Here, we describe universal GC-rich annealing sequences useful for overlap extension PCR and simple in-frame addition of desired sequences.

Key words Fusion PCR, Primer design, Additional sequence, Gene splicing, Recombinant DNA, Homologous recombination, Gene targeting, Gene disruption, Yeast, Mammalian cells

1 Introduction

PCR is a universal method used in almost all the molecular biology laboratories. Many DNA fragments are produced routinely by PCR, and inserted into various vectors, used directly for transformation to host organisms, or further manipulated for desired purposes. PCR amplification requires two primers designed from the target sequences. The primers can be ordered from commercial providers and synthesized as oligonucleotides that are usually 18–30 nucleotides in length in conventional PCR amplification [1]. Generally, primer sequence and length are determined by annealing temperature that indicates the binding ability to their complementary sequences. The calculation formulas, for example, $\{(G+C)\times 4\ ^\circ C\}+\{(A+T)\times 2\ ^\circ C\}$, are used to determine annealing temperatures depending on the primer sequence and length [2]. The PCR cycle has an annealing step, which is usually performed at 50–60 °C. Primer annealing occurs at the annealing steps in PCR cycles, followed by the polymerization extension by DNA polymerase from the 3′ end of an annealed primer.

Svein Valla and Rahmi Lale (eds.), *DNA Cloning and Assembly Methods*, Methods in Molecular Biology, vol. 1116, DOI 10.1007/978-1-62703-764-8_12, © Springer Science+Business Media New York 2014

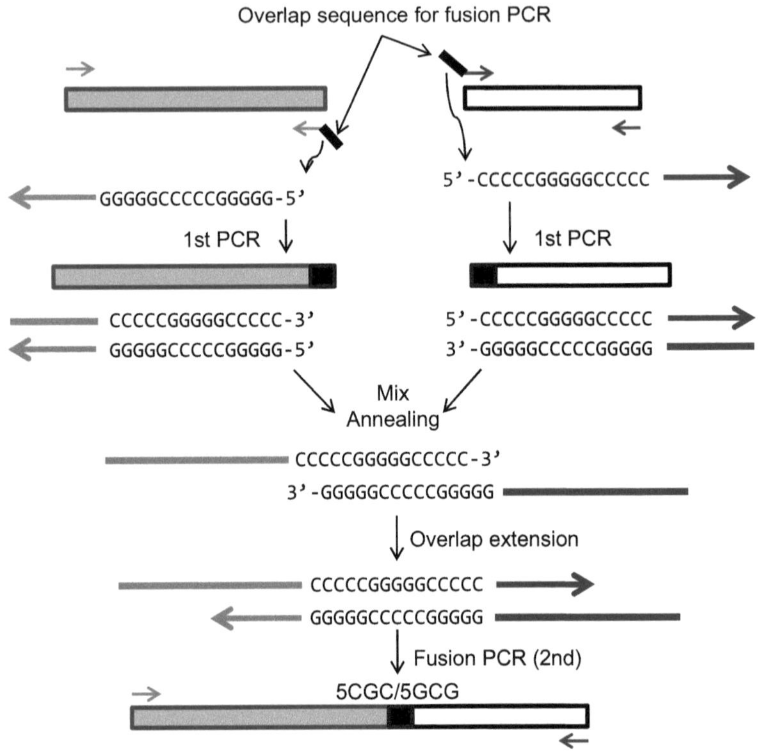

Fig. 1 Schematic representation of overlap extension PCR. Two DNA fragments are amplified with primers containing overlapping sequence. The *left* fragment is amplified with a 5GCG-containing primer and the *right* fragment is with a 5CGC-containing primer. The two DNA fragments are mixed, and PCR is performed with primers that produce fused DNA. 5CGC/5GCG annealing allows efficient overlap extension reaction, resulting in successful fusion DNA construction

If a primer sequence has a high annealing temperature as calculated by the formulas, annealing becomes more specific and results in high yield of PCR products. Since G–C bond is stronger than A–T, GC-rich sequences are thought to be useful for efficient annealing. Based on this concept, we developed novel overlap sequences for fusion PCR (Fig. 1) [3].

Overlap extension PCR or fusion PCR is a traditional method to fuse two DNA fragments [4, 5]. When two fragments contain an overlapping sequence, one with the overlap at the 3′ end and the other at the 5′ end, two fragments can anneal with each other at the overlapping region (Fig. 1). After the annealing of the fragments, annealed DNA sequences can work as primers by themselves for DNA polymerase, from which the DNA extension occurs. This overlap extension produces a fused DNA which becomes a template of the two primers that amplify the fused DNA product. Theoretically, this overlap extension PCR is feasible but practically, it is not reliable [3, 6]. When an AT-rich natural sequence was used

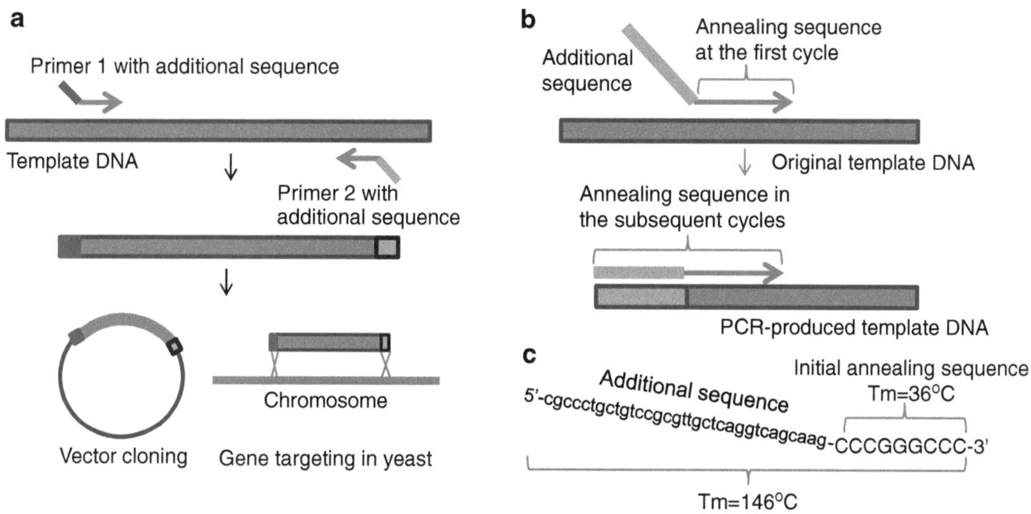

Fig. 2 Application and annealing mode of primers containing additional sequences. (**a**) Primers containing additional sequences are used to attach desired sequences to the 5′ ends of PCR products. These sequences are restriction enzyme sites, recombination targeting sequences, or other sequences to manipulate the genes of interest. (**b**) At the first cycle of PCR, annealing occurs only at the initial annealing sequence. In the subsequent cycles, however, annealing occurs through the entire primer region because the resulting PCR product is used as a template. (**c**) Even if an annealing sequence is as short as 9 nucleotides, the primer can be used for the amplification from a template DNA containing the same annealing sequence. Even if the annealing temperature of the initial annealing region is calculated to 36 °C, Tm of the entire primer sequence becomes higher

as the overlap sequence, only low amount of fusion PCR product was obtained, probably due to the inefficient overlap annealing. We found that GC-rich overlap sequences of only 15 nucleotides in length, such as 15C (5′-CCCCCCCCCCCCCCC-3′) and 5CGC (5′-CCCCCGGGGGCCCCC-3′), were successfully used for fusion PCR [3]. Fusion PCR using these overlapping sequences is reliable without any special attention to PCR conditions and reagents.

PCR primers often contain additional sequences in addition to their annealing sequences, such as 15C and 5CGC in the case of fusion PCR. There are many cases to design additional sequences at their 5′ ends of primers (Fig. 2a). To clone DNA fragments into vectors, restriction enzyme sites or recombination targeting sequences are generated by PCR using primers containing these sequences at their 5′ ends, as flanking sequences [7]. In yeast gene targeting, a transformation selection marker gene is amplified with primers containing homologous recombination sequences of 40–60 nucleotides in length. These DNA fragments can be used for transformation of host yeast to perform gene disruption or targeting [8]. In these PCRs, annealing occurs between the initial annealing sequences of primers and the initial templates at the first cycle but, from the second cycles, annealing can occur through the entire primer sequences because amplified PCR products have entire primer sequences (Fig. 2b). This suggests that the calculation of an annealing temperature from the initial annealing sequence may not be applied in these primers (Fig. 2c). Therefore, if a primer contained an additional sequence, shorter annealing sequence could be designed [9]. This concept led us to develop universal annealing sequences that can be used together with additional sequences. We describe here minimum GC-rich annealing sequences of only 9 nucleotides long for primer annealing without any additional requirements on PCR. These short annealing sequences can also be used for in-frame sequence manipulations. The short universal primer sequences do not only reduce primer cost but also reduce the number of primers to be prepared.

2 Materials

2.1 Oligonucleotide Primers

1. Primers used are listed in Table 1. The naming of the oligonucleotide primers is in accordance with naming conventions and directly related to the oligonucleotide features (*see* **Note 1**).
2. Oligonucleotide primers were dissolved in sterile water to give a final concentration of 10 μM and stored at −20 °C.
3. Oligonucleotides (standard grade) were ordered from commercial suppliers.

2.2 PCR Reagents and Agarose Gel Electrophoresis

1. DNA polymerase: KOD Plus polymerase (Toyobo) and PrimeSTAR GXL DNA polymerase (Takara-Bio).
2. PCR buffer: KOD Plus buffer (10×), 25 mM $MgSO_4$, 2 mM dNTPs, PrimeStar GXL buffer (5×, Mg^{2+} plus), and 2.5 mM dNTPs.
3. TAE buffer (50×): mix 242 g Tris base, 57.1 mL of acetic acid, and 100 mL of 0.5 M EDTA, pH 8.0, and adjusted to 1 L with sterile water. Dilute to 1× TAE for electrophoresis and agarose gel preparation.
4. Agarose gel: 0.8–2.0 % (w/v) agarose was dissolved by heating in 1× TAE buffer with 0.05 μg/mL ethidium bromide.
5. DNA concentration measurement: Qubit fluorometer and Quant-iT dsDNA Assay Kit (Invitrogen).
6. Thermocycler.

2.3 Gene Manipulation in Yeast

1. Yeast plasmids and strains: *Kluyveromyces marxianus* plasmids used were pKM141 containing Sc*URA3-TDH3p*-Nc*Su9*-yEGFP2 [10] maintained in *K. marxianus* strain RAK8961, and pKM382 containing Sc*URA3-TDH3p-yEmRFP* [11] maintained in *K. marxianus* strain RAK9172. The total DNA

Table 1
Oligonucleotide primers

Primer name[a]	Sequence (5′–3′)[b]
6AC-pEGFP-600	acacacacacacgtaatcaattacggggtcatta
pEGFP+699c	GCCGAGAGTGATCCCGGCGGCGGTC
5GCG-pEGFP+699c	GGGGGCCCCCGGGGGgccgagagtgatcccggcggcggt
3GCG-EGFPc	GGGCCCGGGCCCGGGcttgtacagctcgtccatgc
pEGFP+670699-Bax+1	TTCGTGACCGCCGCCGGGATCACTCTCGGCatggacgggtccggggagca
5CGC-Bax+1	CCCCCGGGGGCCCCCatggacgggtccggggagca
3CGC-Bax+1	CCCGGGCCCGGGCCCatggacgggtccggggagca
pEGFP+1018c	tgatgagtttggacaaaccacaact
5CG12-yEGFP2+4	CCCCCGGGGGCCgtctctaagggtgaagaattg
5CG9-yEGFP2+4	CCCCCGGGGgtctctaagggtgaagaattg
5CG6-yEGFP2+4	CCCCCGgtctctaagggtgaagaattg
12C-yEGFP2+4	CCCCCCCCCCCCgtctctaagggtgaagaattg
9C-yEGFP2+4	CCCCCCCCCgtctctaagggtgaagaattg
6C-yEGFP2+4	CCCCCCgtctctaagggtgaagaattg
3CG12-yEGFP2+4	CCCGGGCCCGGGgtctctaagggtgaagaattg
3CG9-yEGFP2+4	CCCGGGCCCgtctctaagggtgaagaattg
3CG6-yEGFP2+4	CCCGGGgtctctaagggtgaagaattg
NcSu9(67-99)-5CG12	cgccctgctgtccgcgttgctcaggtcagcaagCCCCCGGGGGCC
NcSu9(67-99)-5CG9	cgccctgctgtccgcgttgctcaggtcagcaagCCCCCGGGG
NcSu9(67-99)-5CG6	cgccctgctgtccgcgttgctcaggtcagcaagCCCCCG
NcSu9(67-99)-12C	cgccctgctgtccgcgttgctcaggtcagcaagCCCCCCCCCCCC
NcSu9(67-99)-9C	cgccctgctgtccgcgttgctcaggtcagcaagCCCCCCCCC
NcSu9(67-99)-6C	cgccctgctgtccgcgttgctcaggtcagcaagCCCCCC
NcSu9(67-99)-3CG12	cgccctgctgtccgcgttgctcaggtcagcaagCCCGGGCCCGGG
NcSu9(67-99)-3CG9	cgccctgctgtccgcgttgctcaggtcagcaagCCCGGGCCC
NcSu9(67-99)-3CG6	cgccctgctgtccgcgttgctcaggtcagcaagCCCGGG
yEGFP2+720c	TTACTTGTACAATTCGTCCATACCG
SKLc-yEmRFP+687c	TTATAATTTGGAaccagttgaatgtctaccttcagc
3CG9-yEmRFP+687c	CCCGGGCCCaccagttgaatgtctaccttcagc
5CG9-yEmRFP+687c	CCCCCGGGGaccagttgaatgtctaccttcagc

(continued)

Table 1
(continued)

Primer name[a]	Sequence (5′–3′)[b]
9C-yEmRFP+687c	CCCCCCCCCaccagttgaatgtctaccttcagc
SKLc-3CG9	ttataatttggaCCCGGGCCC
SKLc-5CG9	ttataatttggaCCCCCGGGG
SKLc-9C	ttataatttggaCCCCCCCCC
ScTDH3+1000	tgaatttactttaaatcttgcatt
3CG9-pEGFP+699c	CCCGGGCCCgccgagagtgatcccggcggcggtc

[a] *See* **Notes 1** and **2** for primer naming

[b] Overlap and GC-rich annealing sequences are indicated by capital letters

was isolated from the strains according to a standard method [12] and used as a template. *K. marxianus* strain RAK3605 (*ura3*⁻) [13, 14] was used as a host for transformation.

2. Yeast extract-peptone-dextrose (YPD) medium: 1 % yeast extract, 2 % polypeptone, 2 % glucose, and 2 % agar (if necessary).
3. Synthetic dextrose (SD) drop out medium: 0.17 % yeast nitrogen base without amino acids and without ammonium sulfate, 0.5 % ammonium sulfate, 2 % glucose, and required nutrients [12].
4. Transformation mixture (TM): 40 % w/v polyethyleneglycol 3350 (Sigma-Aldrich), 200 mM lithium acetate, and 100 mM dithiothreitol. PEG3350 was dissolved in hot sterile water (60–80 °C), sterilized by autoclaving, and adjusted to final concentration of 60 % (w/v). 4 M lithium acetate solution was prepared in water and autoclaved. 1 M dithiothreitol was prepared in water and filter-sterilized. Fresh TM was prepared for each transformation experiment just before the procedure by mixing the above stock solutions.
5. Yeast transformation [14]: Yeast cells were inoculated with a sterile toothpick from a fresh YPD plate into 50 mL YPD liquid medium in a 250 mL Erlenmeyer flask. The cells were incubated at 30 °C and grown to the stationery phase (18–24 h) with shaking at 150 rpm. The cells were collected by centrifugation and suspended in 2 mL TM by vortexing. Then the cell suspension was transferred to a tube and centrifuged. The pellet was suspended in 2.5 mL TM by vortexing and pipetting. Aliquot of the cell suspension (50 μL) was transferred to a microtube and 2–3 μL of PCR-amplified DNA fragment was added. The mixture was incubated at 42 °C for 30 min. Then, 150 μL of SD was added, and the cell suspension was

spread onto a selection plate. The plates were incubated at 30 °C. After 48 h incubation, cells appeared on the plates were picked and directly observed using a fluorescence microscope.

2.4 Gene Manipulation for Mammalian Cells

1. Plasmid template: pEGFP-C1 (Clontech) and pEGFP-Bax64 [15]. pEGFP-Bax64 was constructed by cloning a mouse Bax mutant in pEGFP-C1.
2. HEK293 cell culture: The human embryonic kidney 293 cells were cultured in RPMI1640 supplemented with 10 % fetal bovine serum containing 1 % penicillin/streptomycin and grown at 37 °C and 5 % CO_2.
3. Transfection of mammalian cells: Fugene HD reagent (Promega) was used for transfection according to the manufacturer's protocol. Cells were seeded at 12,000 cells/300 μL/well in 8-well chamber slides and incubated for 20–24 h prior to transfection. 100 ng of PCR product and 0.2 μL of Fugene HD were mixed with water to give a total of 20 μL, and incubated for 30 min at room temperature. The mixture was mixed with the prepared cells. After 24–48 h, cells were observed using a fluorescence microscope.

3 Methods

3.1 Fusion PCR Via GC-Rich Overlapping Sequences

DNA fragments that are to be fused by PCR are amplified with primers containing overlap sequences, one with the overlap at the 3′ end and the other at the 5′ end. Then, two fragments are mixed and joined by PCR. As an example in this section, the mammalian EGFP green fluorescence protein marker is fused with mouse apoptosis protein Bax. EGFP is amplified from pEGFP-C1 plasmid and a mouse Bax mutant gene is amplified from pEGFP-Bax64 plasmid.

1. For EGFP amplification, 5GCG-pEGFP+699c, pEGFP+699c, and 3GCG-EGFPc reverse primers are used together with a forward primer 6AC-pEGFP-600 to amplify CMV promoter-EGFP region from pEGFP-C1 plasmid (*see* **Note 2**).
2. For Bax amplification, 5CGC-Bax+1, pEGFP+670699-Bax+1, and 3CGC-Bax+1 forward primers are used together with a reverse primer pEGFP+1018c.
3. Mix 5.4 μL of sterile water, 2 μL of 5× PrimeSTAR GXL buffer, 0.8 μL of 2.5 mM dNTPs, 0.3 μL of 10 μM forward primer, 0.3 μL of 10 μM reverse primer, 1 μL of template DNA (0.5 ng/μL), and 0.2 μL of GXL DNA polymerase. Total volume: 10 μL.
4. Run a PCR program comprising an initial denaturation step at 98 °C for 4 min, followed by 30 cycles of 98 °C for 10 s, 55 °C for 15 s, and 68 °C for 1.5 min.

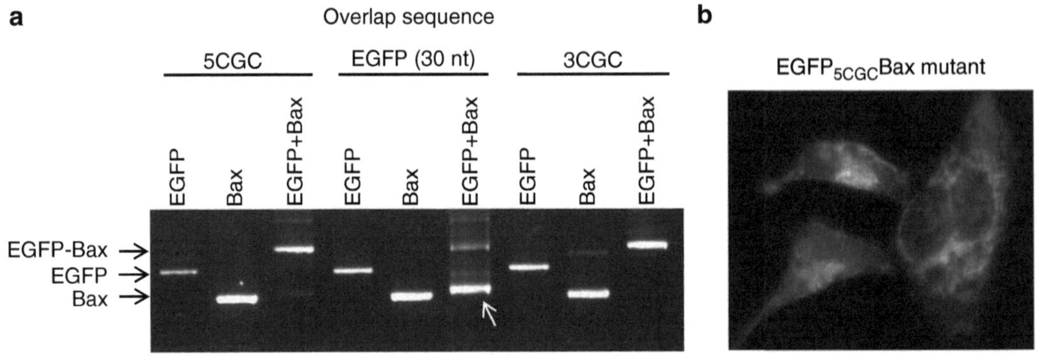

Fig. 3 Construction of an EGFP-Bax mutant fusion with primers containing GC-rich overlap sequence. (**a**) Results of fusion PCR through three types of overlap sequences. CMV promoter-EGFP sequence was fused with a mouse Bax mutant gene using 5CGC, EGFP C-terminal 30-nucleotide (30 nt), or 3CGC overlap sequence. The fusion PCR with the EGFP sequence showed a faint band of EGFP-Bax DNA and a strong unexpected band (*white arrow*) together with the other bands. The fusion PCR with 5CGC and 3CGC overlap sequences showed single strong fusion bands. (**b**) The constructed fusion gene via 5CGC overlap was introduced into HEK293 cells. The expected mitochondrial localization of a Bax mutant was observed, indicating that in-frame fusion between EGFP and Bax was successful

5. Check the PCR products by agarose gel electrophoresis (*see* **Note 3**, Fig. 3a).
6. Measure the DNA concentrations by Qubit fluorometer and Quant-iT dsDNA Assay Kit as recommended by the supplier (Invitrogen) (*see* **Note 4**).
7. Adjust DNA concentration to 0.1 ng/μL (*see* **Note 5**).
8. Use 0.1 ng of each DNA product in a 10-μL reaction mixture for fusion PCR (*see* **Note 6**). Mix combinations are as follows, a 5GCG-pEGFP+699c product with a 5CGC-Bax+1 product, a pEGFP+699c product with a pEGFP+670699-Bax+1 product, and a 3GCG-pEGFP+699c product with a 3CGC-Bax+1 product.
9. Mix 1 μL of an EGFP DNA product, 1 μL of a Bax DNA product, 2 μL of 5× PrimeSTAR GXL buffer, 0.8 μL of 2.5 mM dNTPs, 0.3 μL of 10 μM forward primer (6AC-pEGFP-600), 0.3 μL of 10 μM reverse primer (pEGFP+1018c), 0.2 μL GXL DNA polymerase, and 4.4 μL sterile water. Total volume: 10 μL.
10. Run a PCR program comprising an initial denaturation step at 98 °C for 4 min, followed by 30 cycles of 98 °C for 10 s, 60 °C for 15 s, and 68 °C for 3 min (*see* **Note 7**).
11. Check the PCR products by agarose gel electrophoresis (Fig. 3a, *see* **Note 8**).
12. Fused EGFP-Bax64 constructs are used for transfection to HEK293 cells. The fusion constructs showed mitochondrial

localization of a Bax mutant (Fig. 3b, *see* **Note 9**), indicating that in-frame fusion occurred efficiently by the used overlap sequences.

3.2 Minimum Annealing Sequences for Primer Design When Used with an Additional Sequence

When a primer contains an additional sequence, only 9–12 nucleotides long annealing sequence can be used for PCR without any special attention [9]. As the minimum GC-rich annealing sequences, we propose 9 nucleotides long 5CG9 (5′CCCCCGGGG3′), 9C (5′CCCCCCCCC3′), and 3CG9 (5′CCCGGGCCC3′) for designing primers when it is used with additional sequences. For the preparation of templates containing GC-rich annealing sequences, first PCR is performed with primers containing these GC-rich sequences. Then, the templates are used for the second PCR with the primers containing the minimum GC-rich annealing sequences and desired additional sequences.

1. Use the following forward primers; 5CG12-yEGFP2+4, 5CG9-yEGFP2+4, 5CG6-yEGFP2+4, 12C-yEGFP2+4, 9C-yEGFP2+4, 6C-yEGFP2+4, 3CG12-yEGFP2+4, 3CG9-yEGFP2+4, and 3CG6-yEGFP2+4 (*see* **Note 2** for sequence nomenclature) with a reverse primer yEGFP2+720c to prepare yEGFP2 (yeast-codon-optimized EGFP) templates by PCR.
2. Mix 6.6 μL of sterile water, 1 μL of 10× KOD Plus buffer, 1 μL of 2 mM dNTPs, 0.4 μL of 25 mM $MgSO_4$, 0.2 μL of 10 μM forward primer, 0.2 μL of 10 μM reverse primer yEGFP+720c, 0.4 μL of *K. marxianus* total DNA (RAK8961) containing pKM141 plasmid (0.5 ng/μL) (*see* **Note 10**), and 0.2 μL of KOD Plus polymerase. Total volume: 10 μL.
3. Run a PCR program comprising an initial denaturation step at 94 °C for 1 min, followed by 30 cycles of 94 °C for 20 s, 60 °C for 30 s, and 68 °C for 2 min.
4. Check the PCR products by agarose gel electrophoresis. Each PCR product should yield single bands (0.7 kb) upon gel analysis.
5. Adjust the DNA concentration to 0.5 ng/μL. These PCR products contain GC-rich sequences at the ends of the forward primer side.
6. Perform second PCR with the following forward primers; NcSu9(67-99)-5CG12, NcSu9(67-99)-5CG9, NcSu9(67-99)-5CG6, NcSu9(67-99)-12C, NcSu9(67-99)-9C, NcSu9(67-99)-6C, NcSu9(67-99)-3CG12, NcSu9(67-99)-3CG9, and NcSu9(67-99)-3CG6 with a reverse primer yEGFP2+720c, respectively (*see* **Note 11**).
7. Mix 6.6 μL of sterile water, 1 μL of 10× KOD Plus buffer, 1 μL of 2 mM dNTPs, 0.4 μL of 25 mM $MgSO_4$, 0.2 μL of 10 μM forward primer, 0.2 μL of 10 μM yEGFP2+720c primer, 0.4 μL of template DNA (0.5 ng/μL), and 0.2 μL of KOD Plus polymerase. Total volume: 10 μL.

Primer name	Primer sequence	Template sequence	Result (0.7 kb)
NcSu9(67-99)-5CG12	33nt-CCCCCGGGGGCC	CCCCCGGGGGCC-yEGFP2	Succeeded
NcSu9(67-99)-5CG9	33nt-CCCCCGGGG	CCCCCGGGG-yEGFP2	Succeeded
NcSu9(67-99)-5CG6	33nt-CCCCCG	CCCCCG-yEGFP2	Weak
NcSu9(67-99)-12C	33nt-CCCCCCCCCCCC	CCCCCCCCCCCC-yEGFP2	Succeeded
NcSu9(67-99)-9C	33nt-CCCCCCCCC	CCCCCCCCC-yEGFP2	Succeeded
NcSu9(67-99)-6C	33nt-CCCCCC	CCCCCC-yEGFP2	Weak
NcSu9(67-99)-3CG12	33nt-CCCGGGCCCGGG	CCCGGGCCCGGG-yEGFP2	Failed
NcSu9(67-99)-3CG9	33nt-CCCGGGCCC	CCCGGGCCC-yEGFP2	Succeeded
NcSu9(67-99)-3CG6	33nt-CCCGGG	CCCGGG-yEGFP2	Failed

Fig. 4 Primers with short annealing sequences and a long additional sequence can amplify DNA templates with the same annealing sequences. A template DNA (yEGFP2 gene) was first amplified with the primers containing short annealing sequences (template sequence). These amplified DNAs were used as templates for the second PCR with the primers containing short annealing sequences (primer sequence). Even if annealing sequences were only 9 nucleotides long, such as 5CG9, 9C, and 3CG9, PCR amplification was successful (0.7 kb). 3CG12 showed no amplification probably due to the strong palindromic structure

8. Run a PCR program comprising an initial denaturation step at 94 °C for 1 min, followed by 30 cycles of 94 °C for 20 s, 60 °C for 30 s, and 68 °C for 2 min (*see* **Note 12**).
9. Check the PCR products by agarose gel electrophoresis (Fig. 4).
10. The results demonstrated that the primers with 6-nucleotide long annealing sequences showed only faint or no DNA band but the primers with 9- or 12-nucleotide annealing sequences (*see* **Note 2** for sequence name) showed strong DNA bands except for 3CG12, indicating that the 9-nucleotide long GC-rich sequences can be used for minimum annealing sequences of the primers with additional sequences. The result of failed DNA amplification of 3CG12 may be caused by the palindromic structure within the primer sequence.

3.3 Application of the Minimum GC-Rich Annealing Sequences for the Addition of a Functional Sequence

Nine nucleotides long GC-rich annealing sequences can be used for primer design if the primers contain additional sequences. Sequence addition to the end of DNA fragment is often required but usually it is mediated through the natural sequence of 18–25 nucleotides in length at the end as an annealing sequence. If the annealing sequence becomes 9 nucleotides, primer cost can be reduced. In addition, the minimum annealing sequences can be used with any templates together with the universal primers for the desired sequence addition. For example, restriction enzyme sites, recombination targeting sequences, and epitope tags are often attached to the genes and DNA fragments to be manipulated (Fig. 5). If primers consisting of the GC-rich minimum annealing sequences and the desired additional sequences are prepared, these can be used universally to add the desired sequences

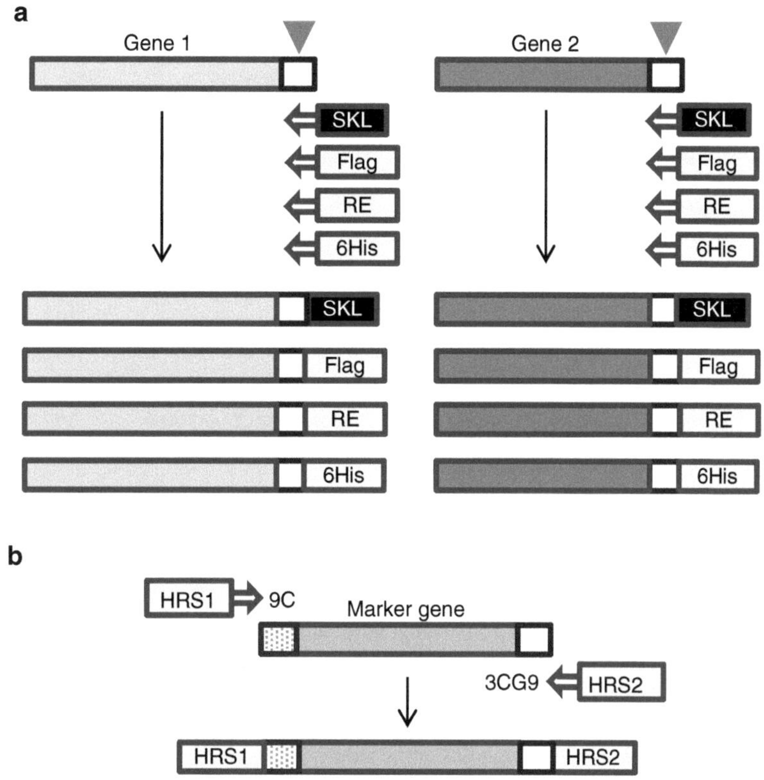

Fig. 5 Application of minimum annealing sequences for the addition of various sequences. (**a**) The minimum annealing sequences are generated to the end of various DNA fragments by PCR with primers containing additional minimum 9-nucleotide sequences (*arrowheads*). The annealing sequence can be used together with a set of primers containing desired additional sequences, such as SKL peroxisomal targeting sequence, FLAG tag, restriction enzyme sites (RE), and 6His affinity purification tag for gene manipulations. These primers or the other additional sequence primers can be used universally for their addition to any genes through the minimum annealing sequences. (**b**) The minimum annealing sequences can be used simultaneously if they are not complementary. For example, a transformation marker gene is amplified with primers containing homologous recombination sequences (HRS) through 9C and 3CG9 annealing sequences for gene targeting in yeast (*see* **Note 13**)

to any templates with the same minimum annealing sequences (Fig. 5, *see* **Note 13**).

To show the efficiency and accuracy of the annealing of the minimum sequences, in-frame addition of a functional peptide coding sequence is demonstrated in this section.

As an example of in-frame addition of a peptide sequence, C-terminal peroxisome-targeting signal SKL (Ser-Lys-Leu) is selected [16]. First, yEmRFP (yeast-codon optimized monomeric RFP) [11] is amplified with primers containing minimum annealing sequences.

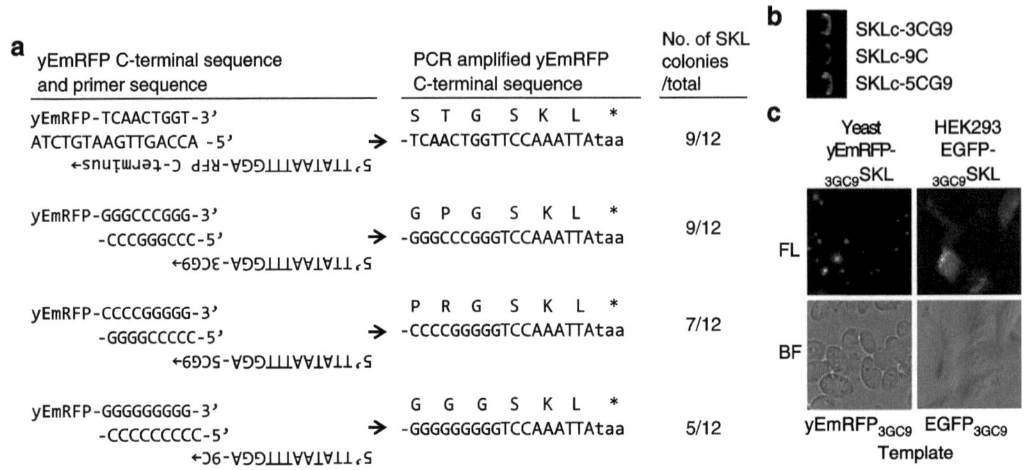

Fig. 6 In-frame addition of a peroxisome-targeting sequence to RFP/GFP through GC-rich minimum annealing sequences. (**a**) *URA3-TDH3* promoter-yEmRFP template was amplified with primers containing indicated minimum annealing sequences (*bold*). The amplified products were used as templates for the second PCR with primers containing minimum annealing sequences (5CG9, 9C, 3CG9, and 3GC9) and a complementary sequence (*italics*) of SKL for peroxisome targeting. The PCR products were used for transformation to the yeast *Kluyveromyces marxianus ura3* mutant strain. Transformants showed punctate distribution of RFP signals. Number of the correct in-frame addition of the SKL sequence was counted and compared to that through the RFP C-terminal sequence. (**b**) Second PCR amplification for the addition of SKL sequence. (**c**) Cellular localization of yEmRFP and EGFP with SKL attached by 3CG9 annealing was shown. *FL* fluorescence, *BF* bright field

These PCR products are used as templates for the addition of SKL sequence to the RFP C-terminus (Fig. 6).

1. For the preparation of templates, use a forward primer ScTDH3+1000 with the following reverse primers: 3CG9-yEmRFP+687c, 5CG9-yEmRFP+687c, and 9C-yEmRFP+687c, to prepare templates of yEmRFP by PCR (*see* **Note 2** for sequence name).
2. Mix 5.4 μL of sterile water, 2 μL of 5× PrimeSTAR GXL buffer, 0.8 μL of 2.5 mM dNTPs, 0.3 μL of 10 μM forward primer, 0.3 μL of 10 μM reverse primer, 1 μL of *K. marxianus* total DNA (RAK9172) containing pKM382 plasmid (0.5 ng/μL), and 0.2 μL of GXL DNA polymerase. Total volume: 10 μL (*see* **Note 10**).
3. Run a PCR program comprising an initial denaturation step at 98 °C for 4 min, followed by 30 cycles of 98 °C for 10 s, 55 °C for 15 s, and 68 °C for 5 min.
4. Check the PCR products by agarose gel electrophoresis. Each product showed a clear DNA band (5.3 kb, data not shown).
5. Adjust the DNA concentration to 0.5 ng/μL (*see* **Note 14**). These PCR products contain GC-rich sequences at the ends of the reverse primer side.

6. Perform second PCR with a forward primer ScTDH3+1000 and the following reverse primers of SKLc-3CG9, SKLc-5CG9, and SKLc-9C, and the templates that had been amplified with the primers of 3CG9-yEmRFP+687c, 5CG9-yEmRFP+687c, and 9C-yEmRFP+687c, respectively.
7. Mix 5.4 μL of sterile water, 2 μL of 5× PrimeSTAR GXL buffer, 0.8 μL of 2.5 mM dNTPs, 0.3 μL of 10 μM forward primer, 0.3 μL of 10 μM reverse primer, 1 μL of template DNA (0.5 ng/μL), and 0.2 μL of GXL DNA polymerase. Total volume: 10 μL (*see* **Note 14**).
8. Run a PCR program comprising an initial denaturation step at 98 °C for 4 min, followed by 30 cycles of 98 °C for 10 s, 55 °C for 15 s, and 68 °C for 5 min.
9. Check the PCR products by agarose gel electrophoresis (Fig. 6b, *see* **Note 14**).
10. The DNA products are used for transformation to the yeast *K. marxianus* strain RAK3605 [14]. Transformant colonies are picked and observed by a fluorescence microscopy (Fig. 6).
11. GC-rich 9-nucleotide annealing sequences showed normal PCR production at 55 °C annealing temperature (Fig. 6b) and higher correct annealing was achieved by 3CG9 and 5CG9 sequences (Fig. 6a). 9C annealing showed slightly low frequency of the correct in-frame addition of SKL (5/12).
12. A CMVp-EGFP$_{3GC9}$ construct was also prepared by PCR with an EGFP C-terminal primer containing 3CG9 (3CG9-EGFP+699c) and pEGFP-C1 as a template. The same SKL primer, SKLc-3CG9, was used for the second PCR and the product was processed for the transfection to HEK293 cells. The peroxisomal localization of EGFP was observed (Fig. 6c, *see* **Notes 15** and **16**).

4 Notes

1. Oligonucleotide primers are designated by rules to make it easy to understand their features [17].

   Rule 1. Oligonucleotide names are given according to the sequence from the 5′ to the 3′ end.

   Rule 2. Many of the oligonucleotides are sequences related to their appearance in genes. These oligonucleotides can be identified by their distance from the ATG start codon of the respective genes. When position A of ATG start codon is counted as +1, the 5′ end of many of the primers can be indicated by the distance from this +1 position. For example, Bax+1 indicates the location that is started from the ATG start codon of Bax gene.

Table 2
Nomenclature of GC-rich sequence

Name	Sequence
15C	5′-ccccccccccccccc-3′
15G	5′-ggggggggggggggg-3′
5CGC	5′-cccccgggggccccc-3′
5GCG	5′-gggggcccccggggg-3′
3CGC	5′-cccgggcccgggccc-3′
3GCG	5′-gggcccgggcccggg-3′
12C	5′-cccccccccccc-3′
9C	5′-ccccccccc-3′
6C	5′-cccccc-3′
5CG12	5′-cccccgggggcc-3′
5CG9	5′-cccccgggg-3′
5CG6	5′-cccccg-3′
3CG12	5′-cccgggcccggg-3′
3CG9	5′-cccgggccc-3′
3CG6	5′-cccggg-3′

Rule 3. Oligonucleotides that relate to genes are also designed to be complementary to these genes. This can be marked by including "c" at the end of their name. For example, pEGFP+699c indicates that the 5′ end is located at position 699-bp downstream from the A of ATG of EGFP gene and that the sequence is complementary to the EGFP coding sequence.

Rule 4. Names can be combined according to the 5′–3′ sequence and joined with a hyphen. For example, 5GCG-pEGFP+699c indicates that this oligonucleotide contains 5GCG at the 5′ end, followed by the pEGFP+699c sequence.

2. The GC-rich overlap or annealing sequences are designated as follows (Table 2).
3. A 0.8–2 % (w/v) agarose gel is prepared in 1× TAE, and 1–2 μL of a DNA product was loaded.
4. Measurement of DNA concentration is important for reliable fusion PCR. We use a fluorometer to measure DNA concentration as it requires small amount of DNA solution and RNA removal is not required. Usually 40–100 ng/μL DNA is obtained by PCR.

5. Adjust to 0.1–0.5 ng/μL. If the DNA concentration cannot be measured, approximately 100 times dilution of the PCR product will give this range.
6. Concentration of the DNA fragments is critical for fusion PCR. If higher concentration of DNA fragment is used, unexpected bands or smear appears but if it is lower or unequal between two fragments, low or no yield of fusion DNA is obtained [3].
7. Fusion PCR requires relatively higher annealing temperatures. We routinely use 60 °C. Lower annealing temperatures will not give better results.
8. Sometimes low PCR amplification from certain templates was experienced, possibly due to strong GC annealing to unexpected region of the template. In such cases, 10CA/10TG and 10CT/10AG overlaps [17] can be applied.
9. The used Bax mutant localizes at mitochondria in mammalian cells without apoptotic signal. 5CGC and 3CGC showed similar results but the fusion construct via an EGFP C-terminal sequence showed cytoplasmic localization in addition to mitochondrial localization. This may be caused by unexpected products shown as smear in Fig. 3a. 5CGC and 3CGC encode PPGAP and PGPGP, respectively, and these linker sequences did not affect Bax localization [3].
10. A replicating plasmid is constructed and maintained in *K. marxianus*. The plasmid region can be amplified directly from yeast total DNA (chromosomal DNA) as a template.
11. NcSu9(67-99) is a 33-nucleotide long sequence of a mitochondrial targeting signal of *Neurospora crassa* ATPase subunit 9 [10] and used as an additional sequence of the primers in this case.
12. Higher annealing temperatures than 60 °C gave lower yield.
13. We routinely use 3CG9 and 9C, and these can be used simultaneously. In yeast gene manipulation, gene disruption is performed by transformation of a selection marker gene containing homologous sequences, usually 40–60 nucleotides in length, at the both ends. We use 9C and 3CG9 minimum annealing sequences for the addition of the homologous sequences. If another marker gene was required for the disruption, the same primers of the homologous sequences can be used for the amplification of the other markers that attached 9C and 3CG9 at the ends. Genome-wide gene manipulations often use thousands of DNA constructs. Even if only 5–10 nucleotides were reduced in primer design, it will greatly save primer cost for genome-wide design [18], and the same set of primers is possibly used for the other projects.

14. Lower DNA concentrations of the templates are sufficient for amplification. We examined until 5 fg/μL of the final template concentration in a PCR mixture. The same amount of DNA product was obtained in all concentrations.
15. The same primer, SKLc-3CG9 in this case, can be used universally for the addition of the SKL sequence. We also examined 5CG9, 9C, and the natural sequence for annealing. 9C annealing showed slightly higher cytoplasmic localization signal in HEK293 cells, indicating that 9C may produce out-of-frame mutations.
16. It happens that identical or similar 9-nucleotide sequences of 5CG9, 3CG9, and 9C are present in the template sequences. For example, pEGFP-C1 and pmCherry-C1 plasmids (Clontech) contain a 3CG9/3GC9 sequence. However, if DNA amount of the target DNA containing the minimum annealing sequence as a template was sufficiently higher than the nontarget DNA that contained similar annealing sequences, contamination of the unexpected DNA amplification is low.

Acknowledgments

The authors thank Yukie Misumi for her technical assistance. This work was supported in part by the Adaptable and Seamless Technology Transfer Program through Target-Driven R & D and the Advanced Low Carbon Technology Research and Development Program (JST, Japan).

References

1. Dieffenbach CW, Lowe TMJ, Dveksler GS (1995) General concepts for PCR primer design. In: Dieffenbach CW, Dveksler GS (eds) PCR primer: a laboratory manual. Cold Spring Harbor Laboratory Press, New York, pp 133–142
2. Suggs SV, Hirose T, Miyake T et al (1981) Use of synthetic oligodeoxyribonucleotides for the isolation of specific cloned DNA sequences. ICN-UCLA Symp Dev Biol 23:683–693
3. Cha-aim K, Fukunaga T, Hoshida H et al (2009) Reliable fusion PCR mediated by GC-rich overlap sequences. Gene 434:43–49
4. Heckman KL, Pease LR (2007) Gene splicing and mutagenesis by PCR-driven overlap extension. Nat Protoc 2:924–932
5. Lu Q (2005) Seamless cloning and gene fusion. Trends Biotechnol 23:199–207
6. Vallejo AN, Pogulis RJ, Pease LR (1994) *In vitro* synthesis of novel genes: mutagenesis and recombination by PCR. Genome Res 4: s123–s130
7. Levis R (1995) Strategies for cloning PCR products. In: Dieffenbach CW, Dveksler GS (eds) PCR primer: a laboratory manual. Cold Spring Harbor Laboratory Press, New York, pp 539–554
8. Baudin A, Ozier-Kalogeropoulos O, Denouel A et al (1993) A simple and efficient method for direct gene deletion in *Saccharomyces cerevisiae*. Nucleic Acids Res 21:3329–3330
9. Kakihara Y, Matsuura Y, Hoshida H et al (2005) Cost-saving design of PCR primers containing additional sequences. ITE Lett 6: 135–139
10. Benedikt W, Walter N (2000) Mitochondria-targeted green fluorescent proteins: convenient tools for the study of organelle biogenesis in *Saccharomyces cerevisiae*. Yeast 16:1421–1427
11. Keppler-Ross S, Douglas L, Konopka JB et al (2010) Recognition of yeast by murine macrophages requires mannan but not glucan. Eukaryot Cell 9:1776–1787

12. Amberg DC, Burke D, Strathern JN (2005) Method in yeast genetics: a Cold Spring Harbor Laboratory course manual. Cold Spring Harbor Laboratory Press, New York
13. Nonklang S, Abdel-Banat BM, Cha-aim K et al (2008) High-temperature ethanol fermentation and transformation with linear DNA in the thermotolerant yeast *Kluyveromyces marxianus* DMKU3-1042. Appl Environ Microbiol 74:7514–7521
14. Abdel-Banat BM, Nonklang S, Hoshida H et al (2010) Random and targeted gene integrations through the control of non-homologous end joining in the yeast *Kluyveromyces marxianus*. Yeast 27:29–39
15. Zha H, Fisk HA, Yaffe MP et al (1996) Structure-function comparisons of the proapoptotic protein Bax in yeast and mammalian cells. Mol Cell Biol 16:6494–6508
16. Monosov EZ, Wenzel TJ, Lüers GH et al (1996) Labeling of peroxisomes with green fluorescent protein in living *P. pastoris* cells. J Histochem Cytochem 44:581–589
17. Cha-Aim K, Hoshida H, Fukunaga T et al (2012) Fusion PCR via novel overlap sequences. Methods Mol Biol 852:97–110
18. Winzeler EA, Shoemaker DD, Astromoff A et al (1999) Functional characterization of the *S. cerevisiae* genome by gene deletion and parallel analysis. Science 285:901–906

Chapter 13

Simple Cloning and DNA Assembly in *Escherichia coli* by Prolonged Overlap Extension PCR

Chun You and Y.-H. Percival Zhang

Abstract

We developed a simple method (Simple Cloning) for subcloning one, two, or three DNA fragments into any location of a targeted vector without the need for restriction enzyme, ligase, exonuclease, or recombinase. This cloning technology can be applied to a few common *Escherichia coli* hosts (e.g., BL21(DE3), DH5α, JM109, TOP10). The protocol includes three steps: (a) linear DNA fragments (i.e., the insert DNA and the vector backbone) with two overlap ends were generated by regular high-fidelity PCR, (b) the DNA multimers were generated based on these equimolar DNA templates by using prolonged overlap extension PCR (POE-PCR) without primers added, and (c) the POE-PCR product was transformed to *E. coli* strains directly. Because positive colony efficiencies are very high, it is not necessary to identify desired clones by using colony PCR. Simple Cloning provides a new cloning and DNA assembly method with great simplicity and flexibility.

Key words Enzyme-free cloning, *Escherichia coli*, DNA assembly, Prolonged overlap extension PCR, Simple Cloning, Subcloning

1 Introduction

Escherichia coli stains have a long history of use in the biotechnology industry and biological science studies. They are the most preferred model microorganism for molecular cloning and recombinant protein expression, mainly due to their genetic simplicity, fast growth rates under laboratory conditions, good transformation ability for hosting foreign DNA, the availability of their genome sequence, and laboratory safety [1–3]. The classic molecular cloning technology in *E. coli* is based on the use of restriction enzymes that digest an inserted DNA fragment and a vector as well as the use of a DNA ligase that connects two fragments to yield the desired chimeric plasmid [4]. However, limited choices of restriction enzymes, relatively low efficiencies in digestion and ligation, and possible self-ligation of the digested vector may result in difficulties in obtaining positive colonies when ligating two fragments,

Svein Valla and Rahmi Lale (eds.), *DNA Cloning and Assembly Methods*, Methods in Molecular Biology, vol. 1116, DOI 10.1007/978-1-62703-764-8_13, © Springer Science+Business Media New York 2014

especially for beginners, not to mention three-fragment ligation. In addition, the use of cutting sites of restriction enzymes usually results in the addition of several amino acids at the N- and C-termini of the target protein, and such addition could cause unexpected effects on the biochemical properties or expression levels of proteins [5].

Several companies have developed recombinase-based cloning technologies, including the Invitrogen Gateway cloning technology (Chap. 14), Clonetech In-Fusion (Chap. 15), BioCat Cold-Fusion, and Red/ET Recombination [6–11]. However, all the recombinase-based cloning technologies heavily rely on special kits containing special vectors, enzymes, or hosts.

Here we present a simple and easy cloning method through DNA multimers generated by prolonged overlap extension PCR (POE-PCR) [12–14]. Different from regular overlap extension PCR, several modifications are made: (a) PCR primers are designed to be complementary for the amplification of a targeted vector and an insert gene and (b) DNA multimers are generated by POE-PCR based on two PCR products featuring the overlap regions at the 5′ and 3′ ends without the presence of any primers. Such DNA multimers can be transformed into common laboratory *E. coli* hosts (e.g., BL21(DE3), DH5α, JM109, TOP10), yielding a chimeric plasmid, where *E. coli* strains are capable of cleaving assimilated DNA multimers into the circular plasmid. By this new sequence-independent cloning method, the desired positive colony can be obtained within 1 day without the verification by colony PCR. In our laboratory, we have used this method for constructing more than 100 plasmids with variable length of insertion fragments from 0.2 to 7.0 kb. Among them, the largest plasmid has a size of 14 kb.

2 Materials

2.1 Biological and Chemical Materials

1. New England Biolabs (NEB) Phusion polymerase (Ipswich, MA). Store at −20 °C (*see* **Note 1**).
2. 5× HF PCR buffer or 5× GC PCR buffer for PCR (NEB, provided with the NEB Phusion polymerase). Store at −20 °C.
3. Deoxynucleotide solution mix (dNTP, 10 mM each). Store at −20 °C.
4. Oligomers which were synthesized by Integrated DNA Technologies (San Jose, CA) or other companies. Make the concentration of oligomers to 100 μM by adding appropriate amount of ultrapure water. Store them at −20 °C till used.
5. Plasmids and/or genomic DNA for PCR amplification template.
6. DNA Clean & Concentrator Kit from Zymo Research (Irvine, CA) or its equivalent.

7. Gel DNA recovery kit from Zymo Research or its equivalent.
8. Plasmid Miniprep isolation kit from Zymo Research or its equivalent.
9. NEB restriction enzymes or their equivalents.
10. 0.8 % agarose in 1× TAE buffer containing 40 mM Tris-acetate and 1 mM EDTA with ethidium bromide.
11. Sterile SOC liquid medium (0.5 % yeast extract, 2 % tryptone, 10 mM NaCl, 2.5 mM KCl, 10 mM $MgCl_2$, 10 mM $MgSO_4$, 20 mM glucose. Autoclave the solution without glucose. Sterilize the glucose solution by passing it through a 0.2 μm filter. Mix two solutions together). SOC medium can be stored at room temperature and is stable for several years.
12. Lysogeny Broth (LB) medium: 1 % bacto-tryptone, 0.5 % yeast extract, and 1 % NaCl in deionized water. For plates, 1.5 % agar is added.

2.2 Equipment

1. PCR thermocycler.
2. Agarose gel running system.
3. Nanodrop ND-1000 Spectrophotometer.

3 Methods

3.1 Primer Design

A pair of primers (IF/IR) is used to amplify the DNA fragment of insertion and the other pair of primers (VF/VR) is used to amplify the vector backbone (Fig. 1). VF, a 50-bp forward primer for vector linearization, contains the last 25 bp of 3′ terminal of insert sequence and the first 25 bp of 5′ terminal of vector sequence. IF, a 50 bp forward primer for amplifying the insert, contains the last 25 bp of 3′ terminal of vector sequence and first 25 bp of 5′ terminal of insert sequence (*see* **Note 2**). IR and VR are the reverse complementary sequences of VF and IF, respectively, so the melting temperatures (T_m) of IR and VF should be the same, as well as VR and IF. These two melting temperatures should be designed to match with each other as closely as possible as the primer design requirement for overlap extension PCR because it will decrease mishybridization.

3.2 Simple Cloning

The flow scheme of Fig. 2 represents the general procedure of Simple Cloning. A DNA fragment encoding the *cherry-cbm* gene (1.3 kb) is subcloned into pET20b by Simple Cloning.

1. The vector backbone is amplified with a pair of primers of VF and VR through regular PCR by using the Phusion DNA polymerase. The PCR system contains dNTP, 200 μM for each; primers, 0.5 μM for each; template, 0.05 ng/μL; 1× HF

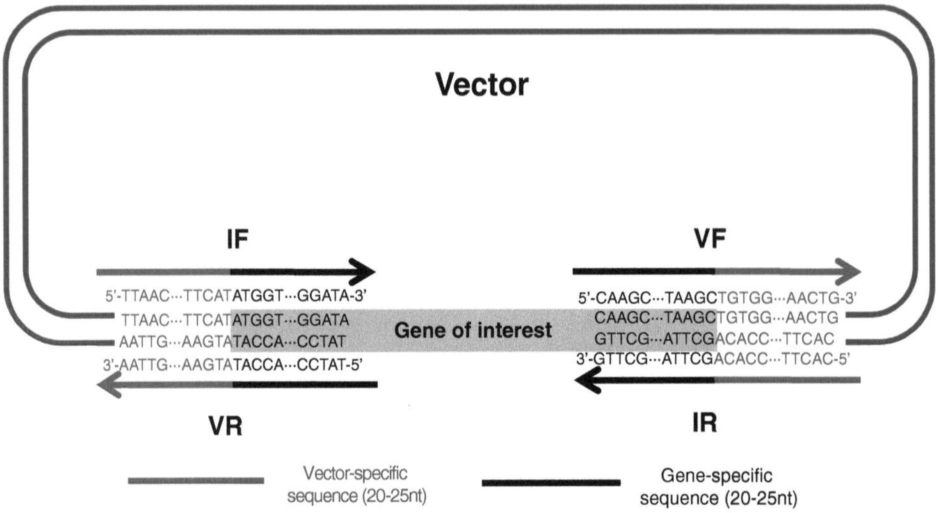

Fig. 1 Primer design for Simple Cloning that can insert one DNA fragment into any place of plasmid

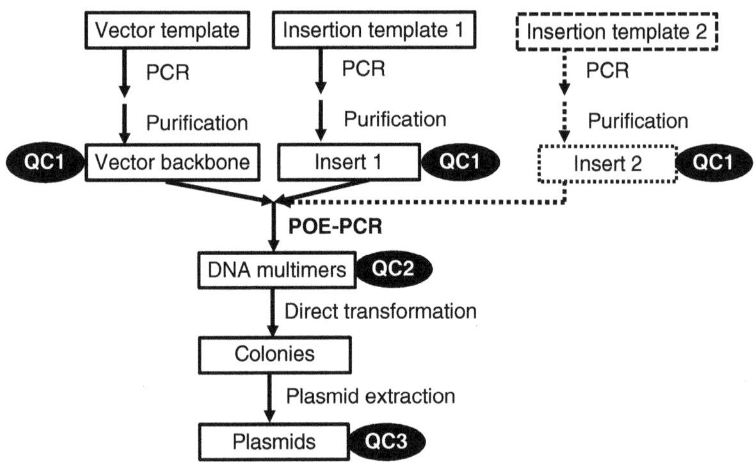

Fig. 2 Flow schemes of Simple Cloning and DNA assembly by POE-PCR. First, two or three or four 3′ and 5′ overlapped insert and vector fragments are generated by regular PCR. Second, DNA multimers are formed in vitro by prolonged overlap extension PCR. Third, *E. coli* strains can cleave DNA multimers to a circular plasmid, the desired chimeric plasmid. QC means quality control

or GC Buffer (*see* **Note 3**); and the Phusion DNA polymerase, 0.02 U/μL. The PCR program is 98 °C denaturation, 30 s; 30 cycles of 98 °C denaturation, 15 s; 55 °C annealing (*see* **Note 4**), 15 s; extension at 72 °C at 3 kb per min for the targeted fragment; and 5 min extension at 72 °C. The quality of the PCR product can be checked by examining 3 μL of the product by 0.8 % agarose gel electrophoresis (Fig. 3a, lane 1).

2. The insert fragment is amplified with a pair of primers of IF and IR through PCR by using the Phusion DNA polymerase.

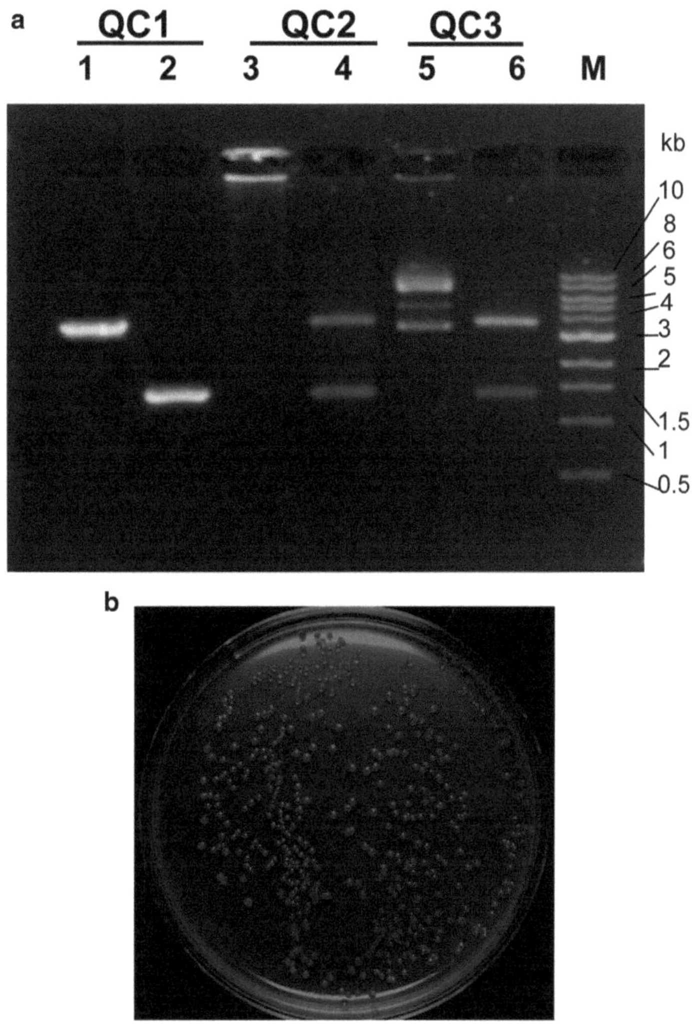

Fig. 3 (**a**) Simple Cloning for one insertion and vector. Lanes: M, DNA markers; *Lane 1* PCR linearized pET20b vector, *Lane 2* PCR linearized *cherry-cbm* insert, *Lane 3* DNA multimers generated by prolonged overlap extension PCR, *Lane 4* PCR products digested with NdeI and XhoI, *Lane 5* a purified plasmid from a randomly selected *E. coli* colony, *Lane 6* a purified plasmid digested with NdeI and XhoI; and M, 1-kb DNA ladder from NEB. (**b**) The *red* transformants on the plate were directly transformed into *E. coli*

Its quality can be checked by examining 3 μL of product in 0.8 % agarose gel electrophoresis (Fig. 3a, lane 2).

3. The two PCR products are cleaned with the DNA Clean & Concentrator Kit (*See* **Note 5**).
4. The DNA concentration of two purified DNA fragments is determined with a Nanodrop ND-1000 Spectrophotometer or roughly estimated by agarose gel electrophoresis.

5. For the generation of DNA multimers, add the below reagents to the tube:

Component	Volume	Final concentration
5× Phusion HF or GC Buffer	10 μL	1×
dNTPs (10 mM each)	1 μL	200 μM each
Vector DNA fragment	Variable	4 ng/μL
Insert DNA fragment	Variable	Equal molar with the vector (*see* **Note 6**)
Water	To 49.5 μL	
Phusion DNA polymerase	0.5 μL	1.0 unit per 50 μL

6. POE-PCR is conducted as the following: 98 °C denaturation, 30 s; 30 cycles of 98 °C denaturation, 15 s, 55 °C annealing 15 s, extension at 72 °C for 2 kb per min for the length of the resulting plasmid; and 10 min extension at 72 °C. Here the extension time was about 1.5-fold of the extension time of typical overlap extension PCR.
7. The quality of the POE-PCR product is examined by adding 3 μL of the product in 0.8 % agarose gel (*see* **Note** 7). The DNA multimers are high molecular weight products so that they cannot migrate into the gel (Fig. 3a, lane 3) (*see* **Note 8**). Also, the quality of the POE-PCR product can be examined by restriction enzyme digestion, where the pattern should be the same as the result of the desired plasmid digested by the same enzyme(s) (Fig. 3a, lane 4).
8. Direct transformation of DNA multimers to *E. coli* cells as the followings. Gently mix 2–5 μL of the POE-PCR product (*see* **Note 9**) with 100 μL component *E. coli* cells in a 1.5-mL polypropylene tube. Place the tube on ice for 20 min, 42 °C for 90 s followed by on ice for 5 min. Add 1 mL of the SOC liquid medium into the tube; incubate the tube at 37 °C for 30 min. After centrifugation at 5,000 rpm for 5 min, discard 1 mL of the supernatant. Re-suspend the cell pellets with the remaining liquid, and then spread the cells on one petri plate containing the LB solid medium supplemented with appropriate antibiotic. Incubate the plate at 37 °C for 12–16 h or until the colonies were easily examined by eyes (Fig. 3b). The transformation efficiencies are ~1–3 × 10^4 per μg of the DNA multimers product by using commercial competent cells with a transformation efficiency of ~3 × 10^9 per μg of plasmid. Since *E. coli* BL21(DE3) host is used, the colonies that can express red fluorescent protein are almost shown as red (Fig. 3b), indicating that more than 99 % of transformants were positive.

9. Pick one of the transformants from the plate and inoculate into 3–5 mL of the LB medium supplemented with appropriate antibiotic and incubate at 37 °C for 10–12 h. Extract plasmid by Plasmid Miniprep isolation kit.
10. Check the plasmid (Fig. 3a, lane 5) and its digestion product by restriction enzymes (Fig. 3a, lane 6).
11. If necessary, the plasmid can be sequenced for further validation.

3.3 DNA Assembly

This method can also be used to assemble three or more fragments in one step (Fig. 4a). For example, three pairs of primers (IF1/IR1, IF2/IR2, and VF/VR) are used to amplify the two DNA insert fragments and one vector backbone. Here fructose-1,6-bisphosphatase (fbp, 0.80 kb) gene from *Thermotoga maritime*, a dockerin module (docRF, 0.25 kb) from *Ruminococcus flavefaciens*, and pET20b backbone are assembled by POE-PCR.

1. The pET20b vector backbone is amplified with a pair of primers of VF and VR through PCR by using the Phusion DNA polymerase (Fig. 4b, lane 1).
2. The *fbp* DNA fragment is amplified with a pair of primers of IF1 and IR1 by using the Phusion DNA polymerase (Fig. 4b, lane 2).
3. The *docRF* DNA fragment is amplified with a pair of primers of IF2 and IR2 by using the Phusion DNA polymerase (Fig. 4b, lane 3).
4. The three PCR products are cleaned with the DNA Clean & Concentrator Kit.
5. Determine the DNA concentration of the three purified DNA fragments with the Nanodrop ND-1000 Spectrophotometer or with agarose gel electrophoresis.
6. For POE-PCR, add the below reagents to the tube:

Component		Final concentration
5× Phusion HF or GC Buffer	10 μL	1×
dNTPs (10 mM each)	1 μL	200 μM each
Vector DNA fragment	Variable	4 ng/μL
Insert DNA fragment 1	Variable	Equal molar with vector
Insert DNA fragment 2	Variable	Equal molar with vector
Water	To 49.5 μL	
Phusion DNA polymerase	0.5 μL	1.0 unit per 50 μL

7. POE-PCR was conducted as described above for the generation of DNA multimers.

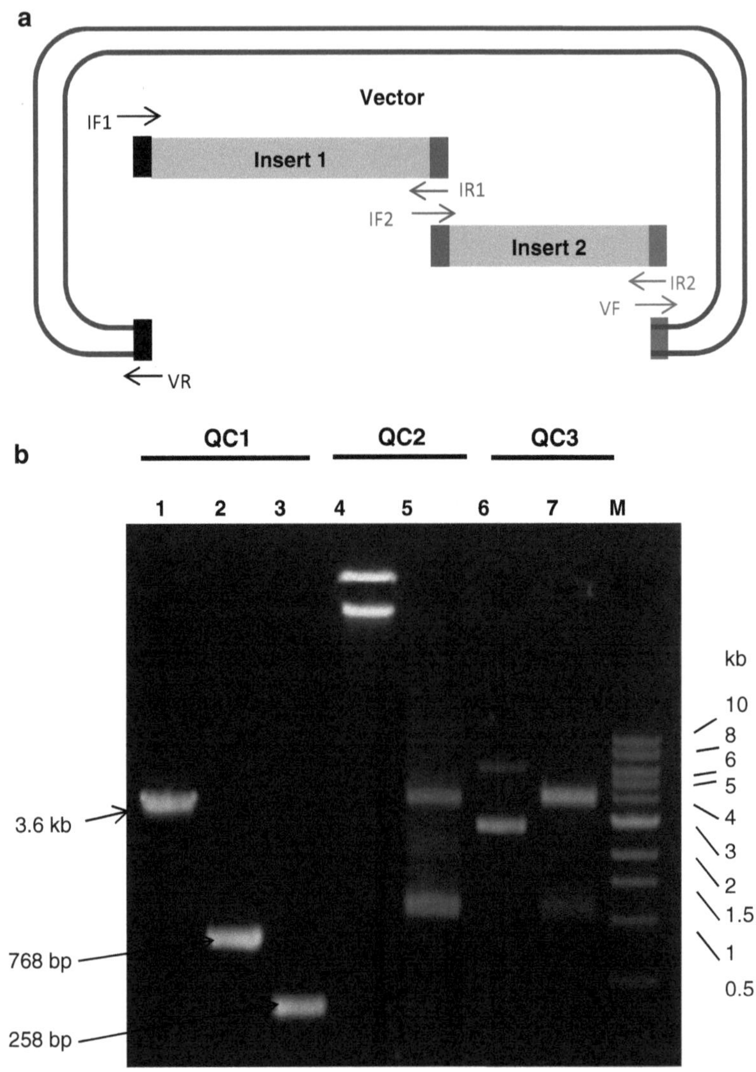

Fig. 4 DNA assembly of three fragments. *Lane 1* PCR linearized pET20b vector, *Lane 2* PCR linearized *fbp* insert, *Lane 3* PCR linearized *rfdoc* insert, *Lane 4* DNA multimers generated by prolonged overlap extension PCR, *Lane 5* DNA multimers digested with NdeI and XhoI, *Lane 6* a purified plasmid from a randomly selected colony, *Lane 7* a purified plasmid digested with NdeI and XhoI; and M, a 1-kb DNA ladder from NEB

8. Three microliters of POE-PCR product was examined by 0.8 % agarose gel electrophoresis (Fig. 4b, lane 4). To check whether the two inserts were successfully incorporated into the plasmid backbone, the POE-PCR product can be digested by the restriction enzymes (Fig. 4b, lane 5).
9. Directly transform the POE-PCR product to competent *E. coli* cells and spread them on plate.

10. Incubate the plate at 37 °C for 12–16 h or until the colonies were easily examined by eyes.
11. Pick one of the transformants from the plate and inoculate into 3–5 mL of the LB medium supplemented with appropriate antibiotic and incubate at 37 °C for 10–12 h. Extract plasmid by Plasmid Miniprep isolation kit.
12. Check the plasmid (Fig. 4b, lane 6) and its digested product by restriction enzymes (Fig. 4b, lane 7).
13. If necessary, the plasmid was sequenced for further validation.

4 Notes

1. To avoid possible mutations during the two-round PCR process, high-fidelity DNA polymerase must be used. High-fidelity DNA polymerases with a feature of long DNA fragment amplification is preferred for DNA assembly.
2. We usually use the overlap region of 25 bp, which means the primers will be 50 bp. Based on our experience, the overlap region should be at least ~20 bp for good performance of the overlap extension process. The overlap region could be longer, but longer primer may result in a higher possibility to get mutations in the synthesized primers and the unit price for primers will be much higher when they exceed 60 bp.
3. HF Buffer was recommended at the first trial. If PCR amplification in the HF Buffer failed, GC Buffer was tried because it can improve the performance of the Phusion DNA polymerase on the templates that were long, GC-rich, or have complex secondary structures.
4. The annealing temperature can be determined by Tm Calculator which is available in NEB website. If PCR amplification failed, optimize the annealing temperature by using gradient temperature PCR.
5. Sometimes the PCR product contained some nonspecific bands; the targeted band based on its immigration rate can be cut from the gel and be purified by using the Gel DNA recovery kit.
6. The amount of insert added in the PCR tube can be calculated based on the length of insert. Here, the insert length is 1.3 kb, the vector length is 3.4 kb, and the final concentration of vector is 4 ng/μL. So the final concentration of equimolar insert with the vector should be $4 \times 1.3/3.4 = 1.5$ ng/μL.
7. If no DNA multimers were generated by POE-PCR, PCR conditions can be optimized by (a) decreasing the annealing temperature, (b) increasing the template amount, (c) increasing the extension time in the POE-PCR, and (d) adding PCR enhancing reagents, such as betaine or DMSO.

8. The POE-PCR product was a kind of sticky solution and can be used to transform competent cells directly. Do not try to clean the POE-PCR product with a DNA cleaning kit. The DNA multimers may be stored at 4 °C for several days for future transformation. Do not freeze the POE-PCR product since it caused severe precipitation. It was frequent to see some high molecular weight DNA at the position of more than 10 kb instead of DNA multimers, which were intermediates generated by POE-PCR.
9. Do not add too much POE-PCR product in competent cells for transformation, since the high-salt PCR solution could decrease the transformation efficiency [15].

Acknowledgments

This work was supported mainly by the Shell GameChanger Program, DOE BioEnergy Science Center, DOE ARPA-E Petro, and the Virginia Tech CALS Biodesign, and NSF SBIR I grant to PZ.

References

1. Baneyx F (1999) Recombinant protein expression in *Escherichia coli*. Curr Opin Biotechnol 10:411–421
2. Blattner FR, Plunkett G 3rd, Bloch CA et al (1997) The complete genome sequence of *Escherichia coli* K-12. Science 277:1453–1462
3. Moxon ER, Higgins CF (1997) *E. coli* genome sequence. A blueprint for life. Nature 389: 120–121
4. Sambrook J, Russel DW (2001) Molecular cloning: a laboratory manual. Cold Spring Harbor Laboratory, New York
5. Liu W, Zhang XZ, Zhang Z et al (2010) Engineering of *Clostridium phytofermentans* Endoglucanase Cel5A for improved thermostability. Appl Environ Microbiol 76:4914–4917
6. Li C, Evans RM (1997) Ligation independent cloning irrespective of restriction site compatibility. Nucleic Acids Res 25:4165–4166
7. Zhu B, Cai G, Hall EO et al (2007) In-fusion assembly: seamless engineering of multidomain fusion proteins, modular vectors, and mutations. Biotechniques 43:354–359
8. Seki T, Seki M, Onodera R et al (1998) Cloning of cDNA encoding a novel mouse DNA topoisomerase III (Topo IIIbeta) possessing negatively supercoiled DNA relaxing activity, whose message is highly expressed in the testis. J Biol Chem 273:28553–28556
9. Walhout AJ, Temple GF, Brasch MA et al (2000) GATEWAY recombinational cloning: application to the cloning of large numbers of open reading frames or ORFeomes. Methods Enzymol 328:575–592
10. Xu R, Li QQ (2008) Protocol: Streamline cloning of genes into binary vectors in *Agrobacterium* via the Gateway(R) TOPO vector system. Plant Methods 4:4
11. Zhang Y, Buchholz F, Muyrers JP et al (1998) A new logic for DNA engineering using recombination in *Escherichia coli*. Nat Genet 20: 123–128
12. You C, Zhang X-Z, Zhang Y-HP (2012) Simple cloning via direct transformation of PCR product (DNA Multimer) to *Escherichia coli* and *Bacillus subtilis*. Appl Environ Microbiol 78:1593–1595
13. Zhang X-Z, Zhang Y-HP (2011) Simple, fast and high-efficiency transformation system for directed evolution of cellulase in *Bacillus subtilis*. Microb Biotechnol 4:98–105
14. You C, Zhang Y-HP (2012) Easy preparation of a large-size random gene mutagenesis library in *Escherichia coli*. Anal Biochem 428:7–12
15. Shafikhani S, Siegel RA, Ferrari E et al (1997) Generation of large libraries of random mutants in *Bacillus subtilis* by PCR-based plasmid multimerization. Biotechniques 23:304–310

Chapter 14

Combinatorial Assembly of Clone Libraries Using Site-Specific Recombination

Vanessa E. Wall, Leslie A. Garvey, Jennifer L. Mehalko, Lauren V. Procter, and Dominic Esposito

Abstract

Generation of DNA clones for use in proteomic and genomic research often requires a significant level of parallel production, as the number of downstream options for these experiments increases. Where a single fluorescently tagged construct may have sufficed before, there is now the need for multiple types of labels for different readouts and different assays. Protein expression, which once utilized a very small set of vectors because of low throughput expression and purification, has now rapidly matured into a high throughput system in which dozens of conditions can be tested in parallel to identify the best candidate clones. This has returned the bottleneck in many of these technologies to the generation of DNA clones, and standard cloning techniques often dramatically limit the throughput and success of such processes. In order to overcome this bottleneck, higher-throughput and more parallel cloning processes need to be developed which would allow rapid, inexpensive production of final clones. In addition, there is a strong need to utilize standardized elements to avoid unnecessarily remaking fragments of clones that could be used in multiple constructs.

The advent of recombinational cloning helped to increase the parallel processing of DNA clones, but was still limited by the need to generate different vector backbones for each specific need. The solution to this problem emerged with the introduction of combinatorial approaches to clone construction, based on either homologous or site-specific recombination processes. In particular, the Gateway Multisite system provides all of the necessary components for a highly parallel, inexpensive, rapid, and diverse platform for clone construction in many areas of proteomic and genomic research. Here we describe our optimized system for combinatorial cloning, including improvements in cloning protocols and construct design that permit users to easily generate libraries of clones which can be combined in parallel to create an unlimited number of final constructs. The system is capable of utilizing the tens of thousands of commercially available Gateway clones already in existence, and allows easy adaptation of most DNA vectors to the system.

Key words Cloning, Gateway, Site-specific recombination, Combinatorics

1 Introduction

The Gateway recombinational cloning system is a robust method for cloning genes into multiple vectors utilizing the power of site-specific DNA recombination [1, 2]. This system eliminates the

Svein Valla and Rahmi Lale (eds.), *DNA Cloning and Assembly Methods*, Methods in Molecular Biology, vol. 1116, DOI 10.1007/978-1-62703-764-8_14, © Springer Science+Business Media New York 2014

Table 1
Gateway *attB* site core sequences (listed 5′–3′)

*attB*1	GTACAAA
*attB*2	GTACAAG
*attB*3	GTATAAT
*attB*4	GTATAGA
*attB*5	GTATACA

need for classical restriction enzyme and ligase (REaL) cloning for the transfer of genes between vectors. Two types of clones are generated using Gateway cloning: Entry clones, which are transcriptionally silent "master" clones which are sequence-verified, and Expression clones, which are the final clones generated by recombination of the Entry clones into a separate construct called a Destination vector. These DNAs carry signals, tags, fusion partners, replication origins, and other specific features required for the specific downstream application. Because the Gateway site-specific recombination reaction does not involve PCR amplification, Expression clones do not need to be resequenced as long as their parent Entry clones have been sequence-verified. For this reason, a large number of Expression clones can be generated easily from a single Entry clone without the need for additional amplification and sequencing. Gateway reactions are driven by recombination between sites called attachment sites (*att* sites), which come in four varieties, *attP*, *attB*, *attL*, and *attR*. All reactions are conservative, directional, and lead to the interconversion of these sites, the types of which can be used to differentiate the vectors. The core recombination site in all of these *att* sites is the same 21 bp DNA sequence which determines the directionality and specificity of the reaction. The protein-coding sequences of basic Gateway clones are always flanked by two slightly different *att* sites, identified with numbers as in *attB*1 and *attB*2. The *att*1 and *att*2 sites differ by a single nucleotide, making them unable to recombine with each other, and thus producing the unique order of the recombination reactions that eliminates the need for directional screening.

An extension of Gateway technology known as Multisite Gateway takes these reactions a step farther [3]. By adding additional *att* site specificities (here identified as *att*3, *att*4, and *att*5, *see* Table 1), it is possible to link multiple Gateway Entry clones together in a single reaction and with a defined order. As with the case of standard Gateway, these subcloning reactions involve only site-specific recombination, and therefore require no additional sequence validation or directional screening. The key to properly applying Multisite technology is to ensure that your fragments are

Table 2
Types of Entry clone/Destination vector combinations for 2-, 3-, and 4-fragment cloning

Fragments	Entry 1	Entry 2	Entry 3	Entry 4	Destination vector
2	*attL4-attR1*	*attL1-attL2*			*attR4-attR2*
3	*attL4-attR1*	*attL1-attL2*	*attR2-attL3*		*attR4-attR3*
3	*attL4-attL5*	*attR5-attR1*	*attL1-attL2*		*attR4-attR2*
4	*attL4-attL5*	*attR5-attR1*	*attL1-attL2*	*attR2-attL3*	*attR4-attR3*

Table 3
Examples of some styles of combinatorial construct designs

Type of construct	First entry clone	Second entry clone	Third entry clone
Basic protein expression	Promoter (*attL4-attR1*)	ORF (*attL1-attL2*)	
Tagged protein expression	Promoter (*attL4-attL5*)	N-terminal fusion (*attR5-attR1*)	ORF (*attL1-attL2*)
Tagged protein expression	Promoter (*attL4-attR1*)	ORF (*attL1-attL2*)	C-terminal fusion (*attR2-attL3*)
Gene expression with reporter	Promoter (*attL4-attR1*)	ORF (*attL1-attL2*)	IRES-GFP (*attR2-attL3*)
shRNA expression	Promoter (*attL4-attR1*)	GFP (*attL1-attL2*)	shRNA (*attR2-attL3*)

designed with a specific *att* site configuration to permit maximum flexibility in combinatorial construction. While several commercial flavors of Multisite Gateway are available, the majority of them are designed to utilize commercially available Destination vectors (*attR1-attR2*) rather than commercially available ORF Entry clones (*attL1-attL2*). As there are tens of thousands of these Entry clones available and only a few hundred Destination vectors, this system is not nearly as universally applicable. Instead, we and others have focused systems that require the construction of new Destination vectors, but can take full advantage of the already available standard Gateway Entry clones [4–7]. This system permits construction of clones with 2, 3, or 4 Entry clone fragments in multiple configurations as shown in Table 2.

In this format, only two different Destination vectors are required: *attR4-attR2* for 2- or 3-fragment cloning, and *attR4-attR3* for 3- or 4-fragment cloning. The choice of which 3-fragment pathway to use likely will depend on your desire to include standard *attL1-attL2* Entry clones. Table 3 shows some examples of how these fragments can be used to make expression vectors in the

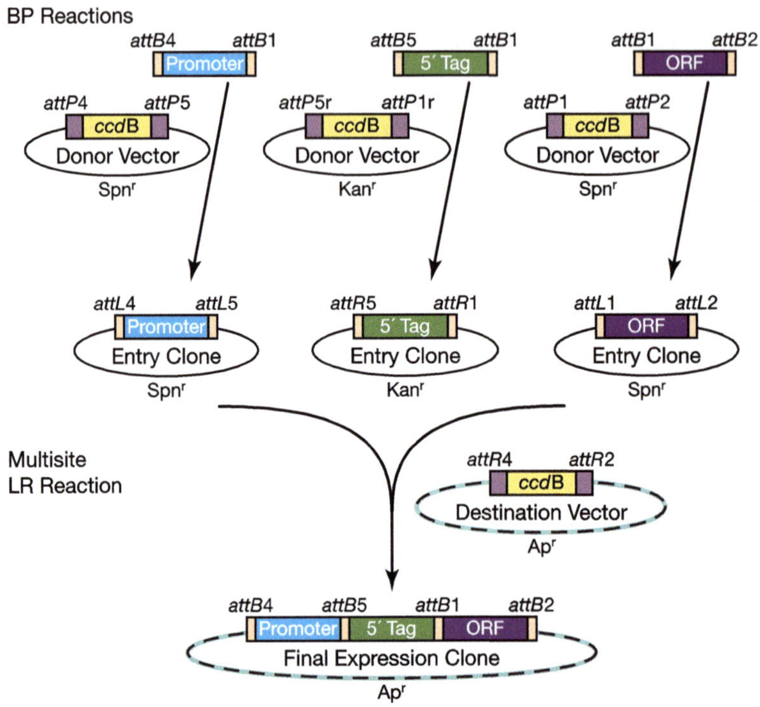

Fig. 1 General Multisite Gateway cloning scheme. This diagram presents a schematic representation of a theoretical 3-fragment Multisite Gateway reaction to link a promoter, N-terminal fusion tag, and gene of interest. The diagram identifies the different plasmid types found in the BP and LR reactions as discussed in the text

various configurations, and Fig. 1 shows a diagram of one of the most common cloning schemes used for protein Expression clones. You can use alternative configurations of the system if necessary as long as you ensure that the leftmost and rightmost *attL* sites from the Entry clones match the *attR* sites in the Destination vector, and that each Entry clone junction contains an identical *attL* and *attR* specificity to allow proper recombination.

DNA sequences of interest can be inserted into Entry clones in several different ways, but to provide the most flexibility in the cloning strategy, we generally use a Gateway recombination reaction (called the BP reaction, *see* Subheading 3.3) to construct Entry clones. Once Entry clones are constructed and sequence-verified, a second recombination reaction (called an LR reaction, *see* Subheading 3.4) is used to carry out the subcloning into the final Destination vector. Destination vectors contain *attR* sites flanking a Gateway "cassette" which encodes a chloramphenicol resistance gene and a toxin called *ccdB*. This region is recombined out during the LR reaction and is replaced by the fragments from the various Entry clones. A combination of positive selection using the marker on the Destination vector and negative selection from *ccdB* to eliminate nonrecombinant vectors permits high efficiency generation of the Expression clone. If you are generating protein Expression clones, careful planning of the original Entry clone sequences also ensures that the reading frame of the cloned gene matches the reading frame of any tags being introduced or that are present in the Destination vector (*see* Subheading 3.2). Because each LR reaction can produce a separate Expression clone, it is clear that a very large number of combinatorial clones can theoretically be made in parallel from a small number of Entry clones, permitting Gateway to function in either low or high throughput modes depending on the choice of the investigator. We will describe methods for the more simple low throughput mode, but additional references offer suggestions for increased throughput cloning [2].

2 Materials

2.1 Oligonucleotide Design

1. Oligonucleotides can be ordered from numerous suppliers, such as Eurofins MWG Operon (Huntsville, AL). We have found that they generally do not require HPLC or gel purification, and for Gateway reactions, the amount of oligonucleotide used is so small that a 50 nmol synthesis scale is more than sufficient.
2. Oligonucleotides for PCR amplification should be resuspended to a concentration of 5 μM in TE (10 mM Tris-Cl, pH 8.0, 0.1 mM EDTA).

2.2 PCR Amplification

1. 2× Phusion Master Mix High Fidelity (New England Biolabs, Beverley, MA).
2. DMSO: dimethyl sulfoxide (provided with the Phusion Master Mix Kit or available via commercial suppliers).
3. 0.2 mL PCR Low Profile strip tubes and caps. QIAquick PCR Purification Kit (Qiagen, Valencia, CA).
4. ReadyLoad 1 kb Plus DNA Ladder (Life Technologies, Carlsbad, CA).
5. 0.8 % agarose gels with TAE buffer w/Ethidium Bromide.
6. Thermal cycler.
7. Gel documentation unit.

2.3 BP Recombination

1. BP Clonase II Kit (Life Technologies, Carlsbad, CA; comes with BP Clonase® II enzyme mix, 2 μg/mL Proteinase K solution).
2. FastPlasmid DNA Kit (Eppendorf, Hamburg, Germany).
3. DH10B chemically competent cells (Life Technologies Corporation, Carlsbad, CA). Store at −80 °C; do not reuse open vials.

4. SB medium: Superior Broth (AthenaES, Baltimore, MD), 40 g/L, autoclave for 20 min.
5. LB agar plates: Lysogeny Broth medium (10 g/L tryptone, 5 g/L yeast extract, 5 g/L NaCl) with 0.7 % agar, autoclave for 20 min, cool to 55 °C, and add antibiotics as needed.
6. Kanamycin (Sigma, St. Louis, MO), 50 μg/mL final concentration; stock solution 50 mg/mL in water, sterile filter, store at 4 °C.
7. Spectinomycin (Sigma, St. Louis, MO), 100 μg/mL final concentration; stock solution 100 mg/mL in water, sterile filter, store at 4 °C.
8. Falcon 2059 culture tubes (Fisher Scientific, Pittsburgh, PA).
9. Supercoiled DNA Ladder (Life Technologies Corporation, Carlsbad, CA). Store at 4 °C.
10. Temperature-controlled water bath.
11. Tabletop centrifuge.

2.4 LR Multisite Recombination

1. LR Clonase II enzyme mix (Life Technologies, Carlsbad, CA; comes with LR Clonase® II enzyme mix, 2 μg/mL Proteinase K solution).
2. Ampicillin (Sigma, St. Louis, MO), 100 μg/mL final concentration; stock solution 100 mg/mL in water, sterile filter, store at 4 °C.
3. QIAprep® Spin Miniprep Kit (Qiagen Inc., Valencia, CA).
4. BsrGI restriction enzyme (New England Biolabs, Beverley, MA).

2.5 Destination Vector Construction

1. *E. coli ccdB* Survival competent cells (Life Technologies, Carlsbad, CA). Store at −80 °C; do not reuse open vials.
2. EcoRV restriction enzyme (New England Biolabs, Beverley, MA).
3. DNA Polymerase I, Klenow fragment (New England Biolabs, Beverley, MA).
4. T4 DNA Polymerase/Quick Blunting Kit (New England Biolabs, Beverley, MA).
5. T4 QuickLigase (New England Biolabs, Beverley, MA).
6. QIAquick PCR Purification Kit (Qiagen, Valencia, CA).
7. ReadyLoad 1 kb DNA ladder (Life Technologies, Carlsbad, CA).
8. Temperature-controlled water bath.
9. Chloramphenicol (15 mg/mL stock in 100 % ethanol).

3 Methods

3.1 Oligonucleotide Design

1. Entry into Gateway cloning using BP recombination requires a PCR amplification step to add the *attB* recombination signal sequences and any other desired sequences on to the gene of interest. For clones containing protein-coding regions, proper design of the oligonucleotides for this step ensures that the correct reading frame is generated in the final Expression clone (*see* **Note 1**).
2. To clone most fragments, 18–21 bp of gene-specific 5′ and 3′ sequences are used for primer annealing. For simple Entry clones which do not contain large amounts of additional features, the recombination signal sequences given in Table 4 can be added directly to the gene-specific primers.
3. Long PCR primers are required for more complicated tagging in Entry clones. Introduction of protease cleavage sites or epitope tags often leads to oligonucleotide lengths in excess of 60 nt. In our experience, such long oligonucleotides often are of reduced quality containing higher numbers of mutations or deletions. This may require more clones be sequenced in order to avoid errors. To avoid this problem, a technique known as adapter PCR can be utilized.
4. Adapter PCR involves the use of multiple nested primers to add long 5′ or 3′ sequences to the gene of interest. First, a primer which contains the gene-specific portion and part or all of a tag sequence (such as an epitope tag) is added to the PCR. After a few rounds of amplification, a second primer is added

Table 4
Oligonucleotide sequences for addition of Gateway sites

*attB*1 (to make *attL*1):	5′-GGGGACAACTTTGTACAAAAAAGTTGGC—gsp
*attB*2 (to make *attL*2)::	5′-GGGGACAACTTTGTACAAGAAAGTTGG—gsp
*attB*3 (to make *attL*3)::	5′-GGGGACAACTTTGTATAATAAAGTTGGC—gsp
*attB*4 (to make *attL*4)::	5′-GGGGACAACTTTGTATAGAAAAGTTGGC—gsp
*attB*5 (to make *attL*5)::	5′-GGGGACAACTTTGTATACAAAAGTTGGC—gsp
*attB*1rev (to make *attR*1):	5′-GGGGCCAACTTTTTTGTACAAAGTTG—gsp
*attB*2rev (to make *attR*2):	5′-GGGGCCAACTTTCTTGTACAAAGTGG—gsp
*attB*5rev (to make *attR*5):	5′-GGGGCCAACTTTTGTATACAAAGTTGA—gsp

gsp, gene-specific primer (should contain 18–22 bp of 5′ or 3′ of gene; in reverse primers this sequence must be the reverse complement of the sense strand of the gene)
Underlined sequences in the *att* sites identify the actual recombination site overlap regions which provide specificity

which contains the appropriate *attB* recombination signal and 12–16 nt of overlapping sequence with the first primer. A mixture of PCR products will be produced, but only the full length product will have the *attB* recombination sites necessary for recombination to occur.

5. Adapter PCR can be used on both ends simultaneously by adding two different adapter primers. One can also nest multiple levels of adapter PCR to insert long 5′ or 3′ sequences if necessary. Often, the same adapter primers can be used for any gene which has a particular 5′ sequence, thus minimizing the cost and number of oligos which need to be generated for library construction.
6. In cases where multiple large genes are to be combined (e.g., a fusion of a protein of interest with a second protein of interest), or when large deletions are desired, overlap PCR can be used [8]. In this process, two separate PCR amplifications are carried out with 20–25 bp of overlapping sequence between the 3′ end of gene 1 and the 5′ end of gene 2. A third PCR is then carried out using the first two PCR products as templates along with flanking primers containing the *attB* sites. Again, the presence of the *attB* sites during only the last round of PCR ensures that no other side-products will be able to be cloned.
7. In addition to oligonucleotides for amplification, oligonucleotides for sequence verification of Entry clones are also required. We generally order 1 primer for every 600 bp of sequence, and an additional primer on the reverse strand that is able to sequence back through the start of the gene. Typical primer lengths are 22–24 nt, and they can be selected manually or with the assistance of many common molecular biology computer programs. Standard Gateway sequencing primers can also be used to sequence into the gene of interest from both directions in the Entry clone.

3.2 PCR Amplification

1. To a 200 μL thin-walled PCR tube, add 1 μL of each 5 μM oligonucleotide primer, 0.75 μL DMSO (*see* **Note 2**), 50–100 ng template DNA (*see* **Note 3**), and water to 12.5 μL final volume.
2. Add 12.5 μL 2× Phusion Master Mix HF, mix well, and carry out the PCR amplification using the following parameters: initial denaturation at 98 °C for 30 s, 20 cycles of (98 °C for 10 s, 55 °C for 30 s, and 72 °C for 30 s per kb of the expected product), followed by a 10 min final elongation at 72 °C, and cooling to 4 °C (*see* **Note 4**).
3. For adapter PCR (*see* Subheading 3.1, **step 4**), after five cycles of amplification, pause the thermal cycler, and add 1 μL of 5 μM adapter primer(s) to the tubes. Continue cycling for additional 20 cycles.

4. If multiple nested adapter primers are used, we suggest increasing the overall cycle time so that there are five cycles between adapter additions, and at least 20 additional cycles after the final adapter is added.
5. If only a small amount of template DNA is available, increase the number of overall cycles from 20 to 30. This will increase the likelihood of PCR errors, but may improve PCR product yield.
6. After cycling, load 5 μL of the PCR product on a 0.8 % agarose gel to verify size by comparison to a linear DNA standard such as the ReadyLoad 1 kb DNA ladder (*see* **Note 5**).
7. Purify the PCR product using the QIAquick PCR Purification Kit following the manufacturer's protocol and elute the DNA in final volume of 50 μL (*see* **Note 6**).

3.3 BP Recombination

1. Add the following reagents to a microcentrifuge tube in the order given (the total reaction volume should be 5 μL): 2 μL H_2O, *attB*-flanked PCR fragment (15–150 ng, *see* **Note 7**), 150 ng Donor vector (*see* **Note 8**), and 1 μL of BP Clonase II. A master mix can be used for all reagents except for BP Clonase II, which must be added last. Mix briefly by gentle vortexing.
2. Incubate the reaction mixture for at least 1 h at 30 °C.
3. Add 1 μL of 2 mg/mL Proteinase K to inactivate the BP Clonase and incubate for 15 min at 37 °C.
4. Add 1 μL of the BP reaction to a microcentrifuge tube containing 20 μL of chemically competent *E. coli* DH10B and incubate on ice for 10–20 min (*see* **Note 9**).
5. Heat shock the cells in 42 °C water bath for 45 s and immediately add 80 μL of LB medium. Shake the reaction for 1 h at 37 °C.
6. Spread 100 μL of the transformation mix on LB agar plates containing the proper antibiotic (*see* **Note 10**) and incubate overnight at 37 °C. A good BP cloning result with a standard length (1 kb) insert should yield greater than 200 colonies per transformation.
7. Pick 2–4 separate Entry clone colonies into Falcon 2059 culture tubes containing 2 mL of SB medium with antibiotic and grow overnight at 37 °C with 200 rpm shaking.
8. Spin 1 mL of the culture in a microcentrifuge to pellet the cells, and isolate plasmid using the FastPlasmid kit, eluting the DNA in 75 μL of elution buffer (*see* **Note 11**).
9. Verify the size of the plasmid using agarose gel electrophoresis. Load 1 μL of the purified Entry clone DNA on a 0.8 % agarose gel, and compare sizes to the Supercoiled DNA ladder.

10. Properly sized Entry clones should be sequence-verified to ensure that no oligonucleotide or PCR-generated errors have been introduced (*see* **Note 12**).
11. Glycerol stocks of the *E. coli* strains containing Entry clones should be made by adding 100 μL sterile filtered 60 % glycerol to 300 μL of culture. After mixing and incubation at room temperature for 5 min, these stocks can be frozen at −80 °C and used to start new cultures if more Entry clone DNA is required in the future.

3.4 Multisite LR Recombination

1. Add the following reagents to a microcentrifuge tube in the order given (the total reaction volume should be 10 μL): 1–5 μL H_2O, each Entry clone DNA (50 ng, *see* **Note 13**), Destination vector DNA (100 ng, *see* **Note 14**), and 2 μL LR Clonase II (*see* **Note 15**).
2. Incubate the reaction mixture overnight at 25 °C (*see* **Note 16**).
3. Add 1 μL of 2 mg/mL Proteinase K to inactivate the LR Clonase II and incubate for 15 min at 37 °C (*see* **Note 17**).
4. Add 1 μL of the LR II reaction to a microcentrifuge tube containing 20 μL of chemically competent *E. coli* DH10B and incubate on ice for 10–20 min (*see* **Note 9**).
5. Heat shock the cells in a 42 °C water bath for 45 s and immediately add 80 μL of LB medium. Shake the reaction for 1 h at 37 °C.
6. Spread 100 μL of the transformation mix onto an LB agar plate containing the correct antibiotic (often ampicillin, but check the Destination vector information) and incubate overnight at 37 °C.
7. Pick two separate colonies into Falcon 2059 culture tubes containing 2 mL of SB medium containing the correct antibiotics and grow overnight at 37 °C with 200 rpm shaking.
8. Spin 1 mL of the culture in a microcentrifuge tube to pellet the cells, and isolate plasmid using the QIAprep Spin Miniprep Kit, eluting the DNA in 50 μL of elution buffer (*see* **Note 18**).
9. Verify the size of the plasmid using agarose gel electrophoresis. Load 1 μL of the purified Expression clone DNA on a 0.8 % agarose gel, and compare sizes to the Supercoiled DNA ladder.
10. Additional confirmation of the Expression clone should be carried out by restriction enzyme analysis (*see* **Note 19**). 1 μL of Expression clone DNA can be digested using BsrGI restriction endonuclease for 1 h at 37 °C.
11. Glycerol stocks of the *E. coli* strains containing Multisite clones should be made by adding 100 μL sterile filtered 60 % glycerol to 300 μL of culture. After mixing and incubating at room temperature for 5 min, these stocks can be frozen at −80 °C and used to start new cultures if more Expression clone DNA is required in the future.

3.5 Destination Vector Construction

1. Cassettes for generating new Destination vectors generally contain two *attR* sites flanking a set of positive and negative selection markers. In the case of standard Gateway vectors, the negative selection is done by the ccdB gene, and the positive selection by the CAT gene for chloramphenicol resistance.
2. To generate new Destination vectors, transfer vectors containing attR4-attR2 (pSpcRFA42) or attR4-attR3 (pSpcRFA43) cassettes flanked by a pair of EcoRV restriction sites can be used. Digest with EcoRV produces a Gateway cassette which can be introduced into any vector with a blunt cutting site, or which has been blunt-ended.
3. Prepare Destination vector cassette by digesting a suitable amount of the parental vector with EcoRV, and purification using the QIAquick PCR Purification Kit. This cut material can be saved for future use, so we suggest doing a single large reaction and storing the cut material at −20 °C for subsequent vector conversion work (*see* **Note 20**).
4. For production of Destination vectors using a blunt cutting site, 1 μg of the vector of interest must first be linearized using a restriction endonuclease that generates blunt ends. After digestion, 2 μL of the linearized product should be confirmed on a 0.8 % agarose gel to verify size by comparison to a linear DNA standard such as the ReadyLoad 1 kb DNA ladder (*see* **Note 21**). Purify the linearized vector using the QIAquick PCR Purification Kit following the manufacturer's protocol and elute the DNA in 50 μL.
5. For production of Destination vectors using a cohesive end cut site, you will require enzymatic blunting to allow correct insertion of the Gateway cassette fragment. The purified linearized vector with cohesive ends must be treated with DNA Polymerase 1, Klenow fragment (for 5′ DNA overhangs), or T4 DNA Polymerase (for 3′ DNA overhangs). Standard manufacturer's protocols (New England Biolabs, Beverly, MA) can be used for both reactions. QiaQuick PCR Purification and subsequent 50 μL elution is used following the Klenow protocol, whereas the product of the NEB Quick Blunting Reaction is used directly for Step 5.
6. Use 1 μL of the blunt-ended linearized vector in a ligation reaction with 3 μL of the digested Gateway cassette of choice. Bring the reaction to 10 μL with water, and add 10 μL of the 10× Ligase Reaction Buffer, followed by 1 μL of T4 DNA Ligase. Incubate the reaction at 25 °C for 10 min (*see* **Note 22**). Based on concentrations of the vector and insert, these amounts may vary, with the goal of a 1:3 vector:insert ratio.
7. Add 1 μL of the ligation reaction to a microcentrifuge tube containing 20 μL of chemically competent *E. coli ccdB* Survival cells and incubate on ice for 10–20 min. (*see* **Note 8**).

8. Heat shock the cells in 42 °C water bath for 45 s and immediately add 80 μL of LB medium. Shake the reaction for 1 h at 37 °C.
9. Spread 100 μL of the transformation mix on LB agar plates containing the proper antibiotics (including 15 μg/mL chloramphenicol for Gateway cassette selection) and incubate overnight at 37 °C.
10. Pick 6–8 separate Destination vector colonies into Falcon 2059 culture tubes containing 2 mL of SB medium with antibiotic and grow overnight at 37 °C with 200 rpm shaking.
11. Spin 1 mL of the culture in a microcentrifuge to pellet the cells, and isolate plasmid using the FastPlasmid kit, eluting the DNA in 50 μL of elution buffer.
12. Verify the size of the plasmid using agarose gel electrophoresis. Load 1 μL of the purified Destination vector DNA on a 0.8 % agarose gel, and compare sizes to the Supercoiled DNA ladder.
13. Additional confirmation of the Destination vector colonies should be carried out by restriction enzyme analysis, using an enzyme that will produce diagnostic bands to provide information on the orientation of the Gateway cassette (*see* **Note 23**). In addition, sequence confirmation across the Gateway cassette boundaries is strongly suggested (*see* **Note 24**).
14. Glycerol stocks of the *E. coli* strains containing Destination vector clones should be made by adding 100 μL sterile filtered 60 % glycerol to 300 μL of culture. After mixing and incubating at room temperature for 5 min, these stocks can be frozen at –80 °C and used to start new cultures if more Destination vector is required in the future.

4 Notes

1. It is a common problem for new Gateway cloners to confuse the reading frames and directions of the "reverse" *att* sites needed to make *attR* containing Entry clones. We strongly recommend the use of in silico modeling of Gateway cloning prior to ordering oligonucleotides. Several products can be used for this purpose, including VectorNTI (Life Technologies, Carlsbad, CA) and Clone Manager (SciEd Software, Cary, NC). Many other programs can also be tricked into carrying out Gateway reactions by pretending that the Gateway *att* sites are restriction sites. In any case, it is always best to validate the reactions in the computer before spending time and money on incorrect sequences.
2. The addition of 3 % DMSO to the PCR can help to reduce the effects of GC-rich primers or template DNA. Often it is not

required, but we have seen no detrimental effect from including it in most PCRs. If templates are very AT-rich, we suggest omitting the DMSO.

3. The use of large amounts of template DNA helps to dramatically reduce PCR errors by forcing the use of original template molecules instead of possibly error-prone PCR products for subsequent PCR cycles. If limited template is available, this amount can be reduced by 10- to 20-fold, but the chances of PCR errors will likely increase.
4. Phusion polymerase has become the standard PCR reagent in our lab due to its robust activity and extreme high fidelity. Other polymerases can also be used, but we recommend using only high fidelity enzymes to limit the introduction of mutation(s) during the PCR procedure, particularly for generating long amplicons. The suggested conditions are optimized for use on BioRad or Applied Biosystems PCR machines; some optimization may be required if PCR machines with slower thermal ramping times are used.
5. The appearance of a properly sized band on an agarose gel does not guarantee the proper addition of adapter primers during adapter PCR. In most cases, the extra length of the adapter primer does not change the overall DNA size sufficiently to detect on a gel. For this reason, if there are failures to BP clone products of an adapter PCR reaction, it is sometimes worth validating that the adapters are working by carrying out two separate PCRs instead of a single adapter PCR.
6. Column purification of PCR products is only successful for products >150 bp in length. For smaller products, a DNA precipitation with polyethylene glycol (PEG) can be carried out as detailed in the Gateway product manuals. Failure to purify PCR products will often lead to a large amount of primer-dimers (small fragments caused by primer misannealing) which will clone very efficiently in the BP reaction. In extreme cases, gel purification may be necessary to eliminate these products entirely.
7. Generally, the more PCR product used in the reaction, the higher the efficiency will be. Particularly with long PCR products (>5 kb), the higher end of the concentration range should be used. Be aware that with adapter PCRs, the effective concentration of PCR product with both *attB* sites may be lower than the total concentration.
8. There are numerous Donor vectors available—these are the *att*P containing vectors which become the backbone of the Entry clone after the BP reaction. pDonr-221 is a common vector for use with *att*1-*att*2 clones. Other vectors are required for different *att* site combinations, and can be obtained from Life Technologies or the authors. Note that Donor vectors

must be propagated in *E. coli ccdB* Survival or another strain which is resistant to the CcdB toxin.

9. DH10B is a recommended *E. coli* strain for Gateway reactions. However, it can be substituted with any other *recA endA* strain (such as TOP10 or DH5α) if necessary. Strains used must not carry the F′ episome (such as in XL1-Blue), as it will interfere with the Gateway negative selection. For good results, be sure that the competent cells have a transformation efficiency of at least 1×10^8 cfu/μg. Electrocompetent cells can also be used instead of chemically competent cells; however, the only advantage would be in the case of a very low efficiency reaction (such as with a very long insert)—usually the number of colonies obtained with standard chemically competent cells is more than sufficient.
10. Most Donor vectors contain either kanamycin or spectinomycin resistance markers, but be sure to verify the correct antibiotic before use.
11. Many commercial kits are available for generating plasmid DNA from *E. coli*. We prefer the FastPlasmid Kit for routine plasmid preps due to its high speed and consistent results. FastPlasmid can only be used for DNA generated in *endA* hosts, as it does not remove nucleases which could affect downstream processes. A standard alkaline lysis plasmid prep will also work in most cases, but these tend to add significant time to the process.
12. Due to the presence of inverted *att* site repeats in some Entry clones, sequencing of very small inserts (<400 bp) can be difficult. For this purpose, one can use specially designed blocking oligos [9] to eliminate the problem, or simply sequence from the inside of the insert using gene-specific primers.
13. In Multisite recombination, the ratio of Entry clones to Destination vector is crucial. Addition of extra Entry clone may improve efficiency, but this will likely also increase the chance of cotransformation of Expression clone DNA and Entry clone DNA into the same cell. We suggest a maximum of 50 ng per Entry clone and 100 ng of Destination vector to avoid cotransformation problems. However, increases in efficiency, if needed, can sometimes be achieved by increasing the amount of Destination vector.
14. Destination vectors can be added as either supercoiled plasmids or plasmids that have been linearized within the Gateway cassette. Contrary to the manufacturer's assertion, using linearized Destination vector will actually improve LR reaction efficiency two- to fivefold. However, due to the high efficiency of the LR reaction, we do not generally find the extra effort required to prepare and purify linearized Destination vector to be worthwhile except in the extreme cases of 4- and 5-fragment cloning.

15. Although a specific product (LR Clonase Plus) is marketed as a requirement for Multisite Gateway reactions, this reagent is not actually necessary for performing Multisite LR reactions. Standard LR Clonase II will work just fine as long as the concentrations of Entry clones and Destination vectors are accurately maintained. Under proper reaction conditions, 95 % of reactions will proceed equally well with the standard Clonase reaction mix.
16. Due to the complexity of the Multisite LR reaction, an overnight incubation at 25 °C is required to increase recombination efficiency and will produce a more than sufficient number of colonies. Shorter incubation times do not produce enough colonies on a consistent basis to be worthwhile, particularly with higher numbers of fragments.
17. Failure to Proteinase K treat the LR reactions will result in dramatically reduced colony counts due to the inability of the DNA to transform while coated with Clonase proteins.
18. Many commercial kits (*see* **Note 11** above) are available for generating plasmid DNA from *E. coli*. The QIAprep Spin Miniprep Kit is used here as many of the LR reaction products we use are destined for transfection into mammalian cells. This kit provides both DNA of a reasonable quality and yield, and also produces DNA with a relatively low level of endotoxin which can be directly transfected into most mammalian cells.
19. Standard Gateway reactions are usually so efficient and accurate that further confirmation of Expression clones is not necessary. However, as the fragment number increases, the slight possibility of unusual recombination byproducts with Multisite reactions exists, and so restriction enzyme analysis is often worthwhile. Some of the Gateway *attB* sites can be cleaved with the restriction enzyme *BsrGI*, which will generally yield diagnostic bands to allow verification of insert sizes. Alternately other restriction sites can be employed within the genes or flanking regions depending on the particular circumstances.
20. Note that the cassette does not need to be gel-purified away from the backbone of the plasmid in many Gateway cassette transfer vectors. The pSpcRFA42 and pSpcRFA43 vectors contain a spectinomycin-resistant backbone which will not be selected for in the final transformation of these vectors, allowing the use of cut but not purified vector for the reactions.
21. It is important to ensure that all of the parental plasmid has been digested in this reaction. Any leftover transfer vector can confer chloramphenicol resistance during cotransformation, and can lead to false positive colonies.
22. The suggested ratio of vector:insert in this ligation reaction is 1:3 on a molar basis. However, we find that it is usually good enough to estimate the amounts visually on a gel rather than performing calculations. In general, addition of an excess of

cassette to vector yields good results. There is a strong selection for the proper clones due to the chloramphenicol selection on the Gateway cassette and the resistance marker on the vector to be converted.

23. With blunt insertion, there is obviously a 50 % chance of obtaining forward or reverse orientation inserts. Several unique restriction sites can be found within the Gateway cassettes at non-symmetric locations—these include EcoRI, NotI, and BamHI. Depending on the restriction map of your vector, one of these may be useful as a diagnostic for directionality of the insert.
24. If allowed to proceed too long or in suboptimal buffer conditions, EcoRV reactions have a tendency to chew back 1–3 nt at the cut sites. While this does not generally affect the function of a Gateway reaction (the EcoRV site is not integral to the site), it can dramatically affect the reading frame if the vector contains a protein-coding region. For this reason, we suggest sequencing with primers internal to the Gateway cassette to ensure that the boundary sequences are accurately known.

Acknowledgments

This work has been funded in whole or in part with federal funds from the National Cancer Institute, National Institutes of Health, under contract HHSN261200800001E. The content of this publication does not necessarily reflect the views or policies of the Department of Health and Human Services, nor does mention of trade names, commercial products, or organizations imply endorsement by the U.S. Government.

References

1. Hartley JL, Temple GF, Brasch MA (2000) DNA cloning using in vitro site-specific recombination. Genome Res 10:1788–1795
2. Esposito D, Garvey LA, Chakiath CS (2009) Gateway cloning for protein expression. Methods Mol Biol 498:31–54
3. Cheo DL, Titus SA, Byrd DR et al (2004) Concerted assembly and cloning of multiple DNA segments using in vitro site-specific recombination: functional analysis of multi-segment expression clones. Genome Res 14: 2111–2120
4. Sasaki Y, Sone T, Yoshida S et al (2004) Evidence for high specificity and efficiency of multiple recombination signals in mixed DNA cloning by the Multisite Gateway system. J Biotechnol 107:233–243
5. Sasaki Y, Sone T, Yahata K et al (2008) Multi-gene gateway clone design for expression of multiple heterologous genes in living cells: eukaryotic clones containing two and three ORF multi-gene cassettes expressed from a single promoter. J Biotechnol 136:103–112
6. Sone T, Imamoto F (2012) Methods for constructing clones for protein expression in mammalian cells. Methods Mol Biol 801:227–250
7. Hopkins RF, Wall VE, Esposito D (2012) Optimizing transient recombinant protein expression in mammalian cells. Methods Mol Biol 801:251–268
8. Horton RM, Cai ZL, Ho SN et al (1990) Gene splicing by overlap extension: tailor-made genes using the polymerase chain reaction. Biotechniques 8:528–535
9. Esposito D, Gillette WK, Hartley JL (2003) Blocking oligonucleotides improve sequencing through inverted repeats. Biotechniques 35: 914–920

Chapter 15

Application of In-Fusion™ Cloning for the Parallel Construction of *E. coli* Expression Vectors

Louise E. Bird, Heather Rada, John Flanagan, Jonathan M. Diprose, Robert J.C. Gilbert, and Raymond J. Owens

Abstract

In-Fusion™ cloning is a flexible DNA ligase-independent cloning technology that has wide-ranging uses in molecular biology. In this chapter we describe the protocols used in the OPPF-UK to design and construct expression vectors using In-Fusion™. Our method for small scale expression screening in *Escherichia coli* of constructs generated by In-Fusion™ is also outlined.

Key words In-Fusion™ enzyme, PCR cloning, Single-stranded annealing, Fusion tags, Co-expression, Periplasmic secretion, *Escherichia coli*

1 Introduction

1.1 In-Fusion™ Enzyme

The advent of structural genomics has involved the construction and expression screening of large numbers of vectors. Over the last 6 years, the strategies employed at many institutes have evolved to a consensus process where only the detailed methodology is different [1–3]. Common to all is the adoption of ligation-independent cloning (LIC) technologies which ensures that the cloning process is independent of the sequence of the target. There are a number of such cloning strategies available which can largely be divided into those that use site-specific recombination and those where a single-stranded region is produced by exonuclease digestion and the subsequent annealing of single-stranded regions.

The In-Fusion™ PCR cloning system is an example of the second approach and employs a commercially available version of Vaccinia DNA polymerase (http://www.clontech.com/GB/Products/Cloning_and_Competent_Cells/Cloning_Kits/In-Fusion_EcoDry?sitex=10020:22372:US#) [4]. The enzyme promotes single-strand annealing reactions (SSA), where DNA molecules that have a short sequence of homologous DNA at their

Svein Valla and Rahmi Lale (eds.), *DNA Cloning and Assembly Methods*, Methods in Molecular Biology, vol. 1116, DOI 10.1007/978-1-62703-764-8_15, © Springer Science+Business Media New York 2014

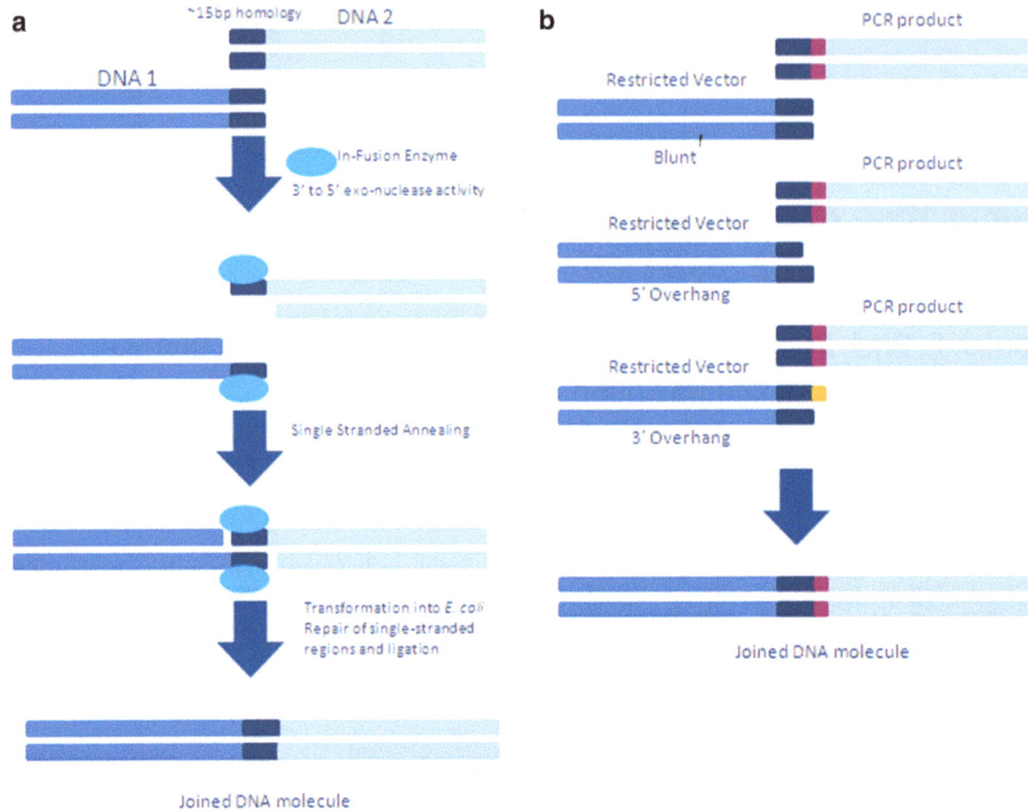

Fig. 1 In-Fusion™ enzyme action. (**a**) Mechanism of In-Fusion™ enzyme action. The In-Fusion™ enzyme has a 3′ to 5′ exonuclease activity which degrades the 3′ ends of two DNA molecules (1 and 2) that have 15 bp of homology at the ends. Single-stranded annealing is then promoted giving rise to joined molecules shown here with small single-stranded gaps. Following transformation into *E. coli* the gaps are repaired and ligated. (**b**) Design of In-Fusion™ primers. 15 bp of homology are standardly used in In-Fusion™ extensions; however, the type of restriction endonuclease used determines which vector sequence is used for the homology extension. *Dark blue boxes* indicate the homology regions on both strands. For blunt digests the 15 bp of homology is to the cleaved vector DNA. For 5′ overhangs the homology should be from the end of the single-stranded overhang. For 3′ overhangs the homology should be to the double-stranded portion of the restricted vector; that is, the single-stranded region shown in *orange* is removed by exonuclease action and the sequence is not in the joined vector. If extra amino acids or a stop codon, etc. are required then these may be added to the 3′ ends of all the extensions (indicated in *magenta*)

ends such as a PCR product and a linearized vector can be joined to make a hybrid molecule ([4, 5]; Fig. 1a). Mechanistically, when the double-stranded insert and linearized vector are incubated with the enzyme in the presence of Mg^{2+} and dNTPs, its exonuclease activity degrades from the 3′ ends of the molecules, thereby exposing complementary sequences. These anneal to create a heterogeneous population of joined molecules that, at the junctions where the annealing has taken place, have nicks, small gaps, and/or short overhangs. The hybrid molecules are relatively resistant to further exonuclease action and thus accumulate with time [5].

On transforming into *E. coli* these non-covalently joined molecules undergo repair resulting in assembly of the vector and insert. By a combination of PCR primer design and the use of restriction endonuclease sites, the junction between vector and inserted DNA fragment can be precisely engineered according to the requirements of the user (Fig. 1b). For example, we frequently use KpnI as a restriction site in our vectors; KpnI has a 4-nucleotide 3′ overhang which is removed by the exonuclease. The homology extension, which is 15 bp in length, is 5′ to this single-stranded region and thus by adding nucleotides 3′ to the extension we can precisely manipulate the junction sequence (Fig. 1b).

There are number of other methods which use an exonuclease to create homologous single-stranded regions on the vector and DNA fragment that subsequently anneal. The prototype of these is LIC in which both the linearized vector and PCR product are treated with T4 polymerase in the presence of one of the four dNTPs to produce 5′ single-stranded ends [6, 7]. The sequences flanking the insert can only be composed of three of the four bases, since the fourth base is required as a "lock" to stop the 3′ to 5′ exonuclease activity of T4 polymerase. A development of LIC which avoids the need for a base lock is SLIC (Sequence and Ligation-Independent Cloning; [8, 9]). SLIC, in common with LIC, is generally a two-stage protocol; the vector and DNA insert/s are processed separately followed by an additional annealing step [9]. However, recent reports do suggest that one-step SLIC is possible [10]. In the Gibson Assembly methodology (Novagen) a 5′ to 3′ exonuclease activity creates single-stranded 3′ overhangs; this facilitates the annealing of fragments that share homology at the ends. A polymerase then fills in the gaps of each annealed fragment and a DNA ligase seals the nicks in the hybrid DNA [11, 12]. The technology is a sequence-independent one-step reaction; however, the directionality of the exonuclease activity means vectors that are designed to utilize the 3′ to 5′ exonuclease activity (e.g., In-Fusion™ and LIC/SLIC) to create precise junctions may not be useful for this methodology [13, 14].

In common with SLIC and Gibson Assembly methodologies, the In-Fusion™ enzyme is sequence-independent with no bases added to the treated sequences [13]. It is a highly efficient one-step process and works over a wide base DNA size (from oligonucleotides to very large PCR fragments) and concentration range and, as such, has a proven record of usefulness in high-throughput experiments to produce precise fusions of vector and fragment [13, 15–20].

1.2 Applications of In-Fusion™ Cloning

In-Fusion™ is a flexible cloning system which is widely applicable to the construction of plasmid vectors and can be used in place of DNA ligase to clone a designed PCR product into a restricted vector [13, 14, 21]. Further, multiple constructs of different targets in one vector may be made in parallel. For example, Zhou et al. used the system to generate recombinant influenza A viruses; genomic sequences were amplified by reverse transcription-PCR (RT-PCR) from human swabs and were cloned into an expression vector allowing generation of viruses for vaccine development [22]. Similarly, Howland et al. have used In-Fusion™ with great efficiency to make a directional random-primed cDNA library using phosphorothioate-modified random primers [17]. Multiple fragments with overlapping ends can also be cloned into a single vector in one step. Zhu et al. constructed an immunoglobulin fusion protein from three fragments [21], and more recently this application of the technology has been used to make BioBrick assemblies [23, 24]. In-Fusion™ can also be used for mutagenesis; mutant primers are used to make a linearized vector with overlapping ends and then the In-Fusion™ protocol is used as normal [21].

In the Oxford Protein Production Facility-UK (OPPF-UK), we use In-Fusion™ cloning for the construction of all expression vectors. Through careful design of expression vectors we have derived subfamilies of the pOPIN suite for different applications [14, 15]. In this chapter we describe our suite of vectors that are used for expression studies in *E. coli*, and standardized protocols for vector construction and expression screening. It is worth emphasizing that PCR In-Fusion™ cloning can be used with any vector which has a unique restriction site(s) and that there is a web tool for designing the appropriate PCR primer extensions available from Clontech (http://www.clontech.com/GB/Products/Cloning_and_Competent_Cells/Cloning_Kits/xxclt_onlineToolsLoad.jsp?citemId=http://bioinfo.clontech.com/infusion/convertPcrPrimersInit.do&xxheight=750).

1.2.1 Vector Families

Soluble expression of recombinant proteins often requires the screening of multiple constructs of a target protein to identify soluble versions [3, 25]. One issue is whether fusion protein tags can improve the expression level and/or solubility of a protein or protein domain [26]. The pOPINF family of vectors has been designed to fuse the N-terminus of a target protein to a range of tags that may be cleaved by treatment with Rhinovirus 3C protease [13–15]. In a recent study we cloned 20 targets into nine different pOPINF family vectors, and showed that the number of targets that gave useful soluble expression could be improved from four using only an N-terminal His-tag to 15 by the addition of a fusion protein, with the SUMO fusion vector being the most effective [15, 27]. The same PCR product can be used to make all of the pOPINF family of fusions as the vector homology extensions which are incorporated into the PCR primers are the same for each vector [14, 15]. The properties of the vectors in the pOPINF family are shown in Table 1. We have also developed a set of C-terminal fusion vectors which share common forward and reverse primers.

Table 1
Vectors designed for fusion at the N-terminus of proteins (pOPINF family)

Vector	Tag	Origin (copy no.)	Resistance	Vector backbone	Vector description	Vector linearization sites	Forward extension	Reverse extension
pOPINF	N-HIS-3C-POI	pUC (50 to >100)	Amp	pTriEx2	T7lacO, CMV enhancer and bb-actin promoter, p10 promoter/lef-2 and 1629 baculo elements	KpnI HindIII	Opti3CInffwd AAGTTCTGTT TCAGGGCCCG	In-Fusion™ 3′ site ATGGTCTA GAAAGCTTTA
pOPINS3C	N-HIS-SUMO-3C-POI	pUC (50 to >100)	Amp	pTriEx2	T7lacO, CMV enhancer and bb-actin promoter, p10 promoter/lef-2 and 1629 baculo elements	KpnI HindIII	Opti3CInffwd AAGTTCTGTT TCAGGGCCCG	In-Fusion™ 3′ site ATGGTCTAG AAAGCTTTA
pOPINTRX	N-HIS-TRX-3C-POI	pUC (50 to >100)	Amp	pTriEx2	T7lacO, CMV enhancer and bb-actin promoter, p10 promoter/lef-2 and 1629 baculo elements	KpnI HindIII	Opti3CInffwd AAGTTCTGTT TCAGGGCCCG	In-Fusion™ 3′ site ATGGTCTA GAAAGCTTTA
pOPINMSYB	N-HIS-MSYB-3C-POI	pUC (50 to >100)	Amp	pTriEx2	T7lacO, CMV enhancer and bb-actin promoter, p10 promoter/lef-2 and 1629 baculo elements	KpnI HindIII	Opti3CInffwd AAGTTCTGTT TCAGGGCCCG	In-Fusion™ 3′ site ATGGTCTA GAAAGCTTTA
pOPINJ	N-HIS-GST-3C-POI	pUC (50 to >100)	Amp	pTriEx2	T7lacO, CMV enhancer and bb-actin promoter, p10 promoter/lef-2 and 1629 baculo elements	KpnI HindIII	Opti3CInffwd AAGTTCTGTT TCAGGGCCCG	In-Fusion™ 3′ site ATGGTCTAG AAAGCTTTA
pOPINN-GFP	N-HIS-eGFP-3C-POI	pUC (50 to >100)	Amp	pTriEx2	T7lacO, CMV enhancer and bb-actin promoter, p10 promoter/lef-2 and 1629 baculo elements	KpnI HindIII	Opti3CInffwd AAGTTCTGTT TCAGGGCCCG	In-Fusion™ 3′ site ATGGTCTAG AAAGCTTTA

(continued)

Table 1
(continued)

Vector	Tag	Origin (copy no.)	Resistance	Vector backbone	Vector description	Vector linearization sites	Forward extension	Reverse extension
pOPINHALO	N-HIS-HALO-3C-POI	pUC (50 to >100)	Amp	pTriEx2	T7lacO, CMV enhancer and bb-actin promoter, p10 promoter/lef-2 and 1629 baculo elements	KpnI HindIII	Opti3CInffwd AAGTTCTGTT TCAGGGCCCG	In-Fusion™ 3′ site ATGGTCTAG AAAGCTTTA
pOPINHALO7	N-HIS-HALO7-3C-POI	pUC (50 to >100)	Amp	pTriEx2	T7lacO, CMV enhancer and bb-actin promoter, p10 promoter/lef-2 and 1629 baculo elements	KpnI HindIII	Opti3CInffwd AAGTTCTGTT TCAGGGCCCG	In-Fusion™ 3′ site ATGGTCTAG AAAGCTTTA
pOPINM	N-HIS-MBP-3C-POI	pUC (50 to >100)	Amp	pTriEx2	T7lacO, CMV enhancer and bb-actin promoter, p10 promoter/lef-2 and 1629 baculo elements	KpnI HindIII	Opti3CInffwd AAGTTCTGTT TCAGGGCCCG	In-Fusion™ 3′ site ATGGTCTAG AAAGCTTTA
pOPINTF	N-HIS-TF-3C-POI	pUC (50 to >100)	Amp	pTriEx2	T7lacO, CMV enhancer and bb-actin promoter, p10 promoter/lef-2 and 1629 baculo elements	KpnI HindIII	Opti3CInffwd AAGTTCTGTT TCAGGGCCCG	In-Fusion™ 3′ site ATGGTCTAG AAAGCTTTA
pOPINNUSA	N-HIS-NUSA-3C-POI	pUC (50 to >100)	Amp	pTriEx2	T7lacO, CMV enhancer and bb-actin promoter, p10 promoter/lef-2 and 1629 baculo elements	KpnI HindIII	Opti3CInffwd AAGTTCTGTT TCAGGGCCCG	In-Fusion™ 3′ site ATGGTCTAG AAAGCTTTA

The pOPINF family of vectors is designed to make a Rhinovirus 3C protease-cleavable fusion at the N-terminus of the target protein. All vectors are linearized with KpnI and HindIII and have common homology extensions. Thus one PCR product may be cloned into all members of the family.

POI protein of interest, *N-HIS* N-terminal His_6 tag, *C-His* C-terminal His_6 tag, *3C* Rhinovirus 3C protease site, *GST* glutathione-S-transferase, *MBP* maltose binding protein, *SUMO* small ubiquitin-like modifier, *MSYB E. coli* MsyB, *E. coli TF* trigger factor, *TRX E. coli* thioredoxin reductase, *HALO* HaloTag® (Promega), *HALO7* HaloTag7® (Promega), *eGFP* enhanced GFP, *NUSA E. coli* NusA

The PCR primer extension of the reverse primer anneals to the sequence of the Rhinovirus 3C cleavage site. Consequently, for the non-cleavable C-his vector (pOPINE) a different reverse primer extension is used (Table 2). In Fig. 2 the gel images of a representative small scale expression trial are shown using the N- and C-terminal fusion vectors fused to the gene encoding *Neisseria meningitidis* MC58 putative cell binding factor (amino acids 21–288).

The backbone of many members of the pOPIN vector suite is derived from pTriEx-2 (Novagen) and these vectors can also be used for heterologous expression in mammalian and insect cells without any sub-cloning required in changing expression systems [13, 14]. In addition, the HaloTag®, HaloTag7®, and eGFP fusions (Tables 1 and 2) have utility in imaging studies [28, 29].

1.2.2 Expression in E. coli of Complexes

Structural and functional studies of protein complexes require the heterologous production of two or more components. In some cases, complexes may be made by preparing the components separately and then mixing them together. However, there are many examples where co-expression of some or all of the components of the complex is required to achieve stoichiometric assembly [30].

There are two options for co-expressing proteins in *E. coli*: (1) expressing more than one open reading frame (ORF) from one vector, transcribed either from the same promoter as a polycistronic mRNA or from more than one promoter in the same vector [31, 32]; (2) co-expression of ORFs from separate plasmids which have different resistance markers and ideally origins of replication [33, 34]. These two approaches are not mutually exclusive and multi-protein complexes can be built up by co-transforming plasmids each containing more than one ORF.

For expressing complexes of two to three proteins, there are benefits in using separate vectors for each component. Firstly, construction of single gene vectors is relatively straightforward compared to multigene formats; secondly, if required, the effect of tag and tag position can be easily studied; and thirdly, there are no issues of gene order which applies to polycistronic or multi-promoter vectors. With this in mind, we have made a series of vectors with different replication origins and selectable markers to enable co-expression of up to three different genes. These vectors were constructed by transferring the cassette containing the fusion tag and the lacZ insert, that enables blue-white screening of recombinant constructs, from members of the pOPIN suite of vectors into plasmids with RSF1030 and CloDF13 origins and kanamycin and spectinomycin resistance markers, respectively [13]. These plasmids and their expression compatibility with other pOPIN vectors are shown in Table 3. PCR products can be transferred between members of the same family; for example, a pOPINE PCR product may be cloned into pOPINRSE and pOPINCDE

Table 2
Vectors designed for fusion at the C-terminus of proteins (pOPINE family)

Vector	Tag	Origin (copy no.)	Resistance	Vector backbone	Vector description	Vector linearization sites	Forward extension	Reverse extension
pOPINE	POI-C-HIS	pUC (50 to >100)	Amp	pTriEX2	T7lacO, CMV enhancer and bb-actin promoter, p10 promoter/ lef-2 and 1629 baculo elements	NcoI PmeI	Kozak(pET/TriEx) Inffwd AGGAGATAT ACCATG	KH6-pTriExInfrev GTGATGGTGA TGTTT
pOPINE-3C-eGFP	POI-3C-eGFP-C-HIS	pUC (50 to >100)	Amp	pTriEX2	T7lacO, CMV enhancer and bb-actin promoter, p10 promoter/ lef-2 and 1629 baculo elements	NcoI PmeI	Kozak(pET/TriEx) Inffwd AGGAGATAT ACCATG	3CFusionRev CAGAACTTCC AGTTT
pOPINE-3C-HALO7	POI-3C-Halo7-C-HIS	pUC (50 to >100)	Amp	pTriEX2	T7lacO, CMV enhancer and bb-actin promoter, p10 promoter/ lef-2 and 1629 baculo elements	NcoI PmeI	Kozak(pET/TriEx) Inffwd AGGAGATAT ACCATG	3CFusionRev CAGAACTTCC AGTTT

pOPINE-3C-FC*	POI-3C-FC-C-HIS	pUC (50 to >100)	Amp	pTriEX2	T7lacO, CMV enhancer and bb-actin promoter, p10 promoter/ lef-2 and 1629 baculo elements	NcoI PmeI	Kozak(pET/TriEx) Inffwd AGGAGATAT ACCATG	3CFusionRev CAGAACTT CCAGTTT
pOPINE-3C-CD4*	POI-3C-CD4-C-HIS	pUC (50 to >100)	Amp	pTriEX2	T7lacO, CMV enhancer and bb-actin promoter, p10 promoter/ lef-2 and 1629 baculo elements	Nco PmeI	Kozak(pET/TriEx) Inffwd AGGAGATAT ACCATG	3CFusionRev CAGAACTT CCAGTTT

The pOPINE family of vectors is designed to make C-terminal fusions. All vectors are linearized with NcoI and PmeI. Since pOPINE has a different 3′ extension to the other members of the family, three primers are needed for PCR amplification to produce two PCR products, one for pOPINE cloning and one for constructing the four C-terminal protein fusion vectors.

POI protein of interest, *C-His* C-terminal His_6 tag, *3C* Rhinovirus 3C protease site, *HALO7* HaloTag7® (Promega), *eGFP* enhanced GFP, *CD4* human CD4 domain, *FC* Human Ig FC domain.

*pOPINE-3C-FC and pOPINE-3C-CD4 are designed for secretion of proteins with a native leader sequence in insect or mammalian cells and are included as members of the family of 3C fusions but are not used in this study.

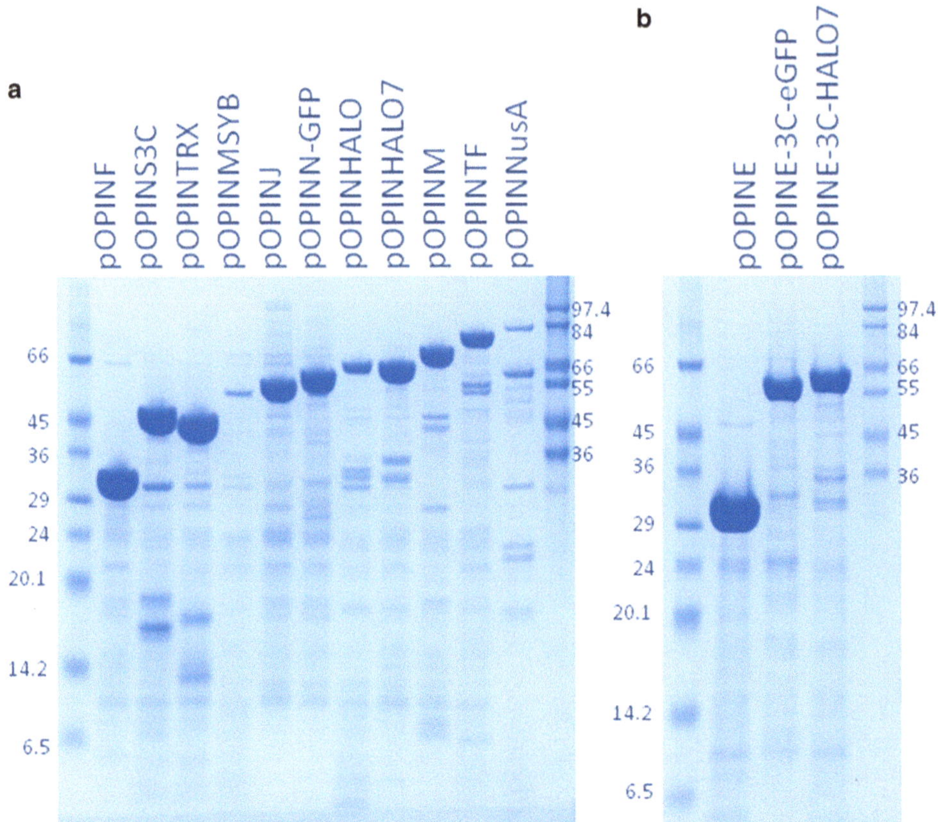

Fig. 2 Expression of *Neisseria meningitidis* MC58 putative cell binding factor (NMB0345; amino acids 21–288) in *E. coli* using pOPIN vectors. (**a**) The pOPINF family of fusion vectors. SDS-PAGE of expression screen results from Rosetta2 (DE3) plysS with IPTG induction of *Neisseria meningitidis* MC58 putative cell binding factor (NMB0345; amino acids 21–288) fused with members of the pOPINF family shown in Table 1. The vector is indicated above the lane and the results are shown in increasing order of size of fusion tag, although as noted previously the SUMO fusion runs with an anomalously high molecular weight [27]. (**b**) Fusion at the C-terminal. SDS-PAGE of expression screen results from Rosetta2 (DE3) plysS with IPTG induction of *Neisseria meningitidis* MC58 putative cell binding factor (NMB0345; amino acids 21–288) fused with C-terminal fusion vectors appropriate for expression in the cytoplasm of *E. coli* shown in Table 2. The vectors are indicated above the lane and they are shown in increasing order of size of fusion tag

since they share the same primer extensions. In Fig. 3 we show an example of the production of two related sub-complexes of *Saccharomyces cerevisiae* eIF3 using these vectors [35]. Each binary complex was assembled by co-expression of one component cloned into pOPINF and the other cloned into pOPINRSF. Using the vectors in Tables 1, 2, and 3 up to three proteins may be co-expressed in a multiple tag format without any deviations from the standard protocol. It is possible to make polycistronic binary and ternary expression vectors by In-Fusion™ cloning in the pOPIN

Table 3
Vectors designed for co-expression in *E. coli*

Vector	Tag	Origin (copy no.)	Resistance	Vector backbone	Vector description	*E. coli* co-expression compatibility	Vector linearization sites	Forward extension	Reverse extension
pOPINCDE	POI-C-HIS	CloDF13 (20–40)	Str	pCDFDuet-1	T7lacO, lacI	pOPINs, pOPINRSs	NcoI PmeI	Kozak(pET/TriEx) Inffwd AGGAGATATA CCATG	KH6-pTriExInfrev GTGATGGTGAT GTTT
pOPINCDF	N-HIS-3C-POI	CloDF13 (20–40)	Str	pCDFDuet-1	T7lacO, lacI	pOPINs, pOPINRSs	KpnI HinDIII	Opti3CInffwd AAGTTCTGTTT CAGGGCCCG	In-Fusion™ 3′ site ATGGTCTAGA AAGCTTTA
pOPINCDJ	N-HIS-GST-3C-POI	CloDF13 (20–40)	Str	pCDFDuet-1	T7lacO, lacI	pOPINs, pOPINRSs	KpnI HinDIII	Opti3CInffwd AAGTTCTGTT TCAGGGCCCG	In-Fusion™ 3′ site ATGGTCTAGA AAGCTTTA
pOPINCDM	N-HIS-MBP-3C-POI	CloDF13 (20–40)	Str	pCDFDuet-1	T7lacO, lacI	pOPINs, pOPINRSs	KpnI HinDIII	Opti3CInffwd AAGTTCTGTT TCAGGGCCCG	In-Fusion™ 3′ site ATGGTCTAGA AAGCTTTA
pOPINRSE	POI-C-HIS	RSF1030 (>100)	Kan	pRSFDuet-1	T7lacO, lacI	pOPINs, pOPINCDs	NcoI PmeI	Kozak(pET/TriEx) Inffwd AGGAGATATA CCATG	KH6-pTriExInfrev GTGATGGTGAT GTTT
pOPINRSF	N-HIS-3C-POI	RSF1030 (>100)	Kan	pRSFDuet-1	T7lacO, lacI	pOPINs, pOPINCDs	KpnI HinDIII	Opti3CInffwd AAG TTCTGTTTCAG GGCCCG	In-Fusion™ 3′ site ATGGTCTAGA AAGCTTTA
pOPINRSJ	N-HIS-GST-3C-POI	RSF1030 (>100)	Kan	pRSFDuet-1	T7lacO, lacI	pOPINs pOPINCDs	KpnI HinDIII	Opti3CInffwd AAGTTCTGTT TCAGGGCCCG	In-Fusion™ 3′ site ATGGTCTAGA AAGCTTTA

The vectors use the same homology extensions as either pOPINE (black) or pOPINF (bold) allowing the same PCR products to be used for either the POPINF or pOPINE vector families. If untagged protein is required the target sequence is cloned in a pOPINE family member with a stop codon at its carboxy-terminus.
POI protein of interest, *N-HIS* N-terminal His_6 tag, *C-His* C-terminal His_6 tag, *3C* Rhinovirus 3C protease site, *GST* glutathione-S-transferase, *MBP* maltose binding protein.

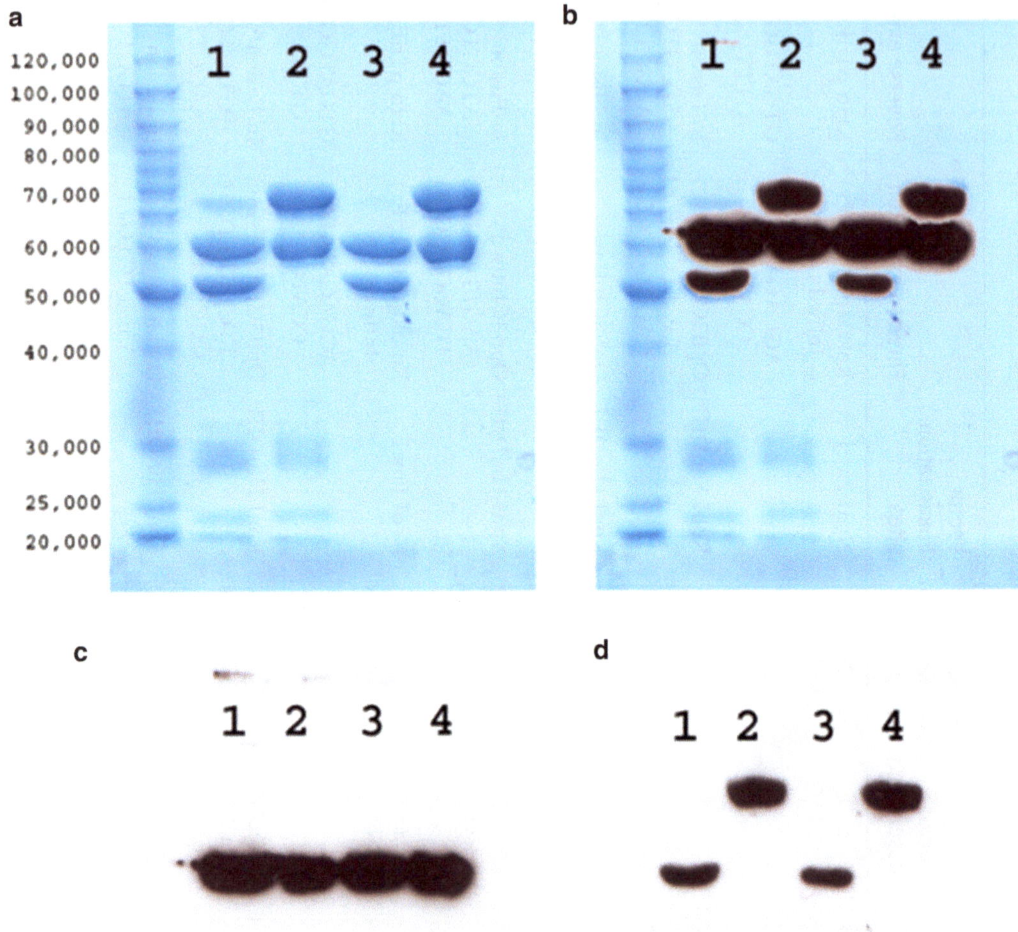

Fig. 3 Purification of yeast eIF5 NIP1/TIP32 sub-complexes. Two related complexes of NIP1 (amino acids 249–806) and TIP32 (amino acids 198–595 or 20–595) were made by cloning StrepII-tagged NIP1 into pOPINRSF (the vector was linearized with NcoI HindIII which removed the vector-derived N-terminal His-tag and a C-terminally StrepII-tagged NIP1 was used as template to provide a StrepII-tagged protein) and the two TIP32 variants into pOPINF (N-hexa-histidine). (**a**) Coomassie-stained polyacrylamide gel showing fractions of a scaled-up purification of the NIP1/TIF32 (amino acids 249–806 and either 198–595 or 20–595) complex. These sub-complexes were purified by Ni-NTA affinity chromatography (lanes 1 and 2, respectively) and gel filtration (lanes 3 and 4, respectively). (**b**) The western blots shown in (**c**) and (**d**) merged onto the gel in (**a**) to show that a His-tagged and Strep-tagged protein have been purified as a complex. (**c**) Anti-strep western blot (NIP-1) against the gel in (**a**). (**d**) Anti-His western blot (TIF32) against the gel in (**a**)

suite of vectors using an inter-cistronic ribosome binding site/s as the homology region (e.g., using pOPINF the 5′ protein would have N-terminal hexa-histidine tag, and the other protein/s are untagged; [36]). Thus by combining multi-fragment In-Fusion™ and the family of pOPIN co-expression vectors larger complexes may be assembled [36].

1.2.3 Secretion into the Periplasmic Space in E. coli

For the production of recombinant proteins that are normally secreted (from either bacterial or eukaryotic sources), there are a number of advantages in targeting them for periplasmic secretion in *E. coli*. Disulphide bond formation is favored in the periplasm, protease activity is lower than in the cytoplasm, and there are fewer host proteins, all of which facilitate recovery of the product. Gram-negative bacteria have a number of different secretory pathways that direct export of proteins to different destinations, such as the periplasmic space and outer membrane [37, 38]. The most commonly used mechanisms for secretion of heterologous proteins in *E. coli* are the Type I and Type II pathways [39, 40]. In Type I secretion, the protein is transported in one step across the inner and outer membranes and is released into the media [41]. By contrast, in Type II secretion the protein is exported via a two-step mechanism. There are three different mechanisms, namely the SecB, signal recognition particle (SRP), and TAT pathways [40]. The SecB and TAT systems translocate proteins post-translationally with the latter system requiring the presence of specific factors in the cytoplasm [40]. By contrast the SRP pathway involves the co-translational translocation of the protein. There is evidence that heterologous proteins may be handled differently by the SRP and SecB pathways. Higher levels of production of recombinant DARPins (small scaffold binding proteins) were observed when the proteins were targeted via the SRP mechanism compared to the SecB mechanism [42].

We have designed a suite of five vectors that target two of the Type II secretory mechanisms. The vectors provide both an N-terminal signal sequence (three are designed to target the SecB mechanism and two the SRP mechanism) and a C-terminal hexahistidine tag (Table 4). Figure 4 shows the results from our standard Ni^{2+}-NTA expression screen for expression of the New Delhi Metallo-β-lactamase 1 (NDM-1) from *Klebsiella pneumoniae*; the cytoplasmically expressed version of the construct with a C-terminal is shown for comparison. Interestingly in Rosetta2 (DE3) plysS we see strong expression from all five signal sequences and only very low levels of the cytoplasmically expressed version of the protein. Moreover in B834 (DE3) we see the same low levels of cytoplasmic expression but only see strong expression from the PelB leader suggesting the choice of screening strain may be an important variable in the periplasmic secretion of heterologous proteins.

2 Materials

2.1 Primers and Vectors

1. The homology regions required for In-Fusion™ cloning are generated by adding extensions to both forward and reverse PCR primers. The sequence and length of the extensions are determined by the type of restriction endonuclease used to

Table 4
Vectors designed to secrete a C-terminal hexa-histidine-tagged protein into the periplasmic space of *E. coli*

Vector	Tag	Signal sequence	Pathway	Origin (copy no.)	Resistance	Vector description	Vector linearization sites	Forward extension	Reverse extension
pOPINP	POI-C-HIS	*E. carotovora* PelB MKYLLPTAAAGL LLLAAQPAMA	Probably Sec	pUC (50 to >100)	Amp	T7lacO, CMV enhancer and bb-actin promoter, p10 promoter/ lef-2 and 1629 baculo elements	KpnI PmeI	pelBleaderfwd CAGCCGGC GATGGCT	KH6-pTriExInfrev GTGATGGT GATGTTT
pOPINO	SS-POI-C-HIS	*E. coli* OmpA MKKTAIAIAVAL AGFATVAQA	Sec	pUC (50 to >100)	Amp	T7lacO, CMV enhancer and bb-actin promoter, p10 promoter/ lef-2 and 1629 baculo elements	KpnI PmeI	ompAleaderfwd CTACCGTAG CGCAAGCT	KH6-pTriExInfrev GTGATGGT GATGTTT
pOPINMalE	SS-POI-C-HIS	*E. coli* MalE MKIKTGARILALSA LTTMMFSASALA	Sec	pUC (50 to >100)	Amp	T7lacO, CMV enhancer and bb-actin promoter, p10 promoter/ lef-2 and 1629 baculo elements	KpnI PmeI	MalEss_InfFwd CCGCCTCGG CTCTCGCC	KH6-pTriExInfrev GTGATGGT GATGTTT

pOPINDsbA	SS-POI-C-HIS	*E. coli* DsbA MKKIWLALAG LVLAFSASA	SRP	pUC (50 to >100)	Amp	T7lacO, CMV enhancer and bb-actin promoter, p10 promoter/ lef-2 and 1629 baculo elements	KpnI PmeI	DsbAss_InfFwd TTTAGCGC ATCGGCG	KH6-pTriExInfrev GTGATGGT GATGTTT
pOPINTolB	SS-POI-C-HIS	*E. coli* TolB MKQALRVAFGFLI LWASVLHA	SRP	pUC (50 to >100)	Amp	T7lacO, CMV enhancer and bb-actin promoter, p10 promoter/ lef-2 and 1629 baculo elements	KpnI PmeI	TolBss_InfFwd CATCAGTTC TGCATGCT	KH6-pTriExInfrev GTGATGGT GATGTTT

All vectors are linearized with KpnI and PmeI and use the same 3′ extension as pOPINE (Table 2; pOPINE can be used to make a vector with a native leader sequence if appropriate) but have signal sequence-specific 5′ extensions. The secretory pathway targeted is indicated in the table.
POI protein of interest, *C-His* C-terminal His_6 tag, *SS* signal sequence.

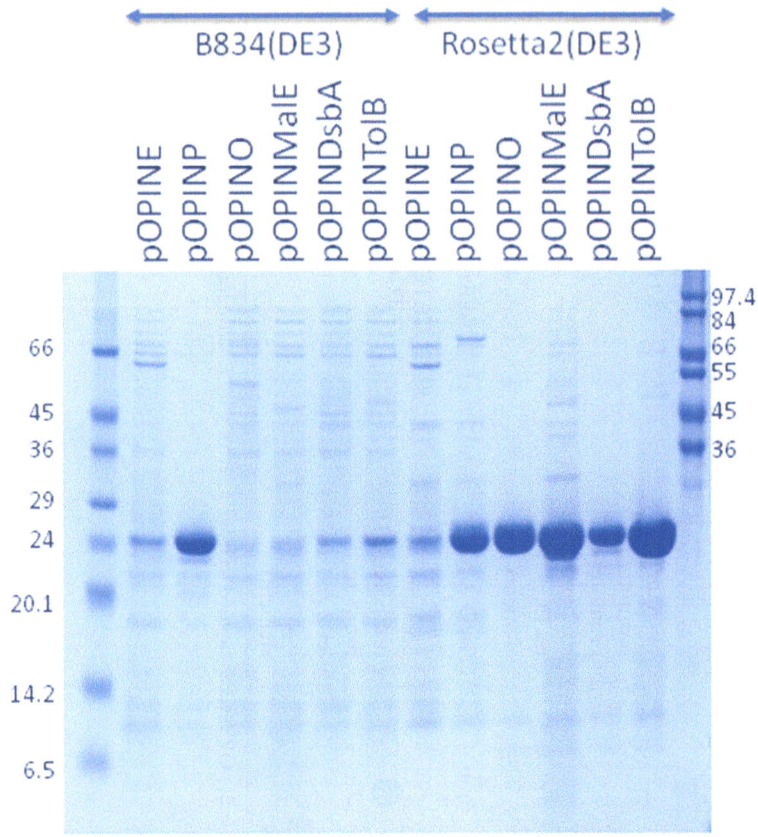

Fig. 4 Expression vectors to secrete proteins in the periplasm of *E. coli*. New Delhi Metallo-β-lactamase 1 (NDM-1) from *Klebsiella pneumoniae* (amino acids 36–270) was cloned into both pOPINE (cytoplasmic expression) and the suite of periplasmic expression vectors (Table 4). SDS-PAGE gel of expression screen results from B834 (DE3) and Rosetta2 (DE3) plysS with IPTG induction. Vectors and strains are indicated above the wells

linearize the vector but typically they are 15 bp long. We use an in-house primer design tool (**Note 1**) and aim for an annealing temperature of >66 °C. Purification of oligonucleotides is not necessary.

2. Any vector can be used with the In-Fusion™ cloning method. The only requirement is that the vector can be linearized using unique restriction site(s). Alternatively, inverse PCR can be used to produce linearized vector, although this may introduce mutations into the vector backbone [43]. OPPF-UK has developed the pOPIN suite of vectors which are described in this chapter; linearization always requires cutting with two restriction enzymes releasing a beta-galactosidase expression cassette [13–15]. Blue-white selection is then used to screen out colonies transformed with non-linearized vector [44]. The pOPIN vectors are available through Addgene http://www.addgene.org/.

2.2 Enzymes and Buffers

1. We use the 96 reaction In-Fusion® HD EcoDry™ Cloning Kit (Clontech; 639691), which is a freeze-dried formulation of enzyme and reaction buffer in a 96-well plate format (8 and 24 reactions are also available). It is critical that the EcoDry™ kit is stored at room temperature on desiccant.
2. TE: 10 mM Tris–HCl pH 8.0, 1 mM EDTA.
3. High fidelity PCR polymerases, e.g., KOD Xtreme™ Hot Start DNA Polymerase (Novagen 71975–3) and Phusion Flash High-Fidelity PCR Master Mix (F-548 L, Thermo Scientific).
4. Restriction endonucleases: DpnI, KpnI, HindIII, NcoI, and PmeI.

2.3 Purification of PCR Products and Linearized Vectors

1. Agencourt® AMPure® XP, 60 mL (A63881 Beckman-Coulter; 5 and 450 mL volumes are also available).
2. 96-Well magnetic separator such as the BILATEST™ magnetic separator M96 for 96-well plates (Bilatest 209601 or Z662429 Sigma).
3. NucleoSpin® Gel and PCR Clean-up; Machery-Nagel.
4. Elution Buffer: 10 mM Tris–HCl pH 8.0 (Qiagen).

2.4 Cloning Strains

1. It is essential to use high competency cloning strains with an efficiency of at least 10^8 cfu/μg circular plasmid DNA, e.g., OmniMax2 cells (Invitrogen) or Stellar competent cells (Clontech; supplied separately or with the In-Fusion™ kit).

2.5 Small Scale E. coli Expression Screen

1. *E. coli expression strains.* Chemically competent expression strains are prepared and stored in 50 μL aliquots at −80 °C [15]. Rosetta2 (DE3) plysS and B834 (DE3) are used routinely but other strains are included where appropriate.
2. *Media*
   (a) Power Broth (MD12-106-1, Molecular Dimensions). Media is premade and autoclaved.
   (b) Overnight Express™ Instant TB Medium (TBONEX; 71491; Novagen). The media is made up immediately prior to use by weighing the granules into sterile water and adding sterile 50 % glycerol (v/v in water) and is NOT autoclaved.
   (c) LB Agar (22700–041; Invitrogen).
   (d) 1 M IPTG (isopropyl-beta-D-thiogalactoside).
3. *Magnetic beads and magnet*
   (e) Ni^{2+} magnetic beads are available from a number of manufacturers, e.g., Qiagen (36113; Ni-NTA Magnetic Agarose Beads: 6 × 1 mL) or GE Healthcare (28-9799-17; His Mag Sepharose Ni: 10 × 1 mL).
   (f) 96-Well magnet such as the 96-Well Magnet Type A (36915; Qiagen).

4. *Buffers*
   (g) NPI-10-Tween: 50 mM NaH_2PO_4, 300 mM NaCl, 10 mM imidazole, and 1 % (v/v) Tween 20. Adjust pH to 8.0 using NaOH and filter before use. Store at 4 °C.
   (h) NPI-20-Tween (Wash Buffer): 50 mM NaH_2PO_4, 300 mM NaCl, 20 mM imidazole, and 0.05 % (v/v) Tween 20. Adjust pH to 8.0 using NaOH and filter before use. Store at 4 °C.
   (i) NPI-250-Tween (Elution Buffer): 50 mM NaH_2PO_4, 300 mM NaCl, 250 mM imidazole, and 0.05 % Tween. Adjust pH to 8.0 using NaOH and filter before use. Store at 4 °C.
   (j) DNAse I Stock Solutions (40,000 units/mL): to a 200,000 unit bottle of DNAse I (Sigma D-4527) add 5 mL of sterile water. Aliquot into 25 μL aliquots and store at −20 °C.
   (k) Lysis Buffer: To 25 mL of NPI-10-Tween buffer add lysozyme (Sigma L-6876) to a final concentration of 1 mg/mL plus one aliquot of DNAseI.

2.6 Plasticware and Seals

1. 96-Well deep well plate (e.g., AB-0932; 2.2 mL Storage Plate Mark II, natural color; AbGene).
2. Gas-permeable adhesive seal (AB-0718; AbGene).
3. 96-Well flat-bottomed microtitre plates (655101, Greiner Bio-One).
4. Foil seals (e.g., #6570; Corning® 96 Well Microplate Aluminum Sealing Tape, these maintain a seal at −80 °C).
5. 24-Well tissue culture plates with lids (e.g., Corning).
6. 24-Well deep well blocks (e.g., 360077; 24 Square well, 10 mL well polypropylene plate; Porvair Sciences Ltd.).

2.7 Equipment

1. Nanodrop (Thermo Scientific).
2. Veriti® 96-Well Fast Thermal Cycler (Invitrogen).
3. 37 °C static incubator.
4. Centrifuge suitable for spinning 96-well deep well blocks (e.g., Avanti J20 XPI with a JS-5.3 rotor; Beckman).
5. Shaking incubator with adapters for shaking 24-well deep well blocks (e.g., Innova® 44/44R; New Brunswick).

3 Methods

The protocols below are intended for high-throughput use in 96-well plates but can be easily adapted for lower throughput (*see* **Notes** for details).

3.1 Preparation of Vector

1. Incubate 15–20 μg of vector (midi-or maxi-prep quality DNA is generally used) with 100 units of each restriction endonuclease (e.g., KpnI and HindIII for the pOPINF family; [15]), in an appropriate buffer in a total reaction volume of 200 μL for 1 h at 37 °C.
2. The linearized vector is purified from solution using a spin column and resuspended in a volume of 30 μL, the concentration is measured using a Nanodrop, and the vector stock is normalized to 100 ng/μL before being stored in aliquots at −20 °C.

3.2 PCR Amplification

Protocols for two different polymerases are shown below (**Note 2**).

1. 10 μM working stocks of oligonucleotides are made in either nuclease-free water or Elution Buffer (**Note 3**).
2. A Master Mix is prepared on ice according to the total number of reactions required:

KOD Xtreme™ Hot Start DNA polymerase				
Number of (50 μL) reactions	1	25	50	100
2× KOD Hot Xtreme Buffer (μL)	25	625	1,250	2,500
dNTP mix (2 mM)	10	250	500	1,000
KOD Xtreme™ (1 U/μL)	1	25	50	100
Sterile water	6	150	300	600
Total volume	42	1,050	2,100	4,200

3. Dispense 42 μL of this Master Mix into each well of the PCR plate/tube.
4. 3 μL each of diluted (10 μM) forward and reverse primers are added to the appropriate wells of the PCR plate/tubes
5. Add 2 μL of template (10–50 ng of plasmid and 0.1–1 μg of genomic DNA) to the appropriate wells of the PCR plate/tube.
6. Perform the thermal cycling using the following parameters:
   (a) 94 °C, 2 min
   (b) 98 °C, 10 s
   (c) 60 °C, 30 s
   (d) 68 °C, *n* min (base this time on the largest PCR product expected and 1 min/kb)
   (e) Go to **step 2** and repeat 29 times
   (f) 68 °C, 2 min
   (g) 4 °C, Hold

7. Alternatively, using Phusion Flash High-Fidelity PCR Master Mix, the reactions are set up using the same volumes of primers and template as used above.

Number of (50 μL) reactions	1	25	50	100
2× Phusion Flash Master Mix (μL)	25	625	1,250	2,500
Sterile water	17	425	850	1,700
Total volume	42	1,050	2,100	4,200

8. Perform the thermal cycling using these parameters:
   (a) 98 °C, 10 s
   (b) 98 °C, 0 or 1 s
   (c) 60 °C, 5 s
   (d) 72 °C, *n* min (base this time on the largest PCR product expected and 15 s/kb)
   (e) Go to **step 2** and repeat 29 times
   (f) 72 °C, 2 min
   (g) 4 °C, Hold
9. Amplified PCR products are treated with DpnI to remove parental templates, 5 units of enzyme are added to the PCRs in 5 μL of 1× restriction buffer, and incubated at 37 °C for 1 h.

3.3 PCR Product Purification

1. Gently shake the AMPure XP bottle to resuspend the magnetic particles. Add 90 μL AMPure to each PCR in the PCR plate.
2. Mix the AMPure XP and PCR thoroughly using a pipette. The color of the mixture should appear homogenous after mixing. Incubate for 3–5 min at room temperature. Place the reaction plate onto a magnet for around 5 min to separate beads from solution. Wait for the solution to clarify before continuing to the next step.
3. With the plate located on the magnet remove the cleared solution and discard. Do not disturb the magnetic beads.
4. Keep the plate on the magnet and dispense 200 μL of 70 % ethanol to each well. Remove the ethanol and discard. Repeat. On the second discard use a fresh tip to remove all of the ethanol from the bottom of the well.
5. The plate should be left to air-dry for 10–20 min on a bench top to allow complete evaporation of any residual ethanol.
6. Take the plate off the magnet and add 30 μL of Elution Buffer to each well and mix by pipetting. Incubate for 1 min at room temperature.
7. Place the plate on the magnet; once the supernatant has clarified, transfer it to a storage plate (**Note 4**). If small amounts of

beads "carry-over" the plate can, if necessary, be placed on a magnet when setting up the In-Fusion™ reaction.

3.4 In-Fusion™ Reaction

1. Add 1 μL (100 ng) of the appropriate linearized vector to each well of a PCR plate.
2. Add x μL of purified insert (ideally it should be in the range of 10–250 ng; **Note 5**) to the appropriate wells of the PCR plate. Add $(9-x)$ μL of water and transfer the 10 μL to a dry-down In-Fusion™ plate (**Notes 6 and 7**). Mix contents briefly by pipetting up and down.
3. Transfer the reactions to a PCR plate and seal.
4. Incubate for 30 min at 42 °C in a thermocycler (**Note 8**).
5. When the In-Fusion™ reaction is complete transfer the reactions to ice and immediately add 40 μL of TE. Immediately freeze or transform into competent cells. For transformation add 3 μL of the diluted In-Fusion™ reaction per 50 μL aliquot of competent cells.
6. Incubate the cells and DNA on ice for 30 min.
7. Heat-shock the cells for 30 s at 42 °C.
8. Return the cells to ice for 2 min. Add 300 μL of SOC per tube.
9. Transfer the cells to a 37 °C static incubator and incubate for 1 h.
10. The LB Agar plates are prepared as follows: 1 mL of LB Agar supplemented with the appropriate antibiotic/s is aliquotted per well of a 24-well tissue culture plate.
11. Plating out 25 μL of a total volume of 350 μL of out-grown cells should give many tens to hundreds of colonies per well of a 24-well plate. A 1 in 5 dilution of the out-grown cells is generally needed to be able to pick single colonies in a 24-well plate.
12. Pick two colonies and mini-prep the vectors. Carry out PCR verification using the construct-specific reverse primer and a vector-specific forward primer (e.g., pOPIN_FOR GAC CGA AAT TAA TAC GAC TCA CTA TAG GG for the pOPIN vectors in this chapter; **Note 9**).

3.5 Small Scale Expression Screening in E. coli: Growth and Harvesting of Cultures

The protocol described below is for high-throughput applications but can be easily adapted for lower throughput experiments.

3.5.1 Transformation into E. coli Expression Strains

1. Thaw the aliquotted *E. coli* on ice.
2. Use 3 μL of mini-prepped expression plasmids to transform the competent cell aliquots.
3. Incubate on ice for 30 min.
4. Heat-shock the cells for 30 s at 42 °C.

5. Return the cells to ice for 2 min.
6. Add 300 μL of Power Broth per tube.
7. Transfer the cells to a 37 °C static incubator and incubate for 1 h.
8. The LB Agar plates are prepared as follows: 1 mL of LB Agar supplemented with the appropriate antibiotic/s is aliquotted per well of a 24-well tissue culture plate.
9. 30 μL of cells is transferred from the transformation mix onto the LB Agar plates (**Note 10**).
10. The plates are shaken in a horizontal plane to spread the cells and dried with the lid off for 10–15 min at 37 °C before replacing the lid and turning upside down.
11. Incubate overnight at 37 °C.

3.5.2 Preparation of Overnight Cultures

1. Add 0.7 mL of Power Broth and the appropriate antibiotic to each well of a 96-well deep well block (the media is supplemented with 1 % (w/v) glucose if the product is likely to be toxic).
2. Individual colonies are picked into each well (**Note 11**).
3. Seal blocks with gas-permeable adhesive seals.
4. Shake the filled blocks at 250 rpm (in a standard incubator) at 37 °C overnight.

3.5.3 Growth and Induction of Cultures

1. For each 24 constructs/per strain prepare two 24-well deep well blocks (**Note 12**).
   (a) Block 1 has 3 mL of Power Broth supplemented with the appropriate antibiotic/s (1 % glucose can also be added) per well.
   (b) Block 2 has of 3 mL of TBONEX supplemented with the appropriate antibiotic/s per well.
2. Dilute the overnight cultures by transferring 150 μL of overnight culture to the 24-well blocks prepared in **step 1**.
3. Shake the blocks at 250 rpm for 3–5 h at 37 °C (until the average O.D. at 595 nm is ~0.5).
   (a) Cool the cultures for IPTG induction by shaking at 250 rpm at 20 °C for 20 min.
   (b) Reduce the temperature to 25 °C for the TBONEX Media. Grow with shaking at 250 rpm for at least 20 h at 25 °C.
4. Induce the IPTG cultures with a final concentration of 1.0 mM IPTG. Grow the cultures overnight (~18 h) by shaking at 250 rpm at 20 °C.

3.5.4 Harvesting of Cultures

1. Transfer 1 mL of culture from each well into a 96-well deep well block.
2. Seal the block and harvest the cells by centrifugation at 6,000 × *g* for 10 min.

3. Decant the media from the cell pellets by inverting the block and drain by tapping onto a paper towel.
4. Seal the plates and store at −80 °C for a minimum of 20 min before Ni^{2+}-NTA screening.

3.6 Ni^{2+}-NTA Miniature Expression Screen Protocol

This manual protocol is similar to the automated protocol used in OPPF-UK (**Note 13**).

1. Resuspend the defrosted cells completely in 210 μL of Lysis Buffer. This cans be done either with a suitable multichannel pipette or on an orbital shaker (~1,000 rpm for 30 min). If resuspending using a multichannel pipette, once resuspended incubate for 30 min at room temperature to allow for the action of the lysozyme and DNAse I.
2. Centrifuge the deep well block at 6,000 × *g* for 30 min at 4 °C.
3. Dispense 20 μL of the Ni-NTA magnetic bead suspension into each well of a flat-bottomed microtitre plate.
4. Transfer the supernatant from **step 1** without disturbing the "insoluble" pellet to each well of the microtitre plate containing the Ni-NTA magnetic beads (**Note 14**).
5. Mix for 30 min at room temperature using either a microtitre plate shaker at 600 rpm or vortex using an adapter for microtitre plates.
6. Place the plate on the 96-well magnet once the supernatant has clarified; remove the supernatant carefully from the beads.
7. Add 200 μL of Wash Buffer to each well, remove from the magnet, and shake (or vortex) for 5 min.
8. Place the plate on the 96-well magnet and once the supernatant has clarified remove the buffer.
9. Repeat **steps 7–8**.
10. Add 60 μL of Elution Buffer (NPI-250) to each well, shake (or vortex) for 1 min, place the plate on the 96-well magnet, and, once it has clarified, transfer the supernatant (eluate) to a fresh plate for analysis on SDS-PAGE gels.

4 Notes

1. The 5′ and 3′ homology extensions for cloning into the pOPIN vectors are indicated in the vector tables. We have developed a software tool for designing the appropriate PCR primers for cloning into the pOPIN suite of vectors. This is available through the following link: https://www.oppf.rc-harwell.ac.uk/Opiner/.
2. Obtaining good quality PCR products is critical. We find the easiest way to optimize PCR is to use an alternative polymerase and would only try other variables, e.g., annealing temperature,

if both polymerases failed to give a product. If a single band cannot be obtained after optimization, then gel purify the correct band or if the extra bands are relatively minor contaminants proceed with the cloning but mini-prep extra colonies from the transformed cells to overcome any possible lowering of cloning efficiency.

3. For high-throughput experiments the working stocks of oligonucleotide are made in a green PCR plate to aid identification when the plate is stored in a freezer.
4. For high-throughput experiments the DpnI-treated purified PCR products are stored in a purple plate PCR to aid identification when the plate is stored in a freezer.
5. For high-throughput experiments, we use one standard volume of PCR product in the In-Fusion™ reaction. However, for lower throughput experiments, variable volumes can be used to achieve a 2:1 molar ratio of fragment to vector.
6. Smaller reaction volumes can be used; for example, we have used reactions as small as 3.3 μL. Alternatively, where a PCR product is being cloned into two (or more) vectors that have the same homology extensions but different selectable markers then both/all vectors may be placed in a single reaction and the transformed cells plated out onto two plates with the appropriate antibiotics.
7. For lower throughput applications the liquid formulation of In-Fusion™ may be more appropriate (In-Fusion® HD cloning kit; Clontech).
8. This protocol is not the manufacturer's current protocol but is the original protocol used in OPPF-UK for many years and as it works efficiently we have not changed it.
9. Picking two mini-preps/construct and following the In-Fusion™ cloning protocol outlined in the text, for 4,400 constructs our cloning efficiency is 95 %.
10. Pipettes that can be expanded and contracted to transfer between 24- and 96-well pitches are useful. We use 8-channel Matrix IMPACT pipettors and 6-channel Rainin Pipet-Lite™ XLS Adjustable Spacer pipettors.
11. Pick colonies using a 200 μL multichannel pipette tip; leave the tip in the well after picking the colony. Once the plate is full the tips can then be removed using a multichannel pipette and discarded.
12. It is a good idea to have different strains in separate blocks as they often grow at different rates and will thus require inducing at different times.
13. This protocol uses Ni^{2+}-NTA magnetic beads; however, protocols that use filter plates and Ni^{2+}-NTA agarose beads (e.g., Qiagen, Ni-NTA Superflow 96 BioRobot Kit (4), 969261 or GE

Healthcare, 50 μL Ni Sepharose High Performance; 4×96-well filter plates, 28-4009-89) are equally effective. For lower throughput screening, spin columns (e.g., Qiagen, Ni-NTA Spin Columns (50), 31014) are a useful alternative.

14. The pellets can be resuspended in 8 M Urea, 300 mM NaCl, 10 mM imidazole, and 10 mM Tris–HCl pH 8.0 for direct analysis of the "insoluble" fraction either by SDS-PAGE or by an Ni^{2+}-NTA pull-down in the presence of Urea using sepharose beads and SDS-PAGE. We use the Qiagen Ni-NTA Superflow 96 BioRobot Kit (969261).

Acknowledgements

The Oxford Protein Production Facility-UK is supported by the UK Medical Research Council and the Biotechnology and Biology Research Council (MRC Grant MR/K018779/1). We thank Jo Nettleship for helpful discussions on *E. coli* secretion vectors.

References

1. Berrow NS, Bussow K, Coutard B et al (2006) Recombinant protein expression and solubility screening in Escherichia coli: a comparative study. Acta Crystallogr D Biol Crystallogr 62: 1218–1226
2. Savitsky P, Bray J, Cooper CD et al (2010) High-throughput production of human proteins for crystallization: the SGC experience. J Struct Biol 172:3–13
3. Graslund S, Nordlund P, Weigelt J et al (2008) Protein production and purification. Nat Methods 5:135–146
4. Irwin CR, Farmer A, Willer DO et al (2012) In-fusion(R) cloning with vaccinia virus DNA polymerase. Methods Mol Biol 890:23–35
5. Hamilton MD, Nuara AA, Gammon DB et al (2007) Duplex strand joining reactions catalyzed by vaccinia virus DNA polymerase. Nucleic Acids Res 35:143–151
6. Aslandis C, de Jong PJ (1990) Ligation-independent cloning of PCR products (LIC-PCR). Nucleic Acids Res 18:6069–6074
7. Haun RS, Serventi IM, Moss J (1992) Rapid, reliable ligation-independent cloning of PCR products using modified plasmid vectors. Biotechniques 13:515–518
8. Li MZ, Elledge SJ (2012) SLIC: a method for sequence- and ligation-independent cloning. Methods Mol Biol 852:51–59
9. Li MZ, Elledge SJ (2007) Harnessing homologous recombination in vitro to generate recombinant DNA via SLIC. Nat Methods 4: 251–256
10. Jeong JY, Yim HS, Ryu JY et al (2012) One-step sequence- and ligation-independent cloning as a rapid and versatile cloning method for functional genomics studies. Appl Environ Microbiol 78:5440–5443
11. Gibson DG, Young L, Chuang RY et al (2009) Enzymatic assembly of DNA molecules up to several hundred kilobases. Nat Methods 6: 343–345
12. Gibson DG, Smith HO, Hutchison CA III et al (2010) Chemical synthesis of the mouse mitochondrial genome. Nat Methods 7:901–903
13. Berrow NS, Alderton D, Sainsbury S et al (2007) A versatile ligation-independent cloning method suitable for high-throughput expression screening applications. Nucleic Acids Res 35:e45
14. Berrow NS, Alderton D, Owens RJ (2009) The precise engineering of expression vectors using high-throughput In-Fusion PCR cloning. Methods Mol Biol 498:75–90
15. Bird LE (2011) High throughput construction and small scale expression screening of multi-tag vectors in Escherichia coli. Methods 55:29–37
16. Chen JH, Jung JW, Wang Y et al (2010) Immunoproteomics profiling of blood stage Plasmodium vivax infection by high-throughput screening assays. J Proteome Res 9:6479–6489
17. Howland SW, Poh CM, Renia L (2011) Directional, seamless, and restriction enzyme-free construction of random-primed complementary

DNA libraries using phosphorothioate-modified primers. Anal Biochem 416:141–143
18. Nettleship JE, Ren J, Rahman N et al (2008) A pipeline for the production of antibody fragments for structural studies using transient expression in HEK 293T cells. Protein Expr Purif 62:83–89
19. Au K, Berrow NS, Blagova E et al (2006) Application of high-throughput technologies to a structural proteomics-type analysis of Bacillus anthracis. Acta Crystallogr D Biol Crystallogr 62:1267–1275
20. Marsischky G, LaBaer J (2004) Many paths to many clones: a comparative look at high-throughput cloning methods. Genome Res 14:2020–2028
21. Zhu B, Cai G, Hall EO et al (2007) In-fusion assembly: seamless engineering of multidomain fusion proteins, modular vectors, and mutations. Biotechniques 43:354–359
22. Zhou B, Donnelly ME, Scholes DT et al (2009) Single-reaction genomic amplification accelerates sequencing and vaccine production for classical and Swine origin human influenza a viruses. J Virol 83:10309–10313
23. Sleight SC, Bartley BA, Lieviant JA et al (2010) Designing and engineering evolutionary robust genetic circuits. J Biol Eng 4:12
24. Sleight SC, Bartley BA, Lieviant JA et al (2010) In-Fusion BioBrick assembly and re-engineering. Nucleic Acids Res 38:2624–2636
25. Alzari PM, Berglund H, Berrow NS et al (2006) Implementation of semi-automated cloning and prokaryotic expression screening: the impact of SPINE. Acta Crystallogr D Biol Crystallogr 62:1103–1113
26. Esposito D, Chatterjee DK (2006) Enhancement of soluble protein expression through the use of fusion tags. Curr Opin Biotechnol 17:353–358
27. Marblestone JG, Edavettal SC, Lim Y et al (2006) Comparison of SUMO fusion technology with traditional gene fusion systems: enhanced expression and solubility with SUMO. Protein Sci 15:182–189
28. Ohana RF, Encell LP, Zhao K et al (2009) HaloTag7: a genetically engineered tag that enhances bacterial expression of soluble proteins and improves protein purification. Protein Expr Purif 68:110–120
29. Chudakov DM, Matz MV, Lukyanov S et al (2010) Fluorescent proteins and their applications in imaging living cells and tissues. Physiol Rev 90:1103–1163
30. Li C, Schwabe JW, Banayo E et al (1997) Coexpression of nuclear receptor partners increases their solubility and biological activities. Proc Natl Acad Sci U S A 94:2278–2283
31. Alexandrov A, Vignali M, LaCount DJ et al (2004) A facile method for high-throughput co-expression of protein pairs. Mol Cell Proteomics 3:934–938
32. Scheich C, Kummel D, Soumailakakis D et al (2007) Vectors for co-expression of an unrestricted number of proteins. Nucleic Acids Res 35:e43
33. Kholod N, Mustelin T (2001) Novel vectors for co-expression of two proteins in E. coli. Biotechniques 31(322–323):326–328
34. Yang W, Zhang L, Lu Z et al (2001) A new method for protein coexpression in Escherichia coli using two incompatible plasmids. Protein Expr Purif 22:472–478
35. Hinnebusch AG (2006) eIF3: a versatile scaffold for translation initiation complexes. Trends Biochem Sci 31:553–562
36. Busso D, Peleg Y, Heidebrecht T et al (2011) Expression of protein complexes using multiple Escherichia coli protein co-expression systems: a benchmarking study. J Struct Biol 175:159–170
37. Economou A, Christie PJ, Fernandez RC et al (2006) Secretion by numbers: protein traffic in prokaryotes. Mol Microbiol 62:308–319
38. Desvaux M, Hebraud M, Talon R et al (2009) Secretion and subcellular localizations of bacterial proteins: a semantic awareness issue. Trends Microbiol 17:139–145
39. Choi JH, Lee SY (2004) Secretory and extracellular production of recombinant proteins using Escherichia coli. Appl Microbiol Biotechnol 64:625–635
40. Mergulhao FJ, Summers DK, Monteiro GA (2005) Recombinant protein secretion in Escherichia coli. Biotechnol Adv 23:177–202
41. Binet R, Letoffe S, Ghigo JM et al (1997) Protein secretion by Gram-negative bacterial ABC exporters—a review. Gene 192:7–11
42. Steiner D, Forrer P, Stumpp MT et al (2006) Signal sequences directing cotranslational translocation expand the range of proteins amenable to phage display. Nat Biotechnol 24:823–831
43. Benoit RM, Wilhelm RN, Scherer-Becker D et al (2006) An improved method for fast, robust, and seamless integration of DNA fragments into multiple plasmids. Protein Expr Purif 45:66–71
44. Ruther U (1980) Construction and properties of a new cloning vehicle, allowing direct screening for recombinant plasmids. Mol Gen Genet 178:475–477

Chapter 16

Seamless Ligation Cloning Extract (SLiCE) Cloning Method

Yongwei Zhang, Uwe Werling, and Winfried Edelmann

Abstract

SLiCE (*S*eamless *Li*gation *C*loning *E*xtract) is a novel cloning method that utilizes easy to generate bacterial cell extracts to assemble multiple DNA fragments into recombinant DNA molecules in a single in vitro recombination reaction. SLiCE overcomes the sequence limitations of traditional cloning methods, facilitates seamless cloning by recombining short end homologies (15–52 bp) with or without flanking heterologous sequences and provides an effective strategy for directional subcloning of DNA fragments from bacterial artificial chromosomes or other sources. SLiCE is highly cost-effective and demonstrates the versatility as a number of standard laboratory bacterial strains can serve as sources for SLiCE extract. We established a DH10B-derived *E. coli* strain expressing an optimized λ prophage Red recombination system, termed PPY, which facilitates SLiCE with very high efficiencies.

Key words SLiCE cloning, In vitro recombination, Seamless cloning, Bacterial cell extract, DNA cloning, Recombinant DNA

1 Introduction

The generation of recombinant DNA molecules is an essential tool in modern molecular biology. The conventional DNA cloning strategies that have been used for several decades typically involve the use of type II restriction enzymes to generate appropriate DNA fragments, the modification of DNA ends to generate blunt or sticky ends, and the ligation of the DNA fragments to generate plasmid or other type DNA vectors [1–3]. However, these procedures depend on the presence of appropriate restriction sites to generate both vector and insert molecules and often leave unwanted sequences at the junction sites. In addition, the restriction enzymes and modifying enzymes required for these manipulations are often expensive making these procedures costly especially in high throughput settings. To circumvent these limitations, we developed a new restriction site-independent cloning method that does not leave any unwanted sequences at the junction sites (seamless) and is based on in vitro recombination between short regions of

Svein Valla and Rahmi Lale (eds.), *DNA Cloning and Assembly Methods*, Methods in Molecular Biology, vol. 1116, DOI 10.1007/978-1-62703-764-8_16,

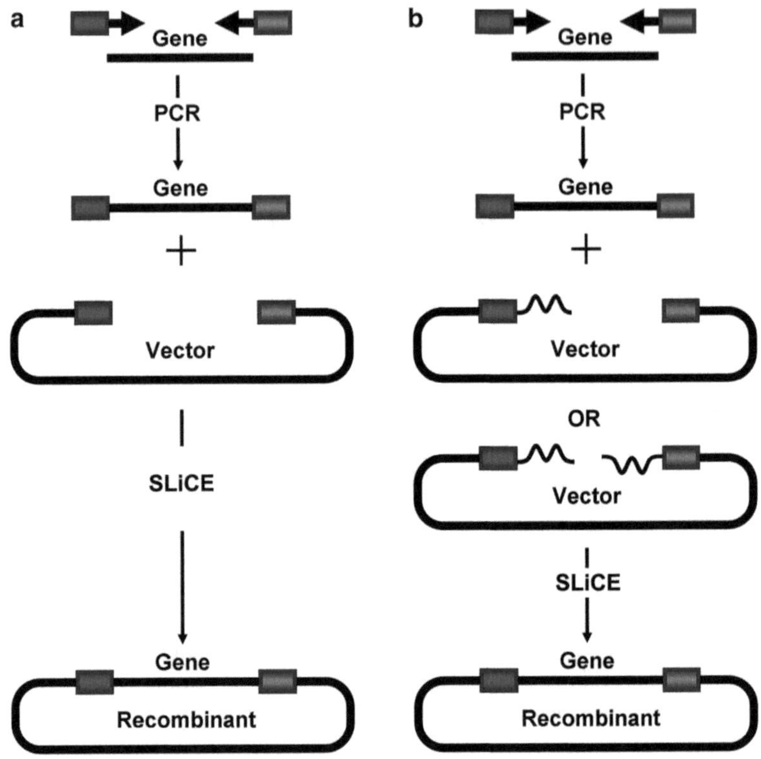

Fig. 1 SLiCE cloning of a single DNA fragment. (**a**) SLiCE cloning without flanking heterologous sequences. (**b**) SLiCE cloning with flanking heterologous sequences (Reproduced from [4], with permission from Oxford University Press)

homologies (15–52 bp) in bacterial cell extracts termed SLiCE (*S*eamless *Li*gation *C*loning *E*xtract) [4]. SLiCE allows for efficient restriction site-independent cloning of DNA fragments generated by restriction digestion or PCR amplification into linearized vectors. In addition, SLiCE does not require the use of enzymes for the modification of vector and insert end sequences (such as Klenow or T4 DNA polymerase) or ligases.

The SLiCE method [4] can be used for virtually any type of cloning approach such as the cloning of a single DNA fragment with or without flanking heterologous sequences (*see* Fig. 1), the cloning of multiple DNA fragments in one step (*see* Fig. 2), and the directional subcloning of large DNA fragments from bacterial artificial chromosomes (BACs) (*see* Fig. 3).

The SLiCE method is based on bacterial extracts that can be derived from a variety of common RecA$^-$ *E. coli* laboratory strains such as DH10B and JM109 [4]. In vivo homologous recombination in *E. coli* can be facilitated by three different recombination pathways: The RecA-dependent pathway, an RecA-independent pathway of unknown nature and an RecA-independent pathway that utilizes prophage Red/ET recombination systems [5–10].

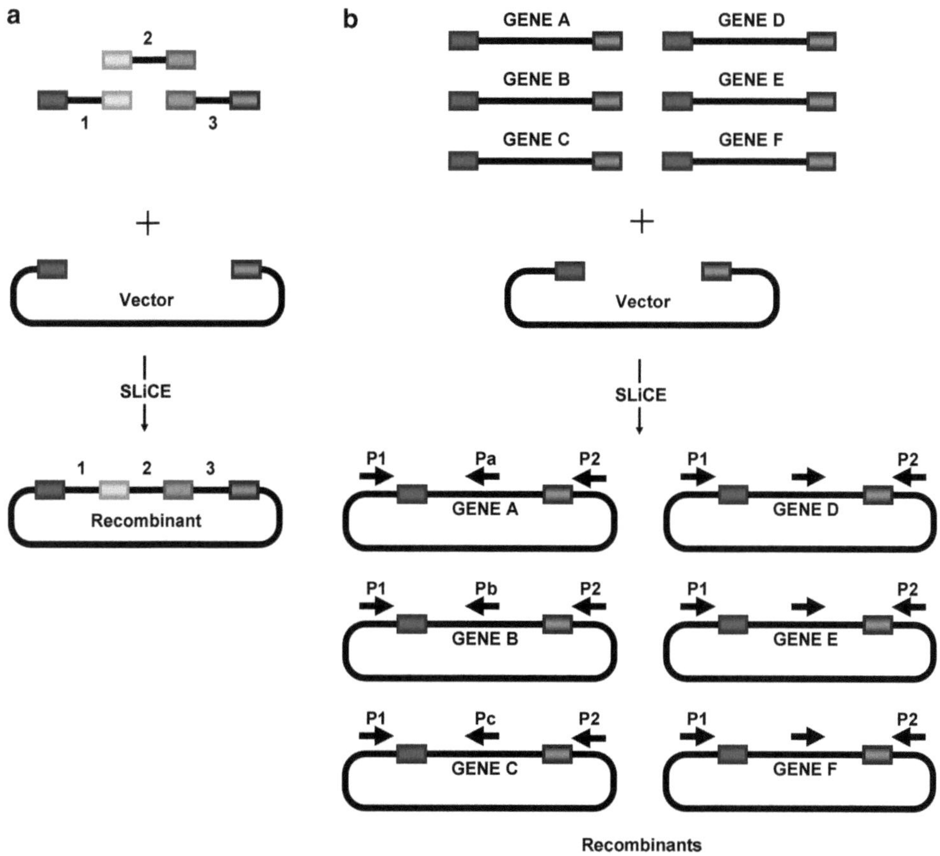

Fig. 2 SLiCE cloning of multiple DNA fragments. (**a**) Schematic illustrating multiple-way SLiCE cloning. A three-way cloning approach is shown. (**b**) Schematic illustrating SLiCE batch cloning (Reproduced from [4], with permission from Oxford University Press)

Our studies indicate that an RecA-independent pathway catalyzes SLiCE. To improve SLiCE cloning efficiencies and capabilities, we established a DH10B-derived strain, termed PPY that was engineered to contain an optimized λ prophage Red recombination system [4–7]. This strain provides the highest cloning efficiencies thus far and can be used for all cloning approaches that are routinely used in the laboratory [4].

SLiCE is a simple and efficient procedure with the entire process consisting of three steps: (1) The preparation of linear vector and insert fragments containing short end homologies introduced by PCR with primers having appropriate 5′ extension sequences; (2) the SLiCE in vitro reaction; and (3) the standard transformation (electroporation or chemical transformation) of recombination products into suitable host bacteria [4].

In this chapter, we describe the preparation of the PPY SLiCE extract, the maintenance of the PPY strain, and the SLiCE cloning procedure.

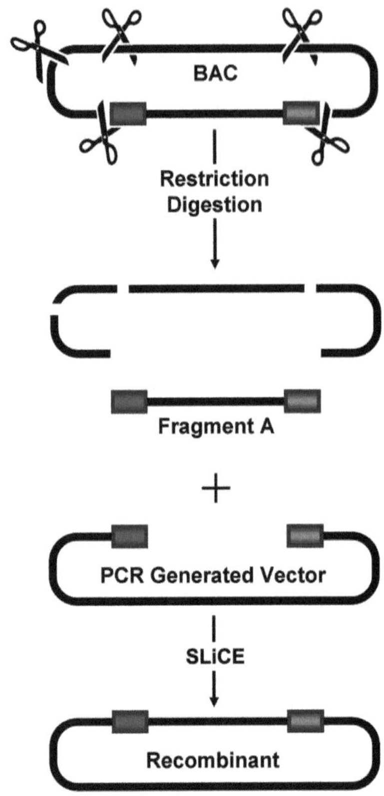

Fig. 3 BAC SLiCE cloning (Reproduced from [4], with permission from Oxford University Press)

2 Materials

2.1 Preparation of PPY SLiCE Extract

1. PPY strain (available from Dr. Yongwei Zhang, yongwei.zhang@einstein.yu.edu) (*see* **Note 1**).
2. 2XYT medium: Dissolve 16 g Bacto-tryptone, 10 g Bacto-yeast extract, and 5 g NaCl in 900 mL ddH_2O. Adjust pH to 7.2 with NaOH. Adjust to 1 L with ddH_2O. Autoclave to sterilize and store at room temperature.
3. Antibiotics: Streptomycin and chloramphenicol.
4. L-(+)-Arabinose (Sigma, A3256).
5. CelLytic™ B Cell Lysis Reagent (Sigma, B7435).
6. 100 % Glycerol.
7. Protein LoBind Tube 1.5 mL (Eppendorf).
8. Protein LoBind Tube 0.5 mL (Eppendorf).
9. 50-mL centrifuge tubes.
10. 250-mL Nalgene Lab Quality Wide-Mouth Bottles.

11. 37 °C shaker.
12. 37 °C incubator.
13. Centrifuge.
14. Spectrophotometer.
15. –20 and –80 °C freezer.

2.2 Maintenance of PPY strain

1. PPY strain.
2. Lysogeny Broth (LB) medium: Dissolve 10 g Bacto tryptone, 5 g Bacto-yeast extract, and 10 g NaCl in 900 mL ddH_2O. Adjust pH to 7.2 with NaOH. Adjust to 1 L with ddH_2O. Autoclave to sterilize and store at room temperature.
3. Antibiotics: Streptomycin and chloramphenicol.
4. 20 % Glycerol.
5. 37 °C shaker.
6. –80 °C freezer.

2.3 SLiCE Cloning

1. PPY SLiCE extract (prepared as described in Subheading 3.1).
2. 10× SLiCE buffer: 500 mM Tris–Cl (pH 7.5 at 25 °C), 100 mM $MgCl_2$, 10 mM ATP, 10 mM DTT, store at –20 °C.
3. Materials and equipment for restriction cutting, PCR, DNA purification and transformation.

3 Methods

3.1 Preparation of PPY SLiCE Extract

1. Streak PPY glycerol stock or fresh culture on an LB agar plate (10 μg/mL streptomycin and 12.5 μg/mL chloramphenicol) and incubate at 37 °C overnight.
2. Inoculate one single colony into a 50-mL centrifuge tube containing 25 mL 2XYT (10 μg/mL streptomycin) and shake at 37 °C and 330 rpm overnight.
3. The next day, measure the OD_{600} (*see* **Note 2**).
4. Dilute the o/n culture to 0.03 OD_{600}, i.e., inoculate appropriate volume of the o/n culture into a 250-mL Nalgene Lab Quality Wide-Mouth Bottle containing 50 mL 2XYT medium (10 μg/mL streptomycin) (*see* **Note 3**).
5. Shake at 37 °C and 330 rpm until the culture reaches an OD_{600} of 5.0–5.5 (*see* **Note 4**).
6. Add 0.2 % L-(+)-arabinose into the culture and continue shaking at 330 rpm for 2 h at 37 °C, to induce expression of λ prophage protein Red. Remove 500 μL from the culture to measure the actual OD_{600}.
7. Transfer 48 mL of the bacterial culture into two 50-mL centrifuge tubes (24 mL each).

8. Pellet the cells by centrifugation at 5,000 × *g* for 20 min at 4 °C.
9. Wash the cell pellet from the 24 mL original culture with 50 mL ddH_2O once.
10. Pellet the cells by centrifugation at 5,000 × *g* for 20 min at 4 °C.
11. Measure the wet weight.
12. Resuspend the cell pellet of about 0.23 g of wet weight or from 24 mL of original culture at $OD_{600} \approx 6$ in 300 μL CelLytic™ B Cell Lysis Reagent (Sigma, B7435).
13. Transfer the resuspended cells into a low binding 1.5 mL tube (Protein LoBind Tube 1.5 mL, Eppendorf) and incubate at room temperature for 10 min to allow lysis to occur.
14. Centrifuge the cell lysate at 20,000 × *g* for 2 min at room temperature to pellet any insoluble material.
15. Remove the resulting supernatant from the cell debris into a low binding 1.5 mL tube (Protein LoBind Tube 1.5 mL, Eppendorf).
16. Mix the cell extracts with equal volume of 100 % glycerol and aliquot into 40–60 μL portions in low binding 0.5 mL tubes (Protein LoBind Tube 0.5 mL, Eppendorf),labeled as "PPY SLiCE extract."
17. Store the PPY SLiCE extract at −20 °C for about 2 months or at −80 °C for long-term storage, which can be thawed on wet ice and refrozen up to ten times without significant loss of activity.

3.2 Maintenance of PPY strain

1. Inoculate 1 single colony of PPY strain from LB agar plate (10 μg/mL streptomycin and 12.5 μg/mL chloramphenicol) into 5 mL LB medium (10 μg/mL streptomycin and 12.5 μg/mL chloramphenicol) and shake at 37 °C and 225–338 rpm overnight.
2. In a sterile tube mix an equal volume of PPY culture with 20 % autoclaved glycerol.
3. Store at −80 °C.

3.3 SLiCE Cloning

1. Preparation of vector and insert DNA.
   (a) PCR primer design and synthesis. The primers for SLiCE have the following characteristics: (1) The 5′-end of the primer contains 15–52 bases that are homologous to the sequence at one end of a DNA fragment for co-assembly or to other appropriate positions within a DNA fragment for co-assembly (i.e., the vector or another insert) (*see* **Note 5**); (2) The 3′-end of the primer contains 15–24 bases that are specific to the DNA fragment for amplification; (3) The primers do not need any further special treatment or modification except for desalting.

Table 1
SLiCE reaction conditions

Components	Standard cloning	Rush cloning	BAC cloning
Vector DNA	Purified, 50–200 ng	Unpurified, 1 μL[a]	Purified, 50–200 ng
Insert DNA	Purified, 1:1–10:1[b]	Unpurified, 1 μL[a]	Phenol/chloroform-purified, 5–10 μg restriction enzyme digested BAC fragments
10× SLiCE buffer	1 μL	1 μL	1 μL
PPY SLiCE extract	1 μL	1 μL	1 μL
ddH_2O	to 10 μL	to 10 μL	to 10 μL
Reaction conditions			
Incubation temperature	37 °C	37 °C	37 °C
Incubation time	15 min[c]–1 h	15 min[c]	1 h

[a]The volume of vector and insert DNAs can be adjusted. Ensure that the volume of unpurified DNA does not exceed 1/5 of total reaction volume
[b]Molar ratio of insert to vector
[c]15 min-incubation is enough for most SLiCE cloning tasks with a slightly reduced efficiency compared with 1 h

(b) Preparation of vector DNA. Linearize vector DNA used for SLiCE by restriction digestion (*see* **Note 6**) or PCR amplification (*see* **Notes 5** and 7). For BAC SLiCE cloning, PCR amplify the vectors using primers containing 5′-end homologies to the target fragments described as above (*see* **Notes 5** and 7).

(c) Preparation of insert DNA. Amplify the cloning inserts using PCR with primers containing 5′-end homologies to the vector or to other inserts for co-assembly described as above (*see* **Notes 5** and 7). For BAC SLiCE cloning, digest the BAC DNA with restriction enzymes.

(d) DNA purification. Purify the vector and insert DNAs by gel extraction, column-based purification methods, phenol/chloroform extraction, or other DNA purification methods and elute in EB (10 mM Tris-Cl, pH 8.5) or ddH_2O (*see* **Note 8**). For BAC SLiCE cloning, purify the restriction-digested BAC DNA fragments by phenol/chloroform extraction.

(e) Alternatively, the DNA purification step can be omitted, especially in simple cloning approaches (*see* **Note 9**).

2. SLiCE Reaction.

(a) Set up the SLiCE reaction (*see* Table 1) in a 0.2 mL microcentrifuge tube and mix well by vortexing (*see* **Notes 10** and **11**).

(b) Incubate the SLiCE reaction mix at 37 °C for 15 min to 1 h using a PCR machine or water bath (*see* Table 1).

3. Transformation (*see* **Note 12**).

(a) Electroporate 1 μL SLiCE reaction mix into 20 μL ElectroMAX DH10B™ Cells (Invitrogen) or chemically transform 1 μL SLiCE reaction mix into 100 μL MAX Efficiency® DH10B™ Competent Cells (Invitrogen) following the manufacturer's instruction. Other competent cells used for DNA cloning can also be used.

(b) Plate the transformed cells onto agar plates containing appropriate antibiotics.

4 Notes

1. The genotype of the PPY strain is as follows: F^- *endA*1 *recA*1 *galE*15 *galK*16 *nupG* *rpsL*Δ*lacX*74 *Φ80lacZ*ΔM15 *araD*139Δ(*ara,leu*)7697 *mcrA* Δ(*mrr-hsdRMS-mcrBC*) *cynX*:: [*araC* pBAD- *red*α EM7-*red*β Tn5-*gam*] λ^-. The PPY strain is recommended as a source of SLiCE extract for its highest efficiency and fidelity in our experiments. Other Rec A and restriction system-deficient strains such as DH10B or JM109 can also be used. For these strains, (1) the cloning efficiency is lower than PPY-derived SLiCE [4]; (2) the preparation of DH10B, JM109, or other bacterial strains cloning extracts follows the same protocol mentioned in Subheading 3.1, except the arabinose induction step (**step 6**) is omitted and appropriate antibiotics need to be used.
2. Centrifuge 500–1,000 μL bacterial cultures at top speed for 2 min and aspirate off all of the medium. Use ddH_2O to resuspend the bacterial pellet and as a black control to measure the absorbance at wavelength = 600 nm using a spectrophotometer. Note that high OD_{600} readings must be calculated by diluting the sample in ddH_2O to enable photometric measurement in the linear range between 0.1 and 0.5 OD_{600}.
3. For preparation of PPY SLiCE extract, avoid using chloramphenicol for drug selection because it inhibits the growth of PPY strain significantly.
4. It typically requires 6–10 h to reach the ideal cell density at 330 rpm and 37 °C. Addition of 0.1 % glycerol to the culture can increase the growth rate. Avoid growing the PPY cells longer than 14 h.
5. In cases where DNA fragments for co-assembly (i.e., vectors linearized with restriction enzymes) have sticky ends, count the end homology length from the 3′-end of DNA fragments. The cloning efficiency of SLiCE increases with increasing end

homology length in a range from 15 to 52 bp, but drops significantly when the end homologies are further increased [4].

6. The nature of vector and insert ends such as blunt ends or 3′ or 5′ sequence overhangs does not influence SLiCE efficiency or accuracy. However, the use of vectors with complementary 5′ or 3′ overhanging ends for SLiCE increases the formation of empty vector background colonies, which is probably due to annealing of the single stranded ends and vector recircularization in the bacterial extracts or in the transformed host cells. It is recommended to avoid linearizing vectors with a single sticky end restriction enzyme or pairs of restriction enzymes generating complementary 5′ or 3′ overhangs. If this cannot be avoided, blunt the vector ends before SLiCE reaction. In addition, the PCR generated vectors yield a 50–100-fold lower cloning efficiency than the restriction linearized vectors isolated from regular bacterial host strains such as DH10B or DH5α. A reason for the lower cloning efficiencies of PCR amplified vectors may be the lack of DNA modifications (such as methylation) in the vector backbone that are introduced during replication in the host bacteria and that might enhance the cloning efficiencies.
7. Digest PCR products using plasmid DNA as templates with DpnI prior to purification or SLiCE reaction to remove residual plasmid template DNA (add 1–2 μL DpnI into a standard 30 μL PCR, at 37 °C for 15–60 min). DpnI works fine in PCR buffers. DpnI treatment can decrease cloning background efficiently in some cases. If unpurified DpnI-treated PCR products are subjected to SLiCE reaction directly, inactivate it at 60 °C for 20 min.
8. For gel purification, it is recommended to use new prepared agarose gels and unused TAE or TBE for electrophoresis to avoid DNA damage caused by inappropriate pH and other factors. Gel purification can be replaced by other DNA purification methods such as column-based purification methods or phenol/chloroform extraction. Note that these methods sometimes lead to higher background caused by uncut vector plasmid or unspecific PCR products.
9. Limited volume of unpurified vector or insert DNA prepared from general PCR or restriction system leads to a significantly reduced but still acceptable SLiCE cloning efficiency, especially for simple cloning. Note that heat inactivation is required when the unpurified DNA contains the restriction enzymes that recognize the restriction sites of recombinant DNA molecules.
10. Before use, thaw 10× SLiCE buffer at 37 °C and vortex vigorously to dissolve any precipitated material. Thaw SLiCE extract stored at −80 °C on ice and vortex.

11. We provide three SLiCE reaction systems (*see* Table 1). Standard cloning is for general cloning of one or multiple DNA fragments into vectors [4]. Rush cloning is an alternative approach mainly for simple cloning. BAC cloning is for directional subcloning of DNA fragments from BAC vectors [4].
12. Standard transformation methods such as electroporation and chemical transformation are all compatible with SLiCE cloning. For cloning of large or complex DNA fragments, electroporation is recommended (yielding 10–100-fold higher transformation efficiencies than that of chemical transformation).

Acknowledgement

This work was supported by National Institute of Health [1R01CA76329 to W.E., 1R01CA93484 to W.E].

References

1. Smith HO, Wilcox KW (1970) A restriction enzyme from *Hemophilus influenzae*. I. Purification and general properties. J Mol Biol 51:379–391
2. Danna K, Nathans D (1971) Specific cleavage of simian virus 40 DNA by restriction endonuclease of *Hemophilus influenzae*. Proc Natl Acad Sci U S A 68:2913–2917
3. Cohen SN, Chang AC, Boyer HW et al (1973) Construction of biologically functional bacterial plasmids in vitro. Proc Natl Acad Sci U S A 70:3240–3244
4. Zhang Y, Werling U, Edelmann W (2012) SLiCE: a novel bacterial cell extract-based DNA cloning method. Nucleic Acids Res 40(8):e55
5. Little JW (1967) An exonuclease induced by bacteriophage lambda. II. Nature of the enzymatic reaction. J Biol Chem 242(4):679–686
6. Radding CM, Carter DM (1971) The role of exonuclease and beta protein of phage lambda in genetic recombination. 3. Binding to deoxyribonucleic acid. J Biol Chem 246(8):2513–2518
7. Carter DM, Radding CM (1971) The role of exonuclease and beta protein of phage lambda in genetic recombination. II. Substrate specificity and the mode of action of lambda exonuclease. J Biol Chem 246(8):2502–2512
8. Lovett ST, Hurley RL, Sutera VA et al (2002) Crossing over between regions of limited homology in *Escherichia coli*. RecA-dependent and RecA-independent pathways. Genetics 160:851–859
9. Dutra BE, Sutera VA, Lovett ST (2007) RecA-independent recombination is efficient but limited by exonucleases. Proc Natl Acad Sci U S A 104:216–221
10. Kuzminov A (2002) Recombinational repair of DNA damage in Escherichia coli and bacteriophage lambda. Microbiol Mol Biol Rev 63: 751–813

Chapter 17

j5 DNA Assembly Design Automation

Nathan J. Hillson

Abstract

Modern standardized methodologies, described in detail in the previous chapters of this book, have enabled the software-automated design of optimized DNA construction protocols. This chapter describes how to design (combinatorial) scar-less DNA assembly protocols using the web-based software j5. j5 assists biomedical and biotechnological researchers construct DNA by automating the design of optimized protocols for flanking homology sequence as well as type IIS endonuclease-mediated DNA assembly methodologies. Unlike any other software tool available today, j5 designs scar-less combinatorial DNA assembly protocols, performs a cost–benefit analysis to identify which portions of an assembly process would be less expensive to outsource to a DNA synthesis service provider, and designs hierarchical DNA assembly strategies to mitigate anticipated poor assembly junction sequence performance. Software integrated with j5 add significant value to the j5 design process through graphical user-interface enhancement and downstream liquid-handling robotic laboratory automation.

Key words DNA assembly, Design automation, BioCAD, Combinatorial library, Synthetic biology

1 Introduction

The preceding chapters of this book, as well as several recent reviews [1, 2], have described, from an experimental point of view, the many advantages of modern DNA construction methodologies over traditional (multiple-cloning site) cloning. What has not been as prominently discussed, however, is how modern standardized methodologies have also enabled the software-automated design of optimized DNA construction (hereafter synonymous with DNA assembly) protocols. Methodologies that are insensitive to the particular DNA sequence being constructed, often referred to as "sequence-independent" methods, inherently lead to the ability to abstract away DNA sequence information details from the protocol design process, and thus facilitate the development of generalized algorithms for optimizing DNA assembly protocols.

DNA construction methodologies that consistently utilize a single validated set of assembly junction sequences (that guide the DNA assembly process) greatly facilitate automated protocol

Svein Valla and Rahmi Lale (eds.), *DNA Cloning and Assembly Methods*, Methods in Molecular Biology, vol. 1116, DOI 10.1007/978-1-62703-764-8_17,

design, since the assembly junctions themselves are fixed and do not need to be designed or optimized. Since these predetermined assembly junction sequences are independent of the DNA sequences to be assembled, they are introduced into the resulting DNA constructs as compulsory insertions. A number of algorithms and automated protocol design processes have been developed for such DNA assembly methodologies. For example, algorithms have been reported [3] to maximize the reuse of DNA assembly intermediates and minimize the number of iterative assembly steps for the BioBrick (Chapter 1) [4] and BglBrick (Chapter 1) [5] DNA assembly methods, and have contributed to the automated execution of these methods utilizing liquid-handling robotics [6, 7]. Efforts are underway (Douglas Densmore, personal communication) to extend these same algorithms to the MoClo assembly method [8], and these algorithms could be further modified for application to the GoldenBraid method (Chapter 10) [9]. Analogous automated protocol design processes have been publicly disclosed by commercial entities, such as Amyris and Ginko BioWorks, for flanking homology sequence DNA assembly methodologies (such as CPEC (Chapter 8) [10] and DNA Assembler (in vivo yeast assembly) [11]) that utilize predefined homology sequence assembly junctions. The major disadvantage to DNA construction methods that exploit predetermined assembly junction sequences, however, is that they introduce the aforementioned compulsory insertions (or "scars") at each assembly junction, namely sequence vestiges of the assembly process that the DNA designer has no power to modify. In certain locations, such as in intergenic regions, these scar sequences may or may not be of significant concern. In other contexts, such as protein coding regions or 5′ untranslated regions of mRNA transcripts, scar sequences must be strictly avoided, as they can have dramatic detrimental effects on the functionality of the resulting constructs (e.g., translated protein sequence mutations; large perturbations to protein translation rates [12–16]).

Many sequence-independent DNA assembly methodologies do not require the use of predetermined assembly junction sequences, and as such do not necessarily result in assembly scar sequences. In contrast with the methods presented immediately above that utilize a single set of assembly junction sequences not related to the DNA sequences being assembled, to prevent the introduction of scar sequences, each scar-less assembly junction sequence must consist solely of a contiguous sequence contained within the concatenation of the corresponding pair of DNA sequences to be assembled. Note that, in general, for each scar-less assembly junction there are many such contiguous sequences to choose from, in terms of length and in terms of location (i.e., containing an equal or skewed number of base pairs from the preceding and subsequent DNA sequences to be assembled). To minimize the input costs of combinatorial scar-less DNA assembly (i.e., for full combinatorial reuse of each DNA input fragment), each scar-less assembly junction sequence must be found within a region of sequence identity across all combinatorial variants to be constructed.

The option to choose between multiple putative scar-less assembly junction sequences presents optimization opportunities as well as challenges. There are a multitude of sensible assembly junction sequence optimization strategies, including but not limited to: avoiding sequences previously resulting in poor assembly efficiencies, avoiding sequences with strong single-stranded DNA secondary structure (e.g., hairpins), optimizing sequence annealing temperature and GC content, avoiding sequences that are significantly homologous to other assembly junction sequences or other sequences found within the DNA to be assembled, achieving 3′ GC-clamps where possible, minimizing sequence length and centering the sequence location between the corresponding pair of DNA sequences to be assembled (to minimize input costs). Furthermore, any given optimization strategy may not be equally effective for all DNA assembly methodologies (e.g., for CPEC, avoiding assembly junction sequences that are significantly homologous to other sequences found within the DNA to be assembled is likely to be important, but this may not be of concern for the Gibson method [2]). The challenges introduced by the option to choose from a set of putative assembly junction sequences, then, are twofold: the additional burden for each new construct of having to select an optimal sequence for each scar-less assembly junction, and the necessity to decide how to prioritize and triage the various optimization strategies for a given DNA assembly methodology.

Design automation software can relieve the burden of selecting and optimizing new scar-less assembly junction sequences, and can predict and potentially mitigate poor assembly junction sequence performance (e.g., via hierarchical DNA assembly design). Several software tools have been developed to help automate the design of protocols for each of the three categories of scar-less DNA assembly methods presented in Table 1. It is very important to point out, however, that the various software tools vary wildly in terms of their sophistication, functionality, flexibility, and especially in terms of their respective capacities to optimize assembly junction sequences and predict and mitigate poor assembly junction sequence performance. As it is beyond the scope of this chapter to review the functionalities of the various software tools presented in Table 1, it is merely suggested here that it is very worthwhile to briefly evaluate each of the tools in turn to determine which is the most appropriate or useful for a given set of DNA assembly protocol design tasks. This is especially prudent, since software tools that appear to be useful on paper are often confounded by factors (as mundane as software inaccessibility, slow web-server response times, or poorly crafted user-interfaces) that limit their suitability in practice.

Table 1
Scar-less DNA assembly methodologies and protocol design software

Category	Representative methodologies	Protocol design software
Flanking homology sequence	Quick and Clean cloning (Chapter 3) [21] In-Fusion® (Chapter 15) SLIC (Chapter 2) [22] Simple cloning (Chapter 13) [23] CPEC (Chapter 8) [10, 22] TPCR (Chapter 7) [25] SLiCE (Chapter 16) [26] Gibson [27, 28] GeneArt® seamless/High-Order DNA assembler [11, 29] Reiterative recombination [30]	j5 (this chapter) [17, 31] Gibthon [32] GeneDesign [33–35] BioCAD [7] In-Fusion® Primer Design [36] GeneArt® OligoDesigner [37]
Type IIS endonuclease	Golden gate (Chapter 9) [38–40] FX cloning (Chapter 11) [41]	j5 GeneDesign BioCAD
Uracil-excision	USER (Chapter 5) [42, 43]	PHUSER (Chapter 5) [44, 45] GeneDesign

This chapter provides a brief tutorial on how to design (combinatorial) scar-less DNA assembly protocols using the web-based software j5 [17]. j5 assists biomedical and biotechnological researchers construct DNA by automating the design of optimized protocols for flanking homology sequence as well as type IIS endonuclease-mediated DNA assembly methodologies (Table 1). Unlike any other software tool available today, j5 designs scar-less combinatorial DNA assembly protocols, performs a cost–benefit analysis to identify which portions of an assembly process would be less expensive to outsource to a DNA synthesis service provider, and designs hierarchical DNA assembly strategies to mitigate anticipated poor assembly junction sequence performance. Software integrated with j5 (e.g., DeviceEditor [18], VectorEditor [19], and PR-PR [20]) add significant value to the j5 design process through graphical user-interface enhancement and downstream liquid-handling robotic laboratory automation.

2 Materials

2.1 j5 and DeviceEditor Software

The j5 and DeviceEditor web-based software described in this chapter is available to academic, nonprofit, and government researchers at no cost on the j5 web-server [31]. The j5 and DeviceEditor software has been exclusively licensed to TeselaGen Biotechnology, Inc. [46] for commercial use and distribution.

2.2 VectorEditor Software

The VectorEditor web-based software described in this chapter is freely available as a stand-alone program [31, 47], as a tool integrated with the public instance of the Joint BioEnergy Institute Inventory of Composable Elements (JBEI-ICE) repository [48], and as source code [49].

2.3 PR-PR Software

The PR-PR web-based software described in this chapter is freely available on the PR-PR web-server [50], and as source code [51].

3 Methods

The process of using j5 to design DNA assembly protocols begins with specifying the DNA sequences that are to be assembled. These specifications are submitted as input to j5, which then outputs the designed DNA assembly protocols. Multiple j5 DNA assembly protocols can be aggregated into a single meta-protocol, facilitating the parallel execution of independent protocols in the same set of multi-well plates. The resulting j5-designed protocols can be performed either by hand or by liquid-handling robot. The following subsections walk through each step in the process, utilizing DeviceEditor [18] on the public j5 web-server [31] to specify the DNA sequences to be assembled and as the graphical front-end interface for j5. The (*see* **Note 1**) below provides a brief synopsis of alternative j5 interfaces.

3.1 Registering for a j5 User Account

The very first step is to register for a j5 user account. Academic, nonprofit, and government researchers who do not already have a j5 user account can register for one at no cost from the main page of the j5 web-server (Fig. 1) [31] by clicking the "Log in/Create account" link at top right, then clicking on the "request one" link, and then filling out and submitting the request account form. It takes approximately one business day for a new account to be approved (if a valid email address and academic, nonprofit, or government professional affiliation credentials are provided). Commercial users can register with TeselaGen [46]. Although the TeselaGen interface for j5 differs from the DeviceEditor interface described below, the overall process of designing DNA assembly protocols with j5 is similar.

3.2 Starting a DeviceEditor Session

After registering for a j5 user account, the next step is to start a DeviceEditor session in which to specify the DNA sequences to be assembled. From the main page of the j5 web-server (Fig. 1) [31], click on the DeviceEditor icon. (Users that are not currently logged-in will be prompted for their j5 account username and password. After logging-in, users will be prompted to continue either to the simplified web-form j5 interface or to DeviceEditor. Click on the DeviceEditor link.) The DeviceEditor interface will open in a new browser tab (Fig. 2). At the top left of the interface,

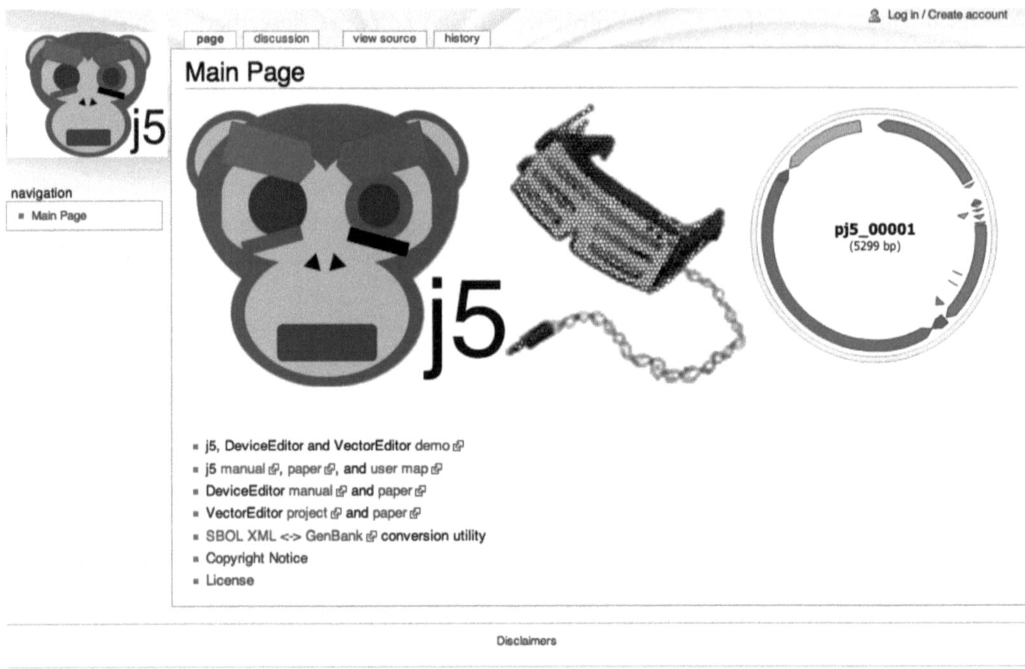

Fig. 1 j5 web-server [31]. (*Top right*) "Create account" link provides access to registering for a j5 user account (Subheading 3.1). (*Middle*) *Left to right*: j5, DeviceEditor, and VectorEditor icons. (*Bottom*) Links to a video demonstration, user's manuals, research papers, and legal information (*see* **Note 2**)

"File" and "Edit" drop-down menus provide access to saving/loading/clearing DeviceEditor designs (Subheading 3.3), importing Eugene design specification rule files (Subheading 3.7), and mapping a GenBank sequence file or information from the system clipboard to a part (Subheading 3.5, *see* **Note 3**); and copying/pasting parts (Subheading 3.8), respectively. Immediately below the "File" and "Edit" drop-down menus, clickable buttons provide access to j5 controls (Subheadings 3.10–3.14), DeviceEditor properties, and the DeviceEditor version, author, and citation information. Directly below the clickable buttons is the design canvas (the largest portion of the interface), below which is the "Part Holding Area" (*see* Subheadings 3.4, 3.6, and 3.8). The top right of the interface displays the name of the current design (*see* Subheading 3.3), immediately below which are the Part and Collection Info panels, which present detailed information about the selected part (*see* Subheadings 3.4–3.7) and the design as a whole (*see* Subheadings 3.4 and 3.9), respectively.

3.3 Saving, Loading, and Clearing Designs in DeviceEditor

To save the current design, click on the "File" drop-down menu at the top left of the interface (*see* Fig. 2) and then select "Save Design." A "Select Location for Download" dialog box will open. Specify the name for DeviceEditor design file (be sure to maintain

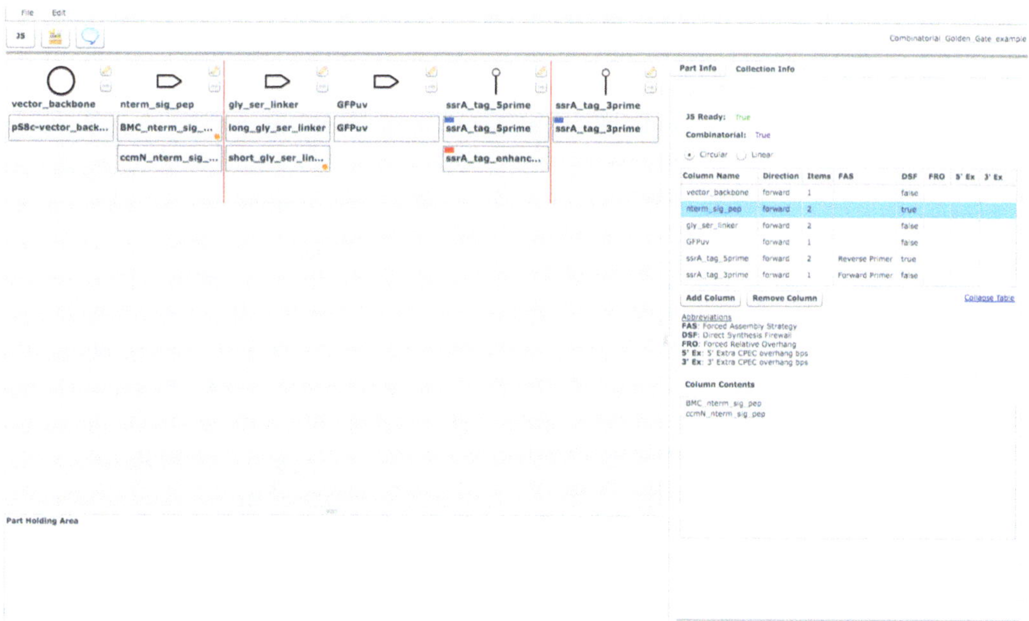

Fig. 2 DeviceEditor interface [64]. (*Top left*) The "J5" button provides access to j5 functionality (Subheadings 3.10–3.14). (*Middle left*) Spreadsheet-like design canvas. DNA fragment columns are ordered left to right from 5′ to 3′. The header row depicts a unique name, an SBOL Visual icon [52], and a directionality (forward or reverse), for each column. Rows below the header contain parts. Parts in the same column are combinatorial variants of each other (*see* Subheading 3.4). Parts with *blue* or *red rectangles* at their *top left* indicate forced assembly strategies (*see* Subheading 3.6). Parts with *orange circles* at their *bottom right* indicate Eugene design specification rules (*see* Subheading 3.7). *Red vertical lines* between columns indicate Direct Synthesis Firewalls (*see* Subheading 3.9). (*Bottom left*) Part Holding Area (*see* Subheadings 3.4, 3.6, and 3.8). (*Right*) Part Info (not shown, *see* Subheadings 3.4–3.7) and Collection Info panels. The Collection Info panel contains controls for toggling the assembly product type (i.e., circular or linear), adding or removing columns, changing column names (*see* Subheading 3.4), and setting additional DNA assembly directives (Subheading 3.9)

the .xml file extension) and where it should be saved to. The name of the saved DeviceEditor design file determines the name of the design displayed at the top right of the interface. Since there is no auto-save or undo functionality (yet) in DeviceEditor, it is important to frequently save designs as they are being developed.

To load a design, click on the "File" drop-down menu, select "Load Design" and then "Design XML." A "Select File to Upload" dialog box will open. Specify which DeviceEditor design file should be loaded. In addition to loading DeviceEditor design files, it is also possible to reconstitute a DeviceEditor design from a set of j5 files. To do so, click on the "File" drop-down menu, select "Load Design" and then "j5 Files." A "J5 File Import" dialog box will open. Specify which "j5 Sequence List," "zipped sequence files," "j5 Part List," and "j5 Target Part Order" files to import, and then click "Done." Since j5 files do not specify Synthetic Biology Open Language (SBOL) Visual icons, they will need to be selected for each of the columns in the design canvas (*see* Subheading 3.4) after importing the j5 files. A subsequent step, if desired, is to import the Eugene rules file (Subheading 3.7) corresponding to the set of imported j5 files. The ability to reconstitute a DeviceEditor design from a set of j5 output files is especially useful when the original DeviceEditor design file is no longer available. Occasionally, it may also be advantageous to import j5 input files that have been prepared with other software tools (*see* **Note 4**). DeviceEditor has several built-in example designs that can also be loaded. To do so, click on the "File" drop-down menu, select "Load Design," "Example Design," and then one of "SLIC/Gibson/CPEC," "Combinatorial SLIC/Gibson/CPEC," "Golden Gate," or "Combinatorial Golden Gate." Figure 2 shows a slightly modified version of the "Combinatorial Golden Gate" built-in example design.

To clear a design in DeviceEditor (i.e., to start with a clean slate), click on the "File" drop-down menu, select "Clear Design," and then click "OK" to confirm that the current design should be cleared.

3.4 Creating a Prototype Design in DeviceEditor

The first step when designing DNA constructs in DeviceEditor is to create a prototype design. Prototype designs do not specify the actual DNA sequences of the fragments to be assembled, but rather specify only the number, names, directionalities (e.g., forward or reverse), and categories (e.g., promoter, origin of replication) of the fragments to be assembled.

The DeviceEditor design canvas (Fig. 2) is organized like a spreadsheet, with each column representing a distinct DNA fragment, and columns organized from left to right so as to correspond with the desired 5′–3′ ordering of the DNA fragments to be assembled. The desired assembly product type may be toggled between "Circular" and "Linear" in the Collection Info panel at the right of the interface. The top row of the design canvas contains the column headers which display a unique name, an SBOL Visual icon [52] (a graphical representation of the category of the parts contained within the column), and a directionality arrow (indicating whether the column should be incorporated in the forward or reverse direction into the resulting constructs) for each column. Below the column header, each row in a given column can be associated with a named part variant for the DNA fragment. Parts in the same column are combinatorial alternatives of each other, and their part-type category and directionality are determined by the column header.

The SBOL Visual icon may be changed for a column header by clicking on the pencil/"Change icon" button and then selecting the desired icon. The directionality of a column header may be toggled by clicking on the arrow/"Change direction" button. The name of the column header may be changed in the Collection Info panel at the right of the interface, by editing the name of the desired column in the "Column Name" column of the Collection Info table. To append a column to the design canvas or remove the final column, click on the "Add Column" or "Remove Column" buttons, respectively, in the Collection Info panel. To insert a column before a particular column, or remove a particular column, click to activate the desired column in the table in the Collection Info panel, and then click on the "Add Column" or "Remove Column" buttons, respectively. Each part that is contained within a removed column is displaced to the Part Holding Area (bottom left of the interface), unless another linked-paste (*see* Subheading 3.8) instance of the part is present elsewhere in the design canvas. To name a part in the design canvas, click on the desired part (column/row), enter the desired part name in the Part Info panel at the top right of the interface, and then click the "Save" button next to the "Part Name" field. New rows are automatically added to the design canvas when bottom-most row of any column is selected.

To create a prototype design in DeviceEditor, then, it is necessary to add a column for each of the design's DNA fragments; select an SBOL Visual icon, unique name, and directionality for each column; and name the parts to be contained within each column. The circular design shown in Fig. 2 contains six columns, all in the forward direction, each with an SBOL Visual icon ("Origin of Replication," "CDS," "CDS," "CDS," "Protein Stability Element," "Protein Stability Element," respectively), each with a unique name ("vector_backbone," "nterm_sig_pep," "gly_ser_linker," "GFPuv," "ssrA_tag_5prime," and "ssrA_tag_3prime," respectively), and each containing named parts ("pS8c-vector_backbone"; "BMC_nterm_sig_pep" and "ccmN_nterm_sig_pep"; "long_gly_ser_linker" and "short_gly_ser_linker"; "GFPuv"; "ssrA_tag_5prime" and "ssrA_tag_enhanced_5prime"; and "ssrA_tag_3prime," respectively).

3.5 Mapping Annotated DNA Sequence Information from VectorEditor to Parts in DeviceEditor

After creating a prototype design in DeviceEditor, the next step is to map (i.e., associate) annotated DNA sequences to the named parts in each column. There are two main methods for mapping an annotated DNA sequence to a part, namely the preferred method of mapping from VectorEditor via the system clipboard (i.e., copy/paste), and mapping directly from a GenBank sequence file (*see* **Note 3**). While the method presented below begins with importing a DNA sequence file into the j5 web-server stand-alone version of VectorEditor, the same methodology largely applies to beginning with opening a JBEI-ICE repository DNA sequence with the integrated version of VectorEditor [19, 48].

To map an annotated DNA sequence from VectorEditor to a part in DeviceEditor, open a new browser tab. From the main page of the j5 web-server (Fig. 1) [31], click on the VectorEditor icon. At the top left of the VectorEditor interface (Fig. 3), click on the

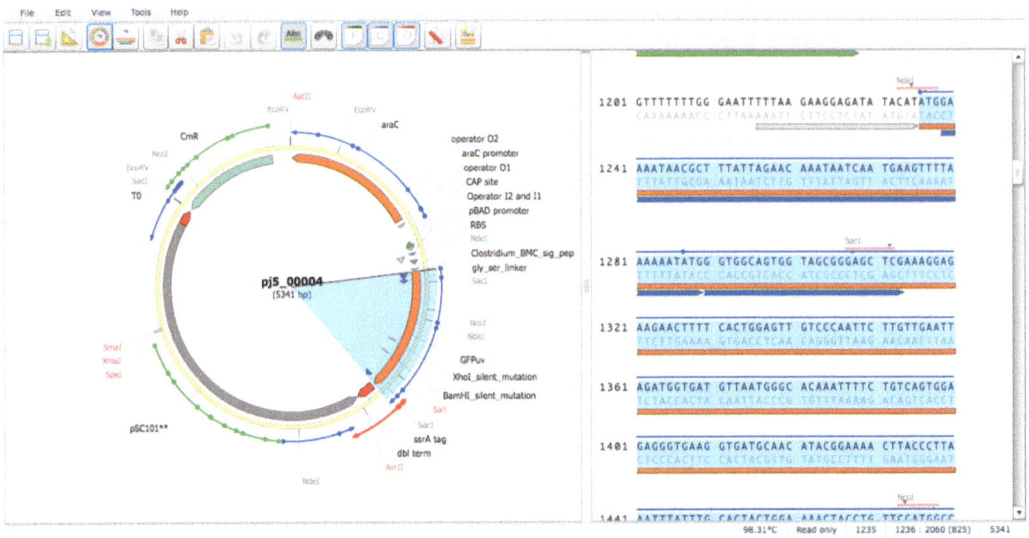

Fig. 3 VectorEditor stand-alone interface [47] (*see* Subheadings 3.5 and 3.11). Activating the ".O" button at *top* displays predicted open reading frames as colored arcs decorated with *arrowheads* indicating directionality and *circles* indicating "ATG" start codons. Activating the ".C" button at *top* displays unique restriction sites in *red* and non-unique restriction sites in *grey*. The utility tool button icon at *top* provides access to configuring the set of restriction sites to be displayed. (*Left*) Plasmid map view. (*Right*) Sequence detail view

"File" drop-down menu and select "Import from File." A "Select File to Upload" dialog box will open. Select the desired FASTA, GenBank, jbei-seq, or SBOL XML/RDF format sequence file (for more information about these sequence formats, refer to the j5 user's manual [53]). Alternatively, plain-text DNA sequence can be copy/pasted into the right panel of the VectorEditor interface. At this point, the DNA sequence can be further annotated as desired by selecting a particular portion of the DNA sequence, right-clicking the selection, selecting "Selected as New Feature," and then filling out the resulting dialog entry form. If desired, the name of the sequence (i.e., the plasmid name) can be modified by clicking on the "File" drop-down menu and selecting "Project Properties." A "Properties" dialog will open, enabling the name of the sequence to be changed. It may be necessary to change the name of the sequence before mapping it to a part in DeviceEditor, if a distinct sequence with the same name has already been mapped to another part (although different portions of the same sequence can be sequentially mapped to multiple parts without requiring a name change). After the sequence has been annotated and named as desired, select the portion of the DNA sequence to map to the part in DeviceEditor, right-click the sequence selection, and then select "Copy".

With the annotated DNA sequence selection copied to the system clipboard, return to the DeviceEditor browser tab. Right-click the desired part and select "Map from Clipboard." A "Paste From

Clipboard" dialog box will open. Right-click in the Paste From Clipboard dialog box, and then select "Paste." A "Specify Part Definition" dialog box will open, which will display the "Part Name," "Part Source," "Source Data," "StartBP," and "StopBP" of the sequence selected in VectorEditor. If the reverse complement of the selected sequence is to be mapped to the part, click to activate the "Reverse Complement" check-box. Clicking the "Done" button concludes the mapping process. The Part Info panel at the right of the interface now displays the Part Source, Reverse Complement status, Start BP, and Stop BP of the resulting mapped part. Clicking on the "Change part definition" button in the Part Info panel re-opens the Specify Part Definition dialog box, enabling changes to be made to the definition of the part. It is possible to re-map an annotated sequence to a previously mapped part by repeating the steps described above.

3.6 Setting Forced DNA Assembly Strategies in DeviceEditor

After an annotated DNA sequence has been mapped to each part in the design canvas, the next step is to set forced DNA assembly strategies, if desired, for each part. To set a forced assembly strategy for a part, click on the part in the design canvas. In the Part Info panel at the right of the interface, in the "Forced Assembly Strategy" section, there is a pull-down menu that includes the following forced assembly strategy options: "None" (the default), "DIGEST," "Direct Synthesis," "PCR," "Embed_in_primer_reverse," "Embed_in_primer_forward," and "Annealed Oligos." "None," the default choice, will allow j5 to determine the cost-optimal assembly strategy for the part. The other options constrain j5 to use the specified strategy for the part when designing the DNA assembly protocol.

A "DIGEST" forced assembly strategy for a part specifies that the part will result from a digest (i.e., restriction enzyme activity). "DIGEST" forced assembly strategies are only allowed for parts located in the first (left-most) column in the design canvas. Any part not in the first column whose forced assembly strategy is set to "DIGEST" is displaced to the Part Holding Area. "Direct Synthesis," "PCR," "Embed_in_primer_reverse," "Embed_in_primer_forward," and "Annealed Oligos" forced assembly strategies specify that the part will result from DNA synthesis, PCR amplification, embedding the part in the reverse primer of a PCR amplification, embedding the part in the forward primer of a PCR amplification, or annealing two DNA oligos together, respectively. For more information, refer to the j5 user's manual [53].

Each part with a forced assembly strategy (other than "None") is indicated by a blue rectangle at its top left in the design canvas (*see* Fig. 2). If two or more parts in the same column in the design canvas have different forced assembly strategies (other than "None"), each part with a strategy differing from the topmost part is indicated by a red (warning) rectangle at its top left (*see* Fig. 2).

Specifying different forced assembly strategies for parts in the same column may significantly increase DNA assembly costs, and is discouraged.

3.7 Creating and Importing Eugene Design Specification Rules in DeviceEditor

After forced assembly strategies have been set as desired for each part in the design canvas, the next step is to create Eugene design specification rules, if desired, for each part. In the design shown in Fig. 2, there are eight exhaustive combinations of parts (2 signal peptides × 2 linkers × 2 degradation tags). However, all eight of the exhaustive combinations may not be desirable. For example, if prior experiments have shown that the BMC signal peptide should not precede the short Gly/Ser linker (because the resulting proteins do not localize properly), then perhaps the two (of the eight total) combinations of parts that contain the BMC signal peptide followed by the short Gly/Ser linker should not be constructed. Eugene design specification rules [54, 55] can be utilized to specify such types of relationships between parts, and prevent certain orders or combinations of parts from being constructed.

To create a new Eugene design specification rule in DeviceEditor, click on the desired part in the design canvas. At the bottom of the Part Info panel at the right of the interface, in the "Eugene Rules for selected part" section, click on the "Add Rule" button. An "Add Eugene Rule" dialog will appear. Enter a name for the rule (e.g., "rule1"), and from the "Operator" pull-down menu, select "AFTER," "NOT AFTER," "BEFORE," "NOT BEFORE," "WITH," "NOT WITH," "THEN," "NOT THEN," "NEXTTO," "NOT NEXTTO," "MORETHAN," or "NOT MORETHAN." For more information about the various operators, refer to the Eugene website [54] and the j5 user's manual [53]. For the most part, the operators are self-explanatory, with the exception of "WITH" and "THEN" (and their "NOT" negated variations). If a Eugene rule specifies *partA* "WITH" *partB*, then both *partA* and *partB* must be present in every combination of parts that will be constructed. If a Eugene rule specifies *partA* "THEN" *partB*, then if *partA* is in a combination of parts then *partB* must also be present for the combination of parts to be constructed. For the example described above, a Eugene rule that specifies BMC signal peptide "THEN" long Gly/Ser linker would exclude the two combinations that have the BMC signal peptide followed by the short Gly/Ser linker, which is the desired behavior. However, a Eugene rule that specifies BMC signal peptide "WITH" long Gly/Ser linker would exclude the six combinations that do not contain both the BMC signal peptide and the long Gly/Ser linker, which is not the desired behavior. The "AFTER" and "BEFORE" (and their "NOT" negated variations) operators are only enforced for linear designs. For the "MORETHAN" or "NOT MORETHAN" operators, enter the desired integral number of times in the "Operand 2" field. For all other operator selections, select the other desired part on the design canvas from the "Operand 2" pull-down menu.

Aside from creating Eugene rules one-at-a-time as described above, it is also possible to import an existing set of Eugene design specification rules from a file. To do so, click on the "File" drop-down menu at the top left of the interface, and select "Import Eugene Rules." A "Select File to Upload" dialog will open. Select the Eugene rules file to import. After the Eugene rules file has been imported, a "Eugene Rules Import" dialog will open, which displays the imported rules in a color-coded fashion. Ignored lines are shown in grey, imported rules in green, rules that are already present in the design in black, and imported rules which have been automatically renamed (so as not to conflict with rules already present in the design) in red.

Each part that is associated with a Eugene design specification rule is indicated by an orange circle at its bottom right in the design canvas (*see* Fig. 2). Detailed information about the Eugene rules associated with a selected part is displayed in the "Eugene Rules for selected part" section at the bottom of the Part Info panel at the right of the interface.

3.8 Cutting, Copying, Pasting, and Deleting Parts in DeviceEditor

At the various stages of developing a design in DeviceEditor, described above in Subheadings 3.4–3.7, it is frequently useful to be able to move a part about the design canvas, replicate a part, or even delete a part all together. In a fashion analogous to spreadsheet software, in DeviceEditor these actions are accomplished through cutting, copying, pasting, and deleting parts.

To move a part from one location in the design canvas to another unoccupied location, right-click the desired part and select "Cut." Right-click the desired destination and select "Paste." This method also applies to moving parts from the Part Holding Area (*see* Subheadings 3.2, 3.4, and 3.6) back to the design canvas.

To replicate a part from one location in the design canvas to another unoccupied location, right-click the desired part and select "Copy." Right-click the desired destination and select "Paste." A "Paste" dialog will appear, offering two different options: "Paste" or "Linked Paste." If "Linked Paste" is selected, the newly replicated part will remain informatically linked to its parent. That is to say, the definition of the part (*see* Subheading 3.5) and all Eugene design specification rules (Subheading 3.7) apply to all linked instances of the part, and any changes made to the definition or Eugene rules of one instance are propagated to all other linked instances. However, each linked instance of a part may have a distinct forced assembly strategy (Subheading 3.6). All linked-parts on the design canvas are indicated by blue outlines. Selecting a part increases the blue outline thickness for all the parts that it is linked to. Linked-parts are useful when the identical part is desired in multiple columns of the design canvas. In contrast with "Linked Paste," if "Paste" is selected, a new unique name for the pasted part must be specified, and the pasted part is not linked to its parent. That is to say, the definitions of the copied and pasted parts are independent, and none of the Eugene design rules for the copied part apply to the pasted part. Pasting a non-linked part can be useful if a slight variant (e.g., different start or stop bp in the source sequence), rather than a carbon-copy, of the copied part is desired. It is also possible to copy/paste parts between DeviceEditor sessions (i.e., across different browser tabs), so as to share parts between designs.

To delete a part from the design canvas, right-click the desired part and select "Delete." In addition to the right-click methods described above to "Cut," "Copy," "Paste," and "Delete" parts, standard keyboard shortcuts (e.g., command/control-c to "Copy") for these methods are also available.

3.9 Setting Additional DNA Assembly Directives in DeviceEditor

Before submitting a design in DeviceEditor to j5 (Subheading 3.10), the final step is to set additional DNA assembly directives, if desired. In the Collection Info panel at the right of the interface, click the "Expand Table" link. This will reveal additional columns in the Collection Info table (*see* Fig. 2), including "DSF" (Direct Synthesis Firewall), "FRO" (Forced Relative Overhang/Overlap), "5′Ex" (5′ Extra Overlap), and "3′Ex" (3′ Extra Overlap). While it is beyond the scope of this chapter to fully address the utility of DSFs in the context of j5, it suffices here to say that they prevent the automatic cost-effective propagation of DNA synthesis from one column (i.e., DNA fragment) to the next (for more details refer to the j5 user's manual [53]). The "DSF" column must be set to either "false" or "true" for each row in the table. If the "DSF" column is set to "true," this directs the presence of a DSF to the right (i.e., 3′) of the corresponding column in the design canvas. DSFs are visually indicated on the design canvas by red vertical lines separating adjacent columns (*see* Fig. 2). Entries in the FRO, 5′ Extra Overlap, and 3′ Extra Overlap columns must either be empty or contain an integral (positive or negative) number of base pairs. In general, users will not need to manually set entries in the FRO, 5′ Extra Overlap, and 3′ Extra Overlap columns, and it is beyond the scope of this chapter to fully address their utility (for more details refer to the j5 user's manual [53]).

3.10 Submitting DeviceEditor Designs to j5 for DNA Assembly Protocol Design

Once a design in DeviceEditor is ready to be submitted to j5, the next step is to click the "j5" button at the top left of the interface (Fig. 2). A "j5 controls" dialog will open. Click the "Run j5 on Server" tab. To edit the j5 design parameters, click the "Edit j5 Parameters" link. By default, DeviceEditor loads the j5 parameters most recently submitted to the j5 server. Each j5 parameter is modifiable through a pull-down menu (e.g., "Output Sequence Format") or a text field. Hover the cursor over a j5 parameter name to display a tool-tip that describes the parameter in additional detail (for more details refer to the j5 user's manual [53]). To return to the default j5 parameter values, click the "Return to Defaults" button.

To return the parameters most recently submitted to the j5 server, click the "Return to Server Values" button. Click "Cancel" to cancel the changes made to the j5 parameters, or "OK" to set the new j5 parameter values.

There are three options for each of the "Master Plasmids," "Master Oligos," and "Master Direct Syntheses" lists. The j5 server maintains a running list of each user's plasmids, DNA oligos, and DNA syntheses. This enables the automatic numbering of new plasmids (e.g., "pj5_00001," "pj5_00002"), oligos, and DNA syntheses; as well as preempts the repurchasing of available oligos and syntheses. Select "Use latest server version" to direct j5 to append to the current list on the server, "Generate empty file" to direct j5 to start afresh with an empty list, or "Choose File" to upload an alternative starting point for j5. The "Choose File" option is valuable for new users who wish to set the naming convention for their plasmids (e.g., "pNJH00001" instead of the default "pj5_00001"), and is an important means to manipulating the lists of DNA plasmids, oligos, or syntheses outside of the j5 design process itself (e.g., to introduce non-j5-designed DNA oligos, remove syntheses which were never ordered). For more information on creating or modifying the master list files, refer to the j5 user's manual [53].

The next step is to select the desired DNA assembly methodology. Select the "(Combinatorial) Mock Assembly" option to request that j5 only assembles the DNA sequences to be constructed, but does not actually design the corresponding assembly protocols; or "(Combinatorial) SLIC/Gibson/CPEC" or "(Combinatorial) Golden Gate" to request that j5 designs the corresponding assembly protocols. "(Combinatorial) Mock Assembly" design is significantly faster than full assembly protocol design, and this option is a prudent preliminary step to rapidly ensure the correctness of the sequences to be constructed (*see* Subheading 3.11) before investing the time required for j5 to design the corresponding assembly protocols. As shown in Table 1, many scar-less DNA assembly methodologies are topologically equivalent (i.e., can utilize the same input DNA fragments, and result in the same DNA products), and only minor design parameter changes are required to apply the same design algorithms across scar-less DNA assembly methods in the same category. With minor changes to the j5 parameters (see above), it is possible to utilize the "(Combinatorial) SLIC/Gibson/CPEC" to design assembly protocols for the GeneArt® Seamless methodology, or "(Combinatorial) Golden Gate" for FX cloning (Chapter 11) [41], for example.

Click the "Run j5" button to submit the design to the j5 server. Once the results are ready, links to each of the sequences to be constructed will be displayed in the table at the bottom of the "j5 controls" dialog (*see* Subheading 3.11), and a "Download Results" button (*see* Subheading 3.12), will appear.

3.11 Visually Assessing j5-Assembled DNA Constructs in VectorEditor

After DeviceEditor has indicated that the j5 DNA assembly protocol design results are ready (*see* Subheading 3.10), click on the corresponding link to open the desired assembled DNA construct in VectorEditor (Fig. 3). In addition to checking that all DNA fragments are present and in the desired order, check that coding sequences spanning multiple DNA fragments are in frame. To do so, click the ".O"/"Show ORF" button icon at the top of the VectorEditor interface. Colored arcs decorated with circles indicating "ATG" start codons and arrowheads indicating directionality provide visual indications of the continuity of reading frames. Check also, if desired, for the existence and locations of restriction sites. To do so, click the ".C"/"Show Cut Sites" button icon at the top of the interface. To control the set of displayed restriction sites, click on the utility tool/"Manage Restriction Enzymes" button icon at the top of the interface. If the assembled DNA construct is as desired, proceed as described in Subheadings 3.10 and 3.12. Otherwise, revise the design in DeviceEditor as described in Subheadings 3.4–3.10.

3.12 Downloading j5 DNA Assembly Protocol Design Output from DeviceEditor and Assessing the Results

After DeviceEditor has indicated that the j5 DNA assembly protocol design results are ready (*see* Subheading 3.10) and the assembled DNA construct sequences have been visually assessed for correctness (Subheading 3.11) (in the "Run j5 on server" tab of the "j5 controls" dialog), click on the "Download Results" button. A "Select Location for Download" dialog will open. Select the j5 output zip file destination. Unzip the j5 output file to create a folder which contains multiple files, including the corresponding set of j5 input files, a combinatorial j5 assembly protocol design comma-separated value (CSV) file (combinatorial designs only), a j5 assembly protocol design CSV file and an annotated DNA sequence file for each assembled construct, and updated master plasmids, oligos, and direct syntheses list CSV files. The formatting and contents of each of these files are documented in great detail in the j5 user's manual [53].

Each j5 assembly protocol design CSV file provides sufficient information to guide the assembly process for the corresponding DNA construct, such as which DNA oligos to purchase, which PCRs to perform, and which DNA assembly pieces to assemble. Open and assess each j5 assembly protocol design CSV file using spreadsheet software (e.g., Excel, OpenOffice). Each j5 assembly protocol design CSV file is broken down into the following sections: header (including the type of assembly protocol, the date j5 designed the assembly protocol, and j5 citation information), Assembly Parameters (the values of the j5 parameters used during the design process), Part Specifications, Target Part Ordering, Assembly Piece Incompatibilities (SLIC/Gibson/CPEC only), Suggested Hierarchical Assembly (SLIC/Gibson/CPEC only), DNA Synthesis, Oligo Synthesis, PCRs, Assembly Pieces, and the Final Assembled Sequence. Check immediately below the Part

Specification section to see if any warning messages resulted from the primer design process. j5 utilizes Primer3 [56] to optimize primer design, and since Primer3's default design constraints are fairly stringent (and a given set of j5 parameters can make them even more so), Primer3 is often not able to find a reasonable set of primers. When Primer3 is not able to identify a reasonable set of primers, j5 progressively relieves Primer3 design constraints until a set of reasonable primers is identified. The warning messages resulting from the primer design process specify which constraints (e.g., minimum Tm) were relieved. Generally speaking, these primer design warning messages should serve merely as starting points for trouble-shooting, and should not be taken as reliable indicators that the PCRs themselves will actually fail. For flanking homology sequence assembly protocols (e.g., SLIC/Gibson/CPEC only), check immediately below the Target Part Ordering section to see if there are any warning messages resulting from the flanking homology overlap design process. Similar to that described above for primer design, j5 also utilizes Primer3 to optimize the design of flanking homology overlap sequences. Generally speaking, these flanking homology overlap design warning messages should serve merely as starting points for trouble-shooting, and should not be taken as reliable indicators that the DNA assembly reaction will fail. However, "mispriming" flanking homology overlap design warning messages should be heeded in earnest, as they identify assembly piece incompatibilities that could result in undesirable off-target mis-assembly products. If j5 detects any such assembly piece incompatibilities, they are presented in the Assembly Piece Incompatibilities section, along with a hierarchical DNA assembly mitigation strategy (in the Suggested Hierarchical Assembly section) designed to avoid off-target mis-assembly products (refer to the j5 user's manual [53] for more information). To supplement j5's current capacity to detect problematic flanking homology overlap sequences, it is prudent to additionally utilize single-stranded DNA secondary structure prediction tools (e.g., the DINAMelt web-server [57]) to identify putative stable secondary structures (e.g., DNA hairpins) within the flanking homology overlap sequences (presented in the Assembly Pieces section) that may inhibit the DNA assembly reaction process. For type IIS endonuclease-mediated assembly protocols (e.g., (Combinatorial) Golden Gate), it is important to check each j5 assembly protocol design CSV file for warning messages presented immediately before the Final Assembled Sequence section concerning the presence of the selected type IIS endonuclease recognition sequence (e.g., for BsaI: "GGTCTC" or "GAGACC") within the assembled DNA construct, which can adversely impact DNA assembly efficiency.

The combinatorial j5 assembly protocol design CSV file provides sufficient information to guide the assembly process for the entire set of combinatorial DNA constructs. For combinatorial

DNA assembly protocol designs, open and assess the combinatorial j5 assembly protocol design CSV file using spreadsheet software (e.g., Excel, OpenOffice). The combinatorial j5 assembly protocol design CSV file aggregates much of the information distributed across the individual j5 assembly protocol design CSV files (described above). While the combinatorial j5 assembly protocol design CSV file does contain hierarchical assembly strategy information, it does not include primer design, flanking homology overlap sequence design, or internal type IIS endonuclease site warning message details. As such, it remains important to open and assess each of the individual design files as described immediately above.

3.13 Condensing j5 Assembly Files in DeviceEditor

Condensed j5 assembly protocol design CSV files facilitate the parallel construction of multiple unrelated designs, enabling a single person with a multichannel pipette (or a liquid-handling robot) to execute in a single set of multi-well plates the DNA construction tasks of an entire research group. Much like the combinatorial j5 assembly protocol design CSV file described at the end of Subheading 3.12, condensed j5 assembly protocol design CSV files provide sufficient information to guide the assembly process for entire sets of (potentially unrelated) DNA constructs.

To condense a set of j5 assembly protocol design CSV files (whether single construct, combinatorial, or condensed) into a single condensed j5 assembly protocol design CSV file, click the "J5" button at the top left of the DeviceEditor interface (Fig. 2). A "j5 controls" dialog will open. Click the "Condense Assembly Files" tab. There are two j5 input files required, namely "Assembly Files to Condense," a CSV file that lists the assembly files to condense, and "Zipped Assembly Files," a zip file that contains the assembly files themselves. The "Assembly Files to Condense" CSV file can be prepared by editing the example CSV file provided in the "Condensation of multiple j5 assembly files" section of j5 user's manual [53] using spreadsheet software (e.g., Excel, OpenOffice). Select the "Assembly Files to Condense" and "Zipped Assembly Files" files to condense, and then click the "Condense Assemblies" button. After DeviceEditor has indicated that the results are ready, click on the "Download Results" button. A "Select Location for Download" dialog will open. Select the j5 output zip file destination. Unzip the j5 output file to create a folder which contains multiple files, including the corresponding set of j5 input files, and the condensed j5 assembly protocol design CSV file. The formatting and contents of each of these files are documented in great detail in the j5 user's manual [53]. Open and assess the condensed j5 assembly protocol design CSV file using spreadsheet software (e.g., Excel, OpenOffice), as described in Subheading 3.12.

3.14 Distributing PCRs in Multi-well Plates Across Thermal Cycler Annealing Temperature Gradients in DeviceEditor

Given a j5 assembly protocol design file (whether single construct, combinatorial library, or condensed), and input files specifying where DNA oligos and DNA templates are located within multi-well plates, j5 can optimize the distribution of PCRs across annealing temperature gradient zones (e.g., in an AB Veriti® Thermal Cycler) and output instructions for liquid-handling robots to prepare the corresponding PCR plates.

To optimize the distribution of PCRs, click the "j5" button at the top left of the DeviceEditor interface (Fig. 2). A "j5 controls" dialog will open. Click the "Downstream Automation" tab. There are two options for the "Downstream Automation Parameters File." Select "Use latest server version" to direct j5 to use the downstream automation parameters most recently submitted to the j5 server, or "Generate file from parameters" to generate a new parameters file. To edit the downstream automation j5 design parameters for the "Generate file from parameters" option, click the "from parameters" link. Each downstream parameter is modifiable through a text field. For more details on the various downstream automation parameters refer to the j5 user's manual [53]. To return to the default downstream automation parameter values, click the "Return to Defaults" button. Click "Cancel" to cancel the changes made to the j5 parameters, or "OK" to set the new downstream automation parameter values.

In addition to the "Downstream Automation Parameters File," there are three additional j5 input files required, namely "Source Plate List," a CSV file that lists the CSV files specifying each multi-well plate, "Zipped Plate Files," a zip file that contains the CSV files specifying the locations and volumes of the DNA oligos and templates within each multi-well plate, and "j5 Assembly File," the j5 assembly protocol design file. The "Source Plate List" CSV file and the multi-well plate CSV files contained within the "Zipped Plate Files" zip file can be prepared by editing the example CSV files provided in the "Distribution of PCR reactions" section of the j5 user's manual [53] using spreadsheet software (e.g., Excel, OpenOffice). Select the desired "Source Plate List," "Zipped Plate Files," and "j5 Assembly File" files, and then click the "Distribute PCR Reactions" button. After DeviceEditor has indicated that the results are ready, click on the "Download Results" button. A "Select Location for Download" dialog will open. Select the j5 output zip file destination. Unzip the j5 output file to create a folder which contains multiple files, including the corresponding set of j5 input files, a distribute PCRs j5 design CSV file, a robotic instruction CSV file for setting up the PCRs with the NextGen/eXeTek liquid-handling robot platform, and a PR-PR script (.par) file for setting up the PCRs with a Tecan Freedom EVO liquid-handling robot platform (*see* Subheading 3.15). The formatting and contents of each of these files are documented in great detail in the j5 user's manual [53]. Open and assess the distribute PCRs j5 design CSV file using spreadsheet software (e.g., Excel, OpenOffice), which is particularly important in that it specifies the optimal annealing temperatures for the various zones of each of the thermal cycler blocks.

3.15 Compiling j5-Output PR-PR Scripts for Execution on the Tecan Evo Liquid-Handling Robotic Platform

As described in the previous Subheading 3.14, j5 can output PR-PR scripts [18] for automating the setup of PCRs using a liquid-handling robot. These PR-PR scripts can be compiled to run on the Tecan Freedom EVO platform.

Open the PR-PR output file (.pr extension, *see* Subheading 3.14) in a text editor (e.g., Notepad, TextEdit) and select/copy all of its contents to the system clipboard. In a new browser tab, open the public PR-PR web-server [50] interface (*see* Fig. 4a). Paste the contents of the PR-PR output file into the PR-PR script section of the PR-PR interface. Either select a built-in robot table file from the "Select table file" pull-down menu, or click the "Browse" button to select a robot table file to upload. The robot table may then be visually previewed by clicking the "Preview table layout button" (Fig. 4b).

Click the "Prepare robot file" button. When ready, the result can be downloaded by clicking on the "Download" button that will be displayed on the right side of the interface. The resulting output file (.esc extension) can be loaded into the Tecan EVOware software, and then executed on the Tecan Freedom EVO robotic platform.

4 Notes

1. Alternative Interfaces for j5. The step-by-step methods presented here utilize DeviceEditor as the graphical front-end interface for j5, but it should be pointed out that, in addition to DeviceEditor, j5 has a simplified web-form interface [61], full j5-functionality is accessible through XML-RPC web-services [53], and a j5 plug-in for the Clotho platform [62, 63] is under development (Avi Robinson-Mosher, personal communication). Depending on a user's particular workflow or protocol design requirements, one of the other interfaces for j5 may be more suitable than DeviceEditor. For the majority of users and DNA assembly protocol design tasks, however, the methods presented here are generally preferred over other current options, largely because DeviceEditor's "correct-by-construction" features [18] preempt users from making common mistakes (that can be tedious and frustrating to identify and correct) when preparing j5 input files.
2. Online Video Demonstrations and User's Manuals. A narrated online video demonstration [58] of j5, DeviceEditor, and VectorEditor serves as a quick-start guide for using these software

Fig. 4 PR-PR web interface [50] (Subheading 3.15). (**a**) Main interface view. The PR-PR version number is presented at *top right*. Links to the PR-PR copyright, web-service disclaimer, and how to guide, and a link to load a sample PR-PR script, are available at *top*. A pull-down menu provides access to built-in robot table files. Alternatively, a robot table file may be selected for upload (not shown). After a robot table file has been selected, the "Preview table layout" button becomes active at the *top right* of the script text field (the largest portion of the interface). The "Prepare robot file" button is located at *bottom right*. (**b**) Preview table layout view for the selected robot table file "Table_JBEI_1.ewt"

tools. Online user's manuals provide additional documentation and specific demonstrations of how to use j5 [53], DeviceEditor [59], and PR-PR [50]. The published j5 [17], DeviceEditor [18], VectorEditor [19], and PR-PR [20] research articles (along with their supplemental materials) provide further information including algorithm and software implementation details. The methods described in this chapter apply specifically to j5 v2.3.6, DeviceEditor v2.1.1, VectorEditor v1.7.3, and PR-PR v1.1. While it is anticipated that these software tools will continue to be developed and that their interfaces and functionalities may diverge from what is described here, the overall workflow for these tools is not anticipated to change significantly in the short term. That said, referring to their online user's manuals (cited immediately above) is the best means to ensuring up-to-date information.

3. Mapping GenBank Files to Parts in DeviceEditor. Mapping annotated DNA sequences to parts in DeviceEditor from VectorEditor via the system clipboard (Subheading 3.5) is generally preferred over mapping directly from GenBank sequence files. Several DNA software tools (such as Sci-Ed Central's Clone Manager software) do not export properly formatted GenBank sequence files. While VectorEditor is able to successfully parse many of these nonstandard GenBank-like sequence files, j5 may not be able to. Passaging DNA sequence information through VectorEditor mitigates the downstream risk of j5 failure due to incorrectly formatted GenBank sequence files. Some DNA software tools (e.g., VectorNTI, ApE [60]) export GenBank sequence files which in general are successfully parsed by j5. Since some users may prefer to map annotated DNA sequences directly from GenBank sequence files, it is important to briefly describe how this is accomplished. To map an annotated DNA sequence directly from a GenBank file to a part in DeviceEditor, right-click the desired part and select "Map from Genbank." A Select File to Upload dialog box will open. Select the desired GenBank sequence file. A "Specify Part Definition" dialog box will open, which will display the "Part Name," the "Part Source" (the "LOCUS" field in the GenBank file), and the "Source Data" (the contents of the GenBank file including all sequence annotation information). As described in Subheading 3.5, DeviceEditor does not allow distinct GenBank sequence files with the same sequence name (i.e., "LOCUS" field) to be sequentially mapped to parts. As such, it may be necessary to rename the various sequences (not the file names, but rather the "LOCUS" fields) in order to make the name of each distinct sequence unique. Either the "Whole Sequence" or a "Specified Sequence" (spanning from "StartBP" to "StopBP") can be mapped to the part. If the reverse complement of the selected sequence is to be mapped to the part, click to activate the "Reverse Complement" check-box. Click the "Done" button, concluding the mapping process.

4. Alternative Methods for Creating Input Files for j5. In addition to DeviceEditor, there are alternative means to creating input files for j5. When designing a DNA assembly protocol, j5 takes as input a set of files that specify the order and sequences of the DNA fragments that are to be assembled. A zipped collection of DNA sequence files provides the annotated sequences of the DNA fragments. CSV files specify which of the zipped DNA sequence files to input; the locations of the DNA fragments within the DNA sequences; the order in which to assemble the DNA fragments; the user's collections of plasmids, DNA oligos, and synthesized DNA fragments; and the parameters (e.g., the cost of DNA synthesis) that tailor the protocol design process. A plain-text Eugene design specifications rule file [54, 55] specifies rules that constrain undesirable DNA part combinations from being constructed (e.g., prevent plasmids with more than one copy of a particular terminator sequence). Aside from DeviceEditor and VectorEditor, various DNA software (e.g., VectorNTI, ApE) can be used to prepare the annotated DNA sequences; spreadsheet software (e.g., Excel, OpenOffice) can be used to prepare the CSV files; and text editor software (e.g., Notepad, TextEdit) can be used to prepare the Eugene design specifications rule file. Manually prepared j5 input files can be either imported into DeviceEditor (*see* Subheadings 3.3 and 3.7) or submitted to j5 via the aforementioned simplified web-form or XML-RPC web-service interfaces.

Acknowledgments

Conflict of Interest Statement: The author declares competing financial interests in the form of pending patent applications related to the j5 software, and equity in TeselaGen Biotechnology, Inc. This work conducted by the Joint BioEnergy Institute and the U.S. Department of Energy Joint Genome Institute was supported by the Office of Science, Office of Biological and Environmental Research, of the U.S. Department of Energy under Contract No. DE-AC02-05CH11231. The author thanks Joanna Chen for constructive comments on the manuscript.

References

1. Ellis T, Adie T, Baldwin GS (2011) DNA assembly for synthetic biology: from parts to pathways and beyond. Integr Biol (Camb) 3:109–118. doi:10.1039/c0ib00070a
2. Hillson NJ (2011) DNA assembly method standardization for synthetic biomolecular circuits and systems. In: Koeppl H, Densmore D, di Bernardo M, Setti G (eds) Design and analysis of bio-molecular circuits, 1st edn. Springer, New York, pp 295–314
3. Densmore D, Hsiau TH, Kittleson JT et al (2010) Algorithms for automated DNA assembly. Nucleic Acids Res 38:2607–2616. doi:10.1093/nar/gkq165

4. Shetty RP, Endy D, Knight TF Jr (2008) Engineering BioBrick vectors from BioBrick parts. J Biol Eng 2:5. doi:10.1186/1754-1611-2-5
5. Anderson JC, Dueber JE, Leguia M et al (2010) BglBricks: a flexible standard for biological part assembly. J Biol Eng 4:1. doi:10.1186/1754-1611-4-1
6. Leguia M, Brophy J, Densmore D et al (2011) Automated assembly of standard biological parts. Methods Enzymol 498:363–397. doi:10.1016/B978-0-12-385120-8.00016-4
7. Beal J, Weiss R, Densmore D et al (2012) An end-to-end workflow for engineering of biological networks from high-level specifications. ACS Synth Biol 1:317. doi:10.1021/sb300030d
8. Weber E, Engler C, Gruetzner R et al (2011) A modular cloning system for standardized assembly of multigene constructs. PLoS One 6:e16765. doi:10.1371/journal.pone.0016765
9. Sarrion-Perdigones A, Falconi EE, Zandalinas SI et al (2011) GoldenBraid: an iterative cloning system for standardized assembly of reusable genetic modules. PLoS One 6:e21622. doi:10.1371/journal.pone.0021622
10. Quan J, Tian J (2009) Circular polymerase extension cloning of complex gene libraries and pathways. PLoS One 4:e6441. doi:10.1371/journal.pone.0006441
11. Shao Z, Luo Y, Zhao H (2011) Rapid characterization and engineering of natural product biosynthetic pathways via DNA assembler. Mol Biosyst 7:1056–1059. doi:10.1039/c0mb00338g
12. Carothers JM, Goler JA, Juminaga D et al (2011) Model-driven engineering of RNA devices to quantitatively program gene expression. Science 334:1716–1719. doi:10.1126/science.1212209
13. Salis HM, Mirsky EA, Voigt CA (2009) Automated design of synthetic ribosome binding sites to control protein expression. Nat Biotechnol 27:946–950. doi:10.1038/nbt.1568
14. Egbert RG, Klavins E (2012) Fine-tuning gene networks using simple sequence repeats. Proc Natl Acad Sci U S A 109:16817–16822. doi:10.1073/pnas.1205693109
15. Mutalik VK, Guimaraes JC, Cambray G et al (2013) Quantitative estimation of activity and quality for collections of functional genetic elements. Nat Methods 10(4):347–353. doi:10.1038/nmeth.2403
16. Mutalik VK, Guimaraes JC, Cambray G et al (2013) Precise and reliable gene expression via standard transcription and translation initiation elements. Nat Methods 10(4):354 –360
17. Hillson NJ, Rosengarten RD, Keasling JD (2012) j5 DNA assembly design automation software. ACS Synth Biol 1:14–21. doi:10.1021/Sb2000116
18. Chen J, Densmore D, Ham TS et al (2012) DeviceEditor visual biological CAD canvas. J Biol Eng 6:1. doi:10.1186/1754-1611-6-1
19. Ham TS, Dmytriv Z, Plahar H et al (2012) Design, implementation and practice of JBEI-ICE: an open source biological part registry platform and tools. Nucleic Acids Res 40:e141. doi:10.1093/nar/gks531
20. Linshiz G, Stawski N, Poust S et al (2012) PaR-PaR laboratory automation platform. ACS Synth Biol 2:216–222. doi:10.1021/sb300075t
21. Thieme F, Engler C, Kandzia R et al (2011) Quick and clean cloning: a ligation-independent cloning strategy for selective cloning of specific PCR products from non-specific mixes. PLoS One 6:e20556. doi:10.1371/journal.pone.0020556
22. Li MZ, Elledge SJ (2007) Harnessing homologous recombination in vitro to generate recombinant DNA via SLIC. Nat Methods 4:251–256. doi:10.1038/nmeth1010
23. You C, Zhang XZ, Zhang YH (2012) Simple cloning via direct transformation of PCR product (DNA Multimer) to *Escherichia coli* and *Bacillus subtilis*. Appl Environ Microbiol 78:1593–1595. doi:10.1128/AEM.07105-11
24. Quan J, Tian J (2011) Circular polymerase extension cloning for high-throughput cloning of complex and combinatorial DNA libraries. Nat Protoc 6:242–251. doi:10.1038/nprot.2010.181
25. Erijman A, Dantes A, Bernheim R et al (2011) Transfer-PCR (TPCR): a highway for DNA cloning and protein engineering. J Struct Biol 175:171–177. doi:10.1016/j.jsb.2011.04.005
26. Zhang Y, Werling U, Edelmann W (2012) SLiCE: a novel bacterial cell extract-based DNA cloning method. Nucleic Acids Res 40:e55. doi:10.1093/nar/gkr1288
27. Gibson DG, Young L, Chuang RY et al (2009) Enzymatic assembly of DNA molecules up to several hundred kilobases. Nat Methods 6:343–345. doi:10.1038/nmeth.1318
28. Ramon A, Smith HO (2011) Single-step linker-based combinatorial assembly of promoter and gene cassettes for pathway engineering. Biotechnol Lett 33:549–555. doi:10.1007/s10529-010-0455-x
29. Shao Z, Zhao H, Zhao H (2009) DNA assembler, an in vivo genetic method for rapid construction of biochemical pathways. Nucleic Acids Res 37:e16. doi:10.1093/nar/gkn991
30. Wingler LM, Cornish VW (2011) Reiterative recombination for the in vivo assembly of librar-

ies of multigene pathways. Proc Natl Acad Sci U S A 108:15135–15140. doi:10.1073/pnas.1100507108
31. j5 website. http://j5.jbei.org
32. Gibthon website. http://gibthon.org
33. Richardson SM, Liu S, Boeke JD et al (2012) Design-A-Gene with GeneDesign. Methods Mol Biol 852:235–247. doi:10.1007/978-1-61779-564-0_18
34. Richardson SM, Nunley PW, Yarrington RM et al (2010) GeneDesign 3.0 is an updated synthetic biology toolkit. Nucleic Acids Res 38:2603–2606. doi:10.1093/nar/gkq143
35. GeneDesign website. http://54.235.254.95/gd/
36. Convert PCR Primers Into In-Fusion® Primers website. http://www.clontech.com/US/Products/Cloning_and_Competent_Cells/Cloning_Kits/xxclt_onlineToolsLoad.jsp?citemId=http://bioinfo.clontech.com/infusion/convertPcrPrimersInit.do&xxheight=750
37. GeneArt® Primer and Construct Design Tool website. http://bioinfo.invitrogen.com/oligoDesigner
38. Engler C, Gruetzner R, Kandzia R et al (2009) Golden gate shuffling: a one-pot DNA shuffling method based on type IIS restriction enzymes. PLoS One 4:e5553. doi:10.1371/journal.pone.0005553
39. Engler C, Kandzia R, Marillonnet S (2008) A one pot, one step, precision cloning method with high throughput capability. PLoS One 3:e3647. doi:10.1371/journal.pone.0003647
40. Engler C, Marillonnet S (2011) Generation of families of construct variants using golden gate shuffling. Methods Mol Biol 729:167–181. doi:10.1007/978-1-61779-065-2_11
41. Geertsma ER, Dutzler R (2011) A versatile and efficient high-throughput cloning tool for structural biology. Biochemistry 50:3272–3278. doi:10.1021/bi200178z
42. Bitinaite J, Rubino M, Varma KH et al (2007) USER friendly DNA engineering and cloning method by uracil excision. Nucleic Acids Res 35:1992. doi:10.1093/nar/gkm041
43. Annaluru N, Muller H, Ramalingam S (2012) Assembling DNA fragments by USER fusion. Methods Mol Biol 852:77–95. doi:10.1007/978-1-61779-564-0_7
44. Olsen LR, Hansen NB, Bonde MT et al (2011) PHUSER (Primer Help for USER): a novel tool for USER fusion primer design. Nucleic Acids Res 39:W61–W67. doi:10.1093/nar/gkr394
45. PHUSER website. http://www.cbs.dtu.dk/services/phuser/
46. TeselaGen website. http://teselagen.com
47. VectorEditor stand-alone software. https://public-registry.jbei.org/static/vesa/VectorEditor.html
48. Public instance of the JBEI Parts Registry. http://public-registry.jbei.org
49. VectorEditor Project. http://code.google.com/p/vectoreditor/
50. PaR-PaR website. http://prpr.jbei.org
51. PaR-PaR source code github repository. https://github.com/jbei/prpr
52. Synthetic Biology Open Language visual standard. http://www.sbolstandard.org/visual
53. j5 user's manual. http://j5.jbei.org/j5manual/index.html
54. Eugene website. http://eugenecad.org/
55. Bilitchenko L, Liu A, Cheung S et al (2011) Eugene—a domain specific language for specifying and constraining synthetic biological parts, devices, and systems. PLoS One 6:e18882. doi:10.1371/journal.pone.0018882
56. Rozen S, Skaletsky H (2000) Primer3 on the WWW for general users and for biologist programmers. Methods Mol Biol 132:365–386
57. DINAMelt web server. http://mfold.rna.albany.edu/?q=DINAMelt/Quickfold
58. j5, DeviceEditor, and VectorEditor demonstration video. http://j5.jbei.org/j5_and_DeviceEditor_Demo_Movie.mov
59. DeviceEditor user's manual. http://j5.jbei.org/DeviceEditor_manual/index.html
60. A plasmid Editor (ApE) software. http://biologylabs.utah.edu/jorgensen/wayned/ape/
61. j5 web-form interface. http://j5.jbei.org/bin/j5_entry_form.pl
62. Clotho website. http://clothocad.org
63. Xia B, Bhatia S, Bubenheim B (2011) Developer's and user's guide to Clotho v2.0 A software platform for the creation of synthetic biological systems. Methods Enzymol 498:97–135. doi:10.1016/B978-0-12-385120-8.00005-X
64. DeviceEditor software. http://j5.jbei.org/bin/deviceeditor.pl

Chapter 18

FastPCR Software for PCR, In Silico PCR, and Oligonucleotide Assembly and Analysis

Ruslan Kalendar, David Lee, and Alan H. Schulman

Abstract

This chapter introduces the software FastPCR as an integrated tools environment for PCR primer and probe design. It also predicts oligonucleotide properties based on experimental studies of PCR efficiency. The software provides comprehensive facilities for designing primers for most PCR applications and their combinations, including standard, multiplex, long-distance, inverse, real-time, group-specific, unique, and overlap extension PCR for multi-fragment assembly in cloning, as well as bisulphite modification assays. It includes a program to design oligonucleotide sets for long sequence assembly by the ligase chain reaction. The in silico PCR primer or probe search includes comprehensive analyses of individual primers and primer pairs. It calculates the melting temperature for standard and degenerate oligonucleotides including LNA and other modifications, provides analyses for a set of primers with prediction of oligonucleotide properties, dimer and G/C-quadruplex detection, and linguistic complexity, and provides a dilution and resuspension calculator. The program includes various bioinformatics tools for analysis of sequences with CG or AT skew, of CG content and purine–pyrimidine skew, and of linguistic sequence complexity. It also permits generation of random DNA sequence and analysis of restriction enzymes of all types. It finds or creates restriction enzyme recognition sites for coding sequences and supports the clustering of sequences. It generates consensus sequences and analyzes sequence conservation. It performs efficient and complete detection of various repeat types and displays them. FastPCR allows for sequence file batch processing, which is essential for automation. The FastPCR software is available for download at http://primerdigital.com/fastpcr.html and online version at http://primerdigital.com/tools/pcr.html.

Key words PCR primer design, Primer linguistic complexity, Sequence assembly, Software probe design, Ligase chain reaction, DNA primers

Abbreviation

OE-PCR	Overlap extension PCR
PCR	Polymerase chain reaction
RT-PCR	Real-time PCR
SSR	Simple sequence repeat

Svein Valla and Rahmi Lale (eds.), *DNA Cloning and Assembly Methods*, Methods in Molecular Biology, vol. 1116, DOI 10.1007/978-1-62703-764-8_18, © Springer Science+Business Media New York 2014

1 Introduction

The polymerase chain reaction (PCR) is fundamental to molecular biology and is the most important practical molecular technique for the DNA research laboratory. However, the utility of the method is dependent on identifying unique primer sequences and designing PCR-efficient primers. Primer design is a critical step in all types of PCR methods to ensure specific and efficient amplification of a target sequence [1–7]. Even though there are currently many online and commercial bioinformatics tools, primer design for PCR is still not as convenient and practical as it might be for routine use. The adaptation of PCR for different applications has made it necessary to develop new criteria for PCR primer and probe design to cover uses such as RT-PCR, real-time PCR, group-specific and unique PCR, combinations of multiple primers in multiplex PCR, overlap extension PCR fort multi-fragments assembling cloning, and bisulphite modification assays. There is a need as well as for a program integrating design of oligonucleotide sets for long sequence assembly by the ligase chain reaction (LCR), discovery of simple sequence repeats (SSRs) and their amplification as diagnostic markers, and for designing TaqMan, molecular beacon, and microarray oligonucleotides [6, 8, 9].

In developing FastPCR and Java web tools (Table 1), our aim was to create practical and easy-to-use software for routine manipulation and analysis of sequences for most PCR applications. The parameters adopted are based on our experimental data for efficient PCR and are translated into algorithms in order to design combinations of primer pairs for optimal amplification. This software, FastPCR, has been successfully used throughout the scientific community in a wide range of PCR and probe applications, and repeat searches and analysis. The first papers describing this software were published in 2001 and since 1999 this software has been applied in various projects, have been cited together over 500 times in scientific journals, patents, PhD theses, and over 1,000 downloads of the installation file per month. The program code is regularly updated.

2 Software, General Information

The FastPCR software (http://primerdigital.com/fastpcr.html) is written in Microsoft Visual Studio 6.0 and compiled to an executive file that, after installation, can be used with any version of Microsoft Windows. For Linux and Mac it requires "Wine" (http://www.winehq.org/) as a compatibility layer for running Windows programs. It is a completely free alternative implementation of the Windows API also for use with native Windows DLLs.

Table 1
Summary of the FastPCR software for PCR, in silico PCR, and oligonucleotide assembly and analysis

Features
PCR tool provides comprehensive facilities for
Design of primers for most PCR applications and their combinations, including standard, multiplex, long-distance, inverse, real-time, unique (specific primers for each from genetically related DNA sequences) or group-specific (universal primers for genetically related DNA sequences), linear-after-the-exponential (LATE)-PCR, bisulphite modification assays, polymerase extension PCR multi-fragment assembly cloning
Design of long oligonucleotides for microarray analyses and dual-labeled oligonucleotides for probes such as molecular beacons
Polymerase chain assembly (PCA) or oligo assembly—for automating the design of oligonucleotide sets for long sequence assembly by ligase chain reaction (LCR) and PCR
In silico (virtual) PCR or multiple primer or probe searches, or in silico PCR against whole genome(s) or a list of predictions by chromosome of probable PCR products, and search for potential mismatching locations of the specified primers or probes
Testing of individual primers, melting temperature calculation for standard and degenerate oligonucleotides including LNA and other modifications
Evaluation of PCR efficiency, linguistic complexity, dimer and G/C-quadruplex detection, dilution and resuspension calculator
Analysis of features of multiple primers simultaneously, including T_m, CG content, linguistic complexity, dimer formation; optimal T_a
Identification of simple sequence repeat (SSR) loci by analyzing the low-complexity regions of input sequences
Restriction digest analyses for Type I, II, and III restriction enzymes and homing endonucleases, finding or creating restriction enzyme recognition sites for coding sequences
Searches for similar sequences (or primers)
Translation of nucleotide (DNA/RNA) sequences to the corresponding peptide sequence in all six frames for standard and degenerate DNA and modifications (inosine, uridine)
Determination of CG:(G–C)/(G+C), AT:(A–T)/(A+T), SW:(S–W)/(S+W), MK:(M–K)/(M+K), purine–pyrimidine (R–Y)/(R+Y) skews, CG% content and the melting temperature, primer quality and linguistic sequence complexity profiles

The online FastPCR (jPCR) software (http://primerdigital.com/tools/) is written in Java with NetBeans IDE (Oracle) and requires the Java Runtime Environment (JRE) on a computer. It can be used with any operating system (64-bit OS preferred for large chromosome files).

3 The Interface

3.1 Inputs to FastPCR

The software contains the menus, the toolbars, and the ribbon and three text editors. The ribbon is designed to help the user quickly find the commands that are needed to complete a task. Commands are organized in logical groups, which are collected together under tabs (Fig. 1). Each tab relates to a type of activity, such as "PCR Primer Design," "in silico PCR," or "Primer Test".

Getting started with a basic project in FastPCR software is as easy as opening a new or existing file and then using copy–paste or starting to type. There are three independent text editors on different tabs within the interface: "General sequence(s)," "Additional sequence(s) or pre-designed primers (probes) list," and "Result report." The first two text editors are necessary for loading sequences for analysis, the text editor "General sequence(s)" is designed for working with the project sequences, and the "Additional sequence(s) or pre-designed primers (probes) list" text editor is used for special and additional sequences such as pre-designed primers, multiple query sequences, or numbers for input.

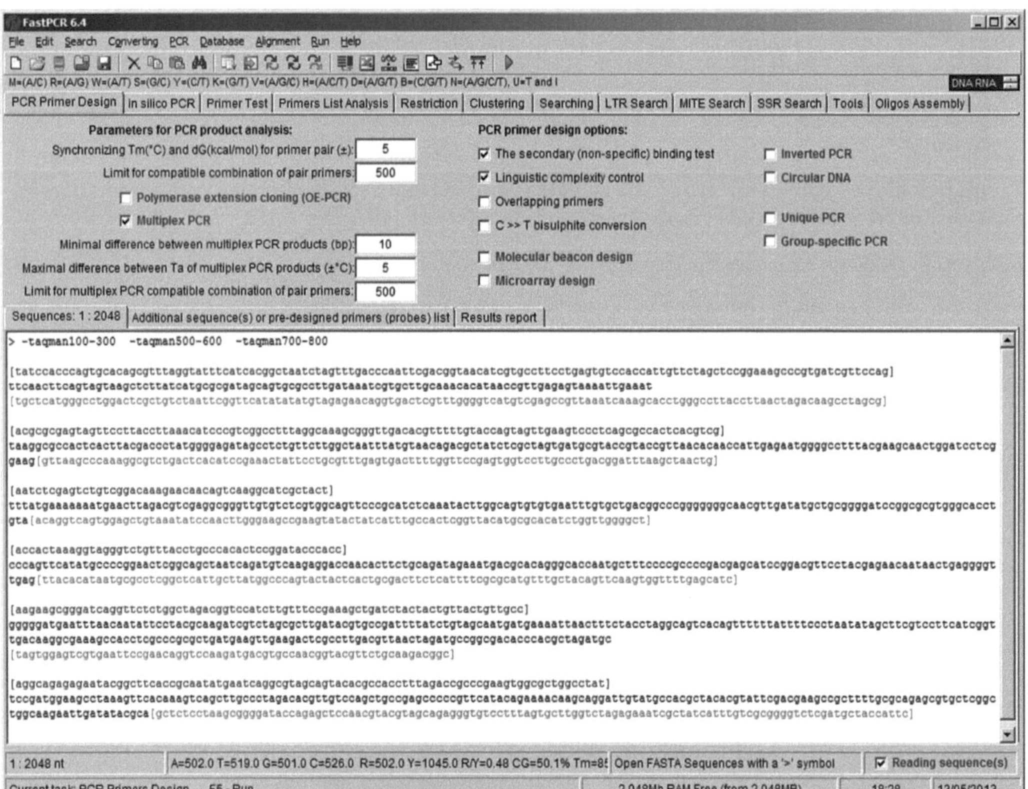

Fig. 1 The FastPCR sequence editor and user interface

3.2 Program Output

FastPCR automatically generates results at a third text editor, "Result report," in tabulated format for transferring to spreadsheet software from the clipboard using copy–paste. Output results are easy to save as Excel worksheet (.xls) or as Rich Text Format (.rtf) text files, compatible with MS Excel or Open Office. The separated output of the primer design is a list of primers, a set of primer pair sequences with their theoretical PCR products, and, for multiplex PCR, the result of the calculation of multiple-PCR primers for given target sequences. In addition, the output shows optimal annealing temperature for each primer pair, the size of PCR product, and complete information for each designed primer and for each multiplex PCR product set.

3.3 Sequence Entry

Sequence data files are prepared using a text editor (Notepad, WordPad, MS Word), and saved in ASCII as text/plain format (.txt) or in .rtf. The program either takes a single sequence or accepts multiple separate DNA sequences in FASTA, tabulated format (two columns from MS Excel sheet or MS Word table), EMBL, MEGA, GenBank, MSF, DIALIGN, simple alignment, or BLAST Queue web alignment formats. The template length is not limited. The FastPCR clipboard allows the user to copy and paste text or tables from MS Word documents or MS Excel worksheets or other programs and to paste them into another Office document. It is important that all target sequences are prepared in the same format. Users can type or import from file(s) into "General Sequence(s)" or "Additional sequence(s) or pre-designed primers (probes) list" editors.

FastPCR allows files to be opened in several ways: the original file can be opened as read-only for editing with text editors; files can be opened to memory without using text editors, which allows larger file(s), up to 200 Mb, to be analyzed; files within a folder can be selected and the files opened during task execution without the use of text editor program. Additionally, the program can open files within a selected folder in order to join all these files in a text editor. For example, this feature can be applied to convert all files from a selected folder into a single file of FASTA sequences. Alternatively this feature allows splitting FASTA sequences to individual files in a particular selected folder.

When a sequence file is opened, FastPCR displays the information about the opened sequence and its format. The information status bar shows the number of sequences, the total sequence length (in nucleotides), the nucleotide composition, and the purine, pyrimidine, CG percentage, and the melting temperature. Files can be saved in various formats including .rtf, .xls, or txt from the text editor in use.

3.4 FASTA Format Description

FastPCR normally expects to read sequence files in FASTA format [10]. FASTA format has the highest priority and is simple, because the raw sequence is merely preceded by a definition line. The definition line begins with a ">" sign and optionally followed immediately by a name for the sequence with no length restriction. Many sequences can be listed in the file, the format indicating a new sequence at each ">" symbol found. It is important to press Enter at the end of each line after ">" to help FastPCR recognize the end and beginning of the sequence and the sequence name. It is important that the first line of each sequence starts with ">".

Degenerate DNA sequences are accepted as IUPAC code [11], an extended vocabulary of 11 letters, which allows the description of ambiguous DNA code. Each letter represents a combination of one or several nucleotides: M = (A/C), R = (A/G), W = (A/T), S = (G/C), Y = (C/T), K = (G/T), V = (A/G/C), H = (A/C/T), D = (A/G/T), B = (C/G/T), N = (A/G/C/T), U = T, and I (Inosine). The program accepts amino acid codes: A (Ala), C (Cys), D (Asp), E (Glu), F (Phe), G (Cly), H (His), I (Ile), K (Lys), L (Leu), M (Met), N (Asn), P (Pro), Q (Gln), R (Arg), S (Ser), T (Thr), U (Sec), V (Val), W (Trp), and Y (Tyr).

3.5 Alignment Format Description

There are many different programs that perform different types of alignment formats. Standardizing on a set of formats enables programs to be written that can read results from many different programs. In all alignment formats, gaps that have been introduced into the sequences to make them align are indicated by the "-" character. The exception to this rule is GCG/MSF format which uses "." as the gap character inside the sequences. The file may begin with as many lines of comment or description as required. The first mandatory line must contain the text "MSF," "Alignment as simple alignment format," "DIALIGN," or "MEGA" to be recognized as alignments from these programs. Following the first line are lines that start with the sequence name, which is separated from the aligned sequence residues by white space.

4 The PCR Primers or Probe Design Analysis Options

4.1 PCR Primer Design Generalities

Primer design is one of the key steps for successful PCR. For PCR applications, primers are usually 18–35 bases in length and should be designed such that they have complete sequence identity to the desired target fragment to be amplified. The parameters, either controllable by the user or selected automatically, are primer length (12–500 nt), melting temperature for short primers calculated by nearest neighbor thermodynamic parameters, theoretical primer PCR efficiency (quality at %) value, primer CG content, 3′ end terminal enforcement, preferable 3′ terminal nucleotide sequence composition in degenerated formulae, and added sequence tags at 5′ termini. The other main parameters used for primer selection are the general nucleotide structure of the primer such as linguistic complexity (nucleotide arrangement and composition), specificity,

Table 2
Default primer design selection criteria

Criteria	Default	Ideal
Length (nt)	20–24	>21
T_m range (°C)[a]	52–68	60–68
T_m[a] 12 bases at 3′ end	30–50	41–47
CG (%)	45–65	50
3′ end composition (5′-NNN-3′)[b]	SWW, SWS, SSW, WSS	SSA, SWS, WSS
Sequence linguistic complexity (*LC*, %)[c]	>75	>90
Sequence quality (*PQ*, %)	>70	>90

[a]Nearest neighbor thermodynamic parameters according to SantaLucia [13]
[b]Ambiguity codes
[c]Sequence linguistic complexity measurement calculated using the alphabet-capacity *L*-gram method

the melting temperature of the whole primer, the melting temperature at the 3′ and 5′ termini, self-complementarity, and secondary (nonspecific) binding.

The software can dynamically optimize the best primer length for the entered parameters. All PCR primer (probe) design parameters are flexible and changeable according to the specifics of the analyzed SEQUENCES and tasks. Primer pairs are analyzed for cross-hybridization and specificity of both primers and, optionally, selected with similar melting temperatures. Primers with balanced melting temperatures (within 1–6 °C of each other) are desirable but not mandatory. The default primer design selection criteria are shown in Table 2. It is possible to use predesigned primers or probes or, alternatively, predesigned primers can act as references for the design of new primers. The program accepts a list of predesigned oligonucleotide sequences and checks the compatibility of each primer with a newly designed primer or probe.

4.2 Melting Temperature Calculation

The melting temperature (T_m) is defined as the temperature at which half the DNA strands are in the double-helical state and half are in the "random-coil" state. The T_m for short oligonucleotides with normal or degenerate (mixed) nucleotide combinations is calculated in the default setting using nearest neighbor thermodynamic parameters [12, 13]. The CG content of an oligonucleotide is the most important factor that influences the T_m value. The melting temperature for mixed bases is calculated by averaging nearest neighbor thermodynamic parameters—enthalpy and entropy values—at each mixed site; extinction coefficient is similarly predicted by averaging nearest neighbor values at mixed sites [2, 3]. Mismatched pairs can be taken into account since the parameters provide for DNA/DNA duplexes and dangling ends, which are unmatched terminal nucleotides [14–16]. The melting temperature for primer (probe) self- or cross-dimers and for in silico PCR experiments with oligonucleotides having mismatches to the target is calculated using values for the thermodynamic parameters for a nucleic acid duplex.

The FastPCR allows the choice of other nearest neighbor thermodynamic parameters or simple non-thermodynamic T_m calculation formulae. For non-thermodynamic T_m calculation, we suggest using simple formulae; the Wallace–Ikatura rule [17] is often used as a rule of thumb when primer T_m is to be estimated at the bench. However, the formula was originally applied to the hybridization of probes in 1 M NaCl and is an estimate of the melting temperature for oligonucleotides shorter than 10 bases:

$$T_m(^\circ\mathrm{C}) = 2(\mathrm{L} + \mathrm{G} + \mathrm{C})$$

For oligonucleotides longer than ten bases the following is used:

$$T_m(^\circ\mathrm{C}) = 64.9 + \frac{41([\mathrm{G}+\mathrm{C}] - 16.4)}{L}$$

or alternatively the formula [18]:

$$T_m(^\circ\mathrm{C}) = 77.1 + 11.7\log_{10}\left[\mathrm{K}^+\right] + \frac{41[\mathrm{G}+\mathrm{C}] - 528}{L}$$

where L is length of primer, $[\mathrm{G}+\mathrm{C}]$ is the number of Gs and Cs, and $[\mathrm{K}^+]$ is salt molar concentration (default value is 50 mM). The two equations above assume that the stabilizing effects of cations are the same on all base pairs. The melting temperature of the PCR product is calculated using the formula [15]:

$$T_m(^\circ\mathrm{C}) = 81.5 + 16.6\log_{10}\left[\mathrm{K}^+\right] + \frac{41[\mathrm{G}+\mathrm{C}] - 675}{L}$$

4.3 Linguistic Complexity of Sequences and Nucleotide-Skew Analysis

The sequence complexity calculation method can be used to search for conserved regions between the compared sequences in order to detect low-complexity regions including SSRs, imperfect direct or inverted repeats, polypurine and polypyrimidine triple-stranded DNA structures, and four-stranded structures (such as G/C-quadruplexes) [19]. Linguistic complexity measurements are performed using the alphabet-capacity L-gram method [20, 21] along the whole sequence length and calculated as the sum of the observed range (x_i), from 1- to L-size words in the sequence, divided by the sum of the expected (E) value for this sequence length. Linguistic complexity (LC) values for sequence length (s) are converted to percentages, in which 100 % means maximal "vocabulary richness" of a sequence:

$$LC(\%) = \frac{100 \times \sum_{i=1}^{L} x_i}{E},$$

where

$$E = \sum_{i=1}^{L} \begin{cases} s - i + 1, s < 4^i - 1 + i \\ 4^i, s \geq 4^i - 1 + i \end{cases},$$

$$L = \left\lceil \log_4 \left(\frac{s}{3} \right) + 1 \right\rceil.$$

For example, the sequence 5′-ACACACACACACACAC, 16 nt ($L = 3$), contains two nucleotides (A, C), but expected $E = 4$ variants; two variants of dinucleotides (AC, CA), but expected $E = (16-1)$ variants; two variants of trinucleotides (ACA, CAC), and expected $E = (16-2)$ variants. The complexity value is $\text{LC} = 100(2+2+2)/(4+16-1+16-2) = 18.2\%$.

The LC tries to describe the "uniqueness" ("vocabulary richness") of a sequence and the likelihood of PCR success of each primer; this value varies from 100 for the best to 5 (e.g., poly(N)) for the worst primers. LC values of 80 and higher allow for the rapid choice of the best primer or probe sequences.

4.4 Primer Quality (Virtual PCR Efficiency) Determination

Our experimental data showed that the primer nucleotide composition and melting temperature of the 12 bases at the 3′ end of the primers are important factors for PCR efficiency. The melting temperature of the 12 bases at the 3′ terminus is calculated preferably by nearest neighbor thermodynamic parameters [13]. The composition of the sequence at the 3′ terminus is important; primers with two terminal C/G bases are recommended for increased PCR efficiency [22]. Nucleotide residues C and G form a strong pairing structure in the duplex DNA strands. Stability at the 3′ end in primer template complexes will improve the polymerization efficiency.

We specify an abstract parameter called primer quality (PQ) that can help to estimate the efficiency of primers for PCR. PQ is calculated by the consecutive summation of the points according to the parameters of total sequence and purine–pyrimidine sequence complexity and of the melting temperatures of the whole primer and of the terminal 12 bases at both the 3′ and 5′ ends. Self-complementarity, which gives rise to possible primer-dimer and hairpin structures, reduces the final value. The PQ tries to describe the likelihood of PCR success of each primer; this value varies from 100 for the best to 0 for the worst primers. To meet multiplexing demands, it is possible in the program to select the best primer with an optimal temperature range, allowing the design of qualified primers or probes for any target sequence with any CG and repeat content. PQ values of 80 and higher allow for the rapid choice of the best PCR primer pair combinations. No adverse effects, due to the modification of the reaction buffer, chosen thermostable polymerases, or variations in annealing temperature, have been observed on the reproducibility of PCR amplification using primers with high PQ.

4.5 Hairpin (Loop) and Dimer Formation

Primer-dimers involving one or two sequences may occur in a PCR. The FastPCR tool eliminates intra- and inter-oligonucleotide reactions before generating a primer list and primer pair candidates. It is very important for PCR efficiency that the production of stable and inhibitory dimers is prevented, especially by avoiding complementarity in the 3′ ends of primers from whence the polymerase will extend. Stable primer-dimer formation is very effective at inhibiting PCR because the dimers formed are amplified efficiently and compete with the intended target.

Primer-dimer prediction is based on analysis of non-gapped local alignments and the stability of both the 3′ end and the central regions of the primers (Fig. 2). Primers will be rejected when they have the potential to form stable dimers, depending on the nucleotide composition and with at least five bases at the 3′ end or seven bases in the central region. Tools to calculate T_m for primer-dimers with mismatches for pure, mixed, or modified (inosine, uridine, or locked nucleic acid (LNA)) bases, using averaged nearest neighbor thermodynamic parameters, are provided for DNA/DNA duplexes [12–14, 23, 24].

In addition to Watson–Crick base-pairing, there is a variety of other hydrogen bonding configurations possible [19, 25–27], including G/C-quadruplexes and wobble base pairs, which the FastPCR software detects. The program provides for the detection of alternative hydrogen bonding during primer-dimer and in silico PCR primer binding site detection. The mismatch stability is examined in order of decreasing stability: **G-C**>**A-T**>G·G>G·T≥G·A>T·T≥A·A>T·C≥A·C≥C·C. Guanine is the most universal base, because it forms the strongest base pair and the strongest mismatches. However, cytosine is the most discriminating base, because it forms the strongest pair and the three weakest mismatches [23, 28]. Therefore, the software also looks for stable guanine mismatches: G·G, G·T and G·A.

G-rich (and C-rich) nucleic acid sequences can fold into four-stranded DNA structures that contain stacks of G-quartets [19]. These quadruplexes can be formed by the intermolecular association of two or four DNA molecules, dimerization of sequences that contain two G bases, or by the intermolecular folding of a single strand containing four blocks of guanines. These are easy to eliminate from primer design because of their low linguistic complexity; LC = 32 % for $(TTAGGG)_4$. The software predicts the presence of putative G- and C-quadruplexes in primer sequences.

Fig. 2 An example of FastPCR duplex formation results

Intermolecular G-quadruplex-forming sequences are detected according to the formula $\ldots G_{m1}X_{n}G_{m2}\ldots$, where m is the number of G residues in each G-tract (m_1, $m_2 \geq 3$); the gap X_n ($n \leq 2$*minimal (m_1:m_2)) can be any combination of residues, including G [16]. The gap sequences (X_n) may have varying lengths, and a relatively stable quadruplex structure may still be formed with a loop more than seven bases long, but in general, increasing the length of the gap leads to a decrease in structure stability. It is also possible for one of the gaps to be zero length when there are long poly-G tracts of >6 bases.

4.6 Calculation of Optimal Annealing Temperature

The optimal annealing temperature (T_a) is the temperature, generally stated as a range, where efficiency of PCR amplification is maximal but nonspecific products minimal. The most important values for estimating the T_a are the primer quality, the T_m of the primers, and the length of PCR fragment. Primers with high T_m (>60 °C) can be used in PCRs with a wide T_a range compared to primers with low T_m (<50 °C). The optimal annealing temperature for PCR is calculated directly as the value for the primer with the lowest T_m (T_m^{min}). However, PCR can work in temperatures up to 10° higher than the T_m of the primer so as to favor primer target duplex formation:

$T_a(°C) = T_m^{min} + \ln L$, where L is the length of the PCR fragment.

4.7 The Secondary Nonspecific Binding Test; Alternative Amplification Products

The specificity of the oligonucleotides is one of the most important factors for good PCR; optimal primers should hybridize only to the target sequence, particularly when complex genomic DNA is used as the template. Amplification problems can arise due to primers annealing to repetitious sequences (retrotransposons, DNA transposons, or tandem repeats). Alternative product amplification can also occur when primers are complementary to inverted repeats and produce multiple bands. This is unlikely when primers have been designed using specific DNA sequences (unique PCR). However, the generation of inverted repeat sequences is exploited in two common generic DNA fingerprinting methods, RAPD and AP-PCR [29, 30]. Because only one primer is used in these PCRs, the ends of the products must be reverse complements and thus can form stem-loops.

The techniques of inter-repeat amplified polymorphism, including inter-retrotransposon amplified polymorphism (IRAP), retrotransposon-microsatellite amplified polymorphism (REMAP), inter-MITE amplification [31, 32], and *Alu*-repeat polymorphism [33, 34], exploit highly abundant, dispersed repeats as markers. However, primers complementary to repetitious DNA may produce many nonspecific bands in single-primer amplification and compromise the performance of unique PCRs. A homology search with the primer as the query sequence, for example, using BLASTn against all sequences in GenBank or EMBL-Bank, will determine whether the primer is likely to interact with dispersed repeats. Alternatively, one can create a small, local, specialized library of repeat sequences based on those in Repbase [35] or TREP [36] and use this for searches.

The mismatches at the 3′ end of the primers affect target amplification much more than mismatches at the 5′ end. A two-base mismatch at the 3′ end of the primer prevents amplification. A single-base mismatch at the 3′ end as well as several mismatches at the 5′ end of the primer permits amplification, albeit with reduced efficiency. However, the presence of multiple primer binding sites does not necessarily lead to alternative amplification products because, for successful amplification, the priming sites for both primers must be both located close to each other, in correct orientation, and sufficiently match the primer sequences.

By default, FastPCR performs a test for nonspecific binding by repeat masking and low-complexity region detection and masking for each given sequence.

Additionally the software allows this test to be performed against a reference sequence or sequences (e.g., a BAC or YAC) or one's own database. Primers that bind to more than one location on given sequences will be rejected. Even though the test for non-specific primer binding is performed as a default for all primers, the user may cancel the operation. Identification of secondary binding sites including mismatched hybridization is normally performed by considering the similarity of the primer to targets along the entire primer sequence. An implicit assumption is that stable hybridization of a primer with the template is a prerequisite for priming by DNA polymerase. FastPCR pays particular attention to the 3′ portion of the primers and calculates the similarity of 3′ end to the target (the length is chosen by user) to determine the stability of any potential interactions.

The secondary nonspecific primer binding test is based on repeat masking using a quick local alignment screen (which allows one mismatch within a hash index of 12-mers) between the reference and input sequences.

5 Methods

Once the input files are selected or sequences copied and pasted to the **General Sequence**(s) text editor, the FastPCR provides various execution features. Figure 3 provides an example for primer design from the user's perspective.

5.1 Execution of the Selected Task

The user selects the ribbon having the task needed. The program will only perform the selected task. The name of the selected executive task is shown on the status bar by "Press F5." The task is executed by using key **F5** or by clicking the arrow on the toolbar using the mouse. Once the executive task is completed, the result is shown in the **Result report** text editor (e.g., *see* Fig. 3).

5.2 PCR Primer Design Options

The "PCR Primer Design" Tab contains various execution options for commonly selected types of PCR and for the most important PCR parameters (Fig. 1). The option panel of "PCR Primers or Probe Design Options" is shown in Fig. 4. Once the user selects any attribute, the option attribute value field shows the default attributes value, which can then be modified. "PCR Primers or Probe Design Options" affects all sequences. PCR primer design options can be customized for each sequence using special commands at the header of the sequence (http://primerdigital.com/soft/pcr_help.html). Typically, it is not necessary to use these commands to manage typical PCR primer design and these are applied to advanced tasks. Default global parameters for primer design will be assigned by typing the help command "/?" in text editor:

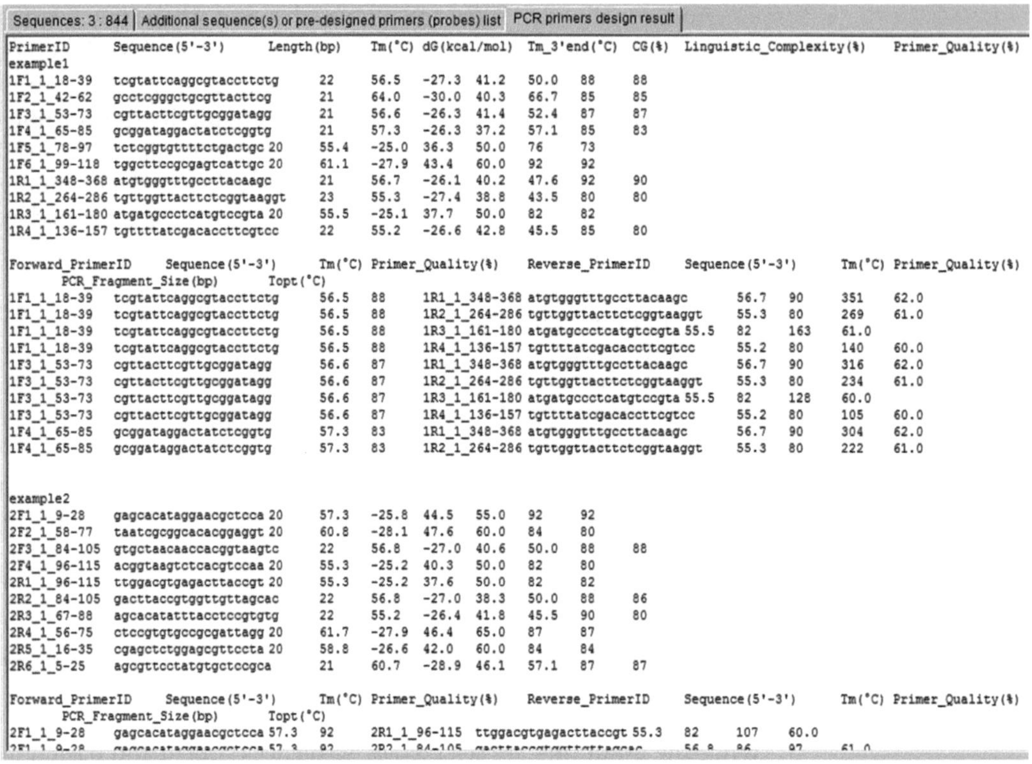

Fig. 3 Primer design result window

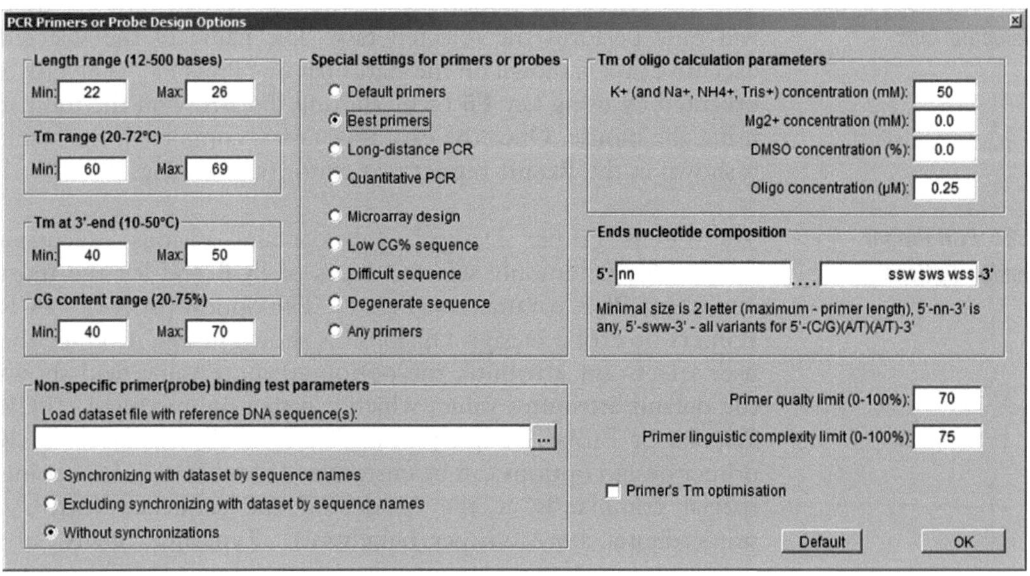

Fig. 4 The "PCR Primers or Probes Design Options" window

"{-ln20-23 -tm55-68 -3tm37-50 -cg41-70 -q70 -lc75 -npr400 -c5[NN] -c3[SWW SSW SWS WSS]}," where "-ln20-23" determines the range of primer length (20–23 bases), "-tm55-68" determines the range of primer T_m (55–68 °C), "-3tm37-50" determines the range of primer T_m at 3′ end (37–50 °C), "-cg41-70" determines the range of primer CG% contents (between 41 and 70 %), "-npr400" shows the maximum number of primers (400) designed to each target, "-c5[NN]" denotes a primer having no specific sequence pattern for 5′ ends, and "-c3[SWW SSW SWS WSS]" specifies primers that conform to particular patterns of ambiguity, such as that shown here for example.

5.3 Examples for Primer Selection Regions

Users can specify, individually for each sequence, multiple locations for both forward and reverse primer placement with the commands: "-FpdN1-N2" for forward primers and "-RpdN1-N2" for reverse primers, where from N1 to N2 are bases from the selected regions; "-pdN1-N2" (see more at: http://primerdigital.com/soft/pcr_help.html). Alternatively, users can specify multiple locations for both forward and reverse primers using ["and"] inside each sequence: the software allows multiple and independent locations for both forward and reverse primers inside each of the sequences. Whilst PCR primer design will be performed independently for different targets, multiplex PCR primer design can be performed simultaneously with multiple amplicons within a single sequence as well as for different sequences, i.e., all possible combinations of ["and"] inside one or more sequences. By default, the software designs primers within the entire sequence length. Optionally, users can specify, individually for each sequence, multiple locations for both forward and reverse primers with the commands:

1. The same location for both forward and reverse primers will be designed in the central [nnnnnnnnnn] part ("[]" is used only once):
   ....[nnnnnn]....
2. Different locations for forward and reverse primers; forward primers will be chosen inside the "[1nnnnnn]" location and reverse primers inside "[2nnnnnn]" location (twice "[]"):
   ....[1nnnnnn]....[2nnnnnn].....
3. Primers must flank the central "]nnnnnn["; forward primers will be chosen from 1 to "A]" bases and reverse primers will be chosen from base "[C" to the end of sequence: ......A] nnnnnn[C.....
4. Forward and reverse primers have an overlapping part "[nnnnnn]"; forward primers will be chosen from "[A to n]" bases and reverse primers will be chosen from "[n base to C]":
   ......[A.....[nnnnnn]......C]....

The software allows the selection of any number of independent PCR primer (or probe) designing tasks for each sequence using multiple combinations of "[..]" and -FpdN1-N2, -RpdN1-N2, or -pdN1-N2 commands. Multiplex PCR can be carried out simultaneously within a single sequence with multiple tasks as well as for different sequences with multiple tasks or a combination of both.

All possible combinations of "[]" (forward) with "[]" (reverse) within the sequence(s):

1. `[]`
2. `] [`
3. `[] []`
4. `[[]]`
5. `([] [])n or/and ([[]])n.`

5.4 User-Defined PCR Product Size

The PCR product size can be specified in a similar way, with the command: "(N1-N2)"; these values can be specified in the form of minimum and maximum values for the product size. For example, the "(400-500)" line defines that the product size ranges from 400 to 500 base pairs. In case a user wants to specify a fixed product size, the command should be a single number, for example, "(500)." FastPCR is flexible and allows PCR product sizes from 50 to 10,000 base pairs in length.

5.5 PCR Set-up Examples with Individual Commands

1. Where primers have already been designed, FastPCR can be used to predict the optimal annealing temperature and PCR product length for one or more predesigned primers (with the "-npd" command, which prohibits the primer design). For example, the command:

   "-fpr[ggagagtagcttacctcgct cggtaaggttct-tcatgc] -npd"

   will analyze the selected sequence between the two primers (5′ ggagagtagcttacctcgct and 5′ cggtaaggttcttcatgc).
2. Design primers with a difference in T_m of about 10°, e.g., for LATE-PCR:

   "-Ftm50-55 -Rtm64-68 -pTMs10."
3. Design primers with a specific restriction enzyme site at the 3′ end ("-z3eNameEnzyme", "-Fz3eNameEnzyme", "-Rz3eNameEnzyme") where "NameEnzyme" is the name of the restriction enzyme: "-z3eXceI." The alternative command ("-c3NN") is also used for a special primer location. For example, "-c3YCATGR" is the same as "-z3eXceI": both commands will design primers with the recognition site for *Xce I* endonucleases 5′-YCATGR-3′ at the 3′ end of the primers.
4. Additional bases can be added to primer ends using the commands "-5eNN" or "-3eNN," where 5 or 3 denotes the end at

which to add the extra bases and "NN" is a given sequence of one to more bases. For example, "`-F5eCGACG -R5eTTTTTT`", means adding the sequence "`CGACG`" to forward primers and "`TTTTTT`" to reverse primers, both at the 5′ ends.

5.6 Bisulphite-Modified DNA

The "C ≫ T bisulphite conversion" option allows the design of specific PCR primers for in silico bisulphite conversion for both strands. Only cytosine not followed by guanidine (CpG methylation) will be replaced by thymine:

```
5'aaCGaagtCC 3'        5'aaCGaagtTT 3'
  ||||||||||     ->      |||||| |
3'ttGCttCagg 5'        3'ttGCttTagg 5'
```

5.7 Uniqueness of Primers

Optionally, the user can synchronize the primer test for secondary, nonspecific binding with a dataset of sequence names. The program recognizes that a given sequence in the screening library dataset (from loading the dataset file) is the same as the sequence for which it is designing primers; this allows sequence selection to be made even if the selection matches the screening sequence perfectly. This allows the same dataset to be used for both primer design and screening without having to make a new screening database for each sequence. In other words, a dataset that contains sequences A, B, C, and D can be used both for choosing primers and for checking primer specificity. Alternatively, the user can input preexisting primers into a second "Additional sequence(s) or pre-designed primers (probes) list" text editor. These primers or probes will be checked for compatibility (inter-primer-dimer formation) with newly designed primers. The number of preexisting primers is not limited to one or two; it can be as many as the user needs.

5.8 PCR Primer Design

The PCR primer design algorithm generates a set of primers having a high likelihood of success in virtually any amplification protocol. All PCR primers designed by FastPCR can be used for PCR or sequencing experiments. The program is able to generate either long oligonucleotides or PCR primers for the amplification of gene-specific DNA fragments of user-defined length. FastPCR provides a flexible approach to designing primers for many applications and for both linear and circular sequences. It will check if either primers or probes have secondary binding sites in the input sequences that may give rise to additional PCR products. The selection of the optimal target region for the design of long oligonucleotides is performed in the same way as for PCR primers. The basic parameters in primer design are also used to measure the oligonucleotide quality and to evaluate the thermodynamic stability of the 3′ and 5′ terminal bases.

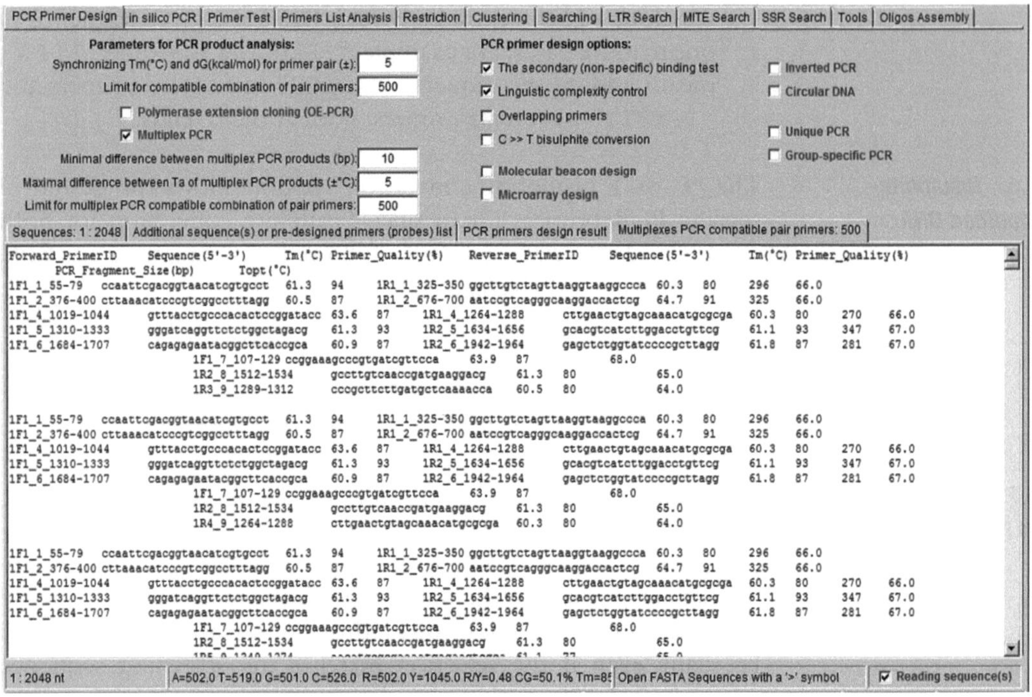

Fig. 5 An example of multiplex PCR results

Both the proposal of primer pairs and the selection of the best pairs are possible. The user can vary the product size or design primer pairs for the whole sequence without specifying parameters by using default or preset parameters. Preset parameters are specified for various situations: for example, for sequences with low CG content, for long-distance PCR, for degenerate sequences, or for manual input. A list of the best primer candidates and all compatible primer pairs that are optimal for PCR is generated. Users can specify, individually for each sequence, multiple locations for both forward and reverse primers within each sequence, whilst PCR design will be performed independently for different targets. Primers for multiplex PCRs can be designed from a single or from multiple targets (Fig. 5).

The program generates primer pairs (and probes) from the input sequences and shows the optimal annealing temperature for each primer pair and the sizes of PCR products together with information for each designed primer. Suggested primers and primer pairs are produced in tabulated format for MS Excel or Open Office (Table 3). The spreadsheets show the following properties: the automatically generated primer name, primer sequence, sequence location, direction, length, melting temperature, CG content (%), molecular weight, molar extinction coefficient, linguistic complexity (%), and *PQ*. For compatible primer pairs, the annealing temperature and PCR product size are also provided.

Table 3
Program output of the primer design

Primer ID	Sequence (5′–3′)	Length (nt)	T_m (°C)	dG, kcal/mol	T_m 3′end (°C)	CG %	*LC*	*PQ*
1F1_1_6-27	gggcggatcacttgaggtcagg	22	63.0	−29.7	40.0	63.6	88	87
1F2_1_107-132	ggcaggagaatcacttcaacctggga	26	62.5	−33.6	40.2	53.8	89	89
1F3_1_133-154	ggcggaggttgcagtgaaccga	22	64.4	−31.3	43.9	63.6	90	90
1F4_1_153-175	gagaccgcgccactgcactccag	23	67.2	−33.7	42.8	69.6	76	76
1F5_1_166-187	tgcactccagcctgggcgacag	22	66.3	−32.1	49.5	68.2	85	85
1F6_1_176-197	cctgggcgacagcctgagactc	22	64.8	−30.8	41.4	68.2	80	80
1F1_2_859-880	cttacggaggccgagatgggca	22	63.4	−30.6	46.9	63.6	88	87
1F2_2_913-934	tggccaacatggtgaaaccctg	22	60.9	−28.9	41.6	54.5	82	80
1F3_2_956-977	aattagctgggcatggtggcac	22	60.9	−29.1	47.8	54.5	80	73
1F4_2_1013-1036	tggagctgaaccactgcactccag	24	63.2	−32.3	42.8	58.3	79	79
1R1_1_412-434	acccagaagagcctgagtgggca	23	63.9	−31.7	46.7	60.9	83	83
1R2_1_309-330	ccttgctcagctctggccatcc	22	63.6	−30.0	44.8	63.6	80	80
1R3_1_299-320	ctctggccatcccagttcaagc	22	61.5	−28.9	42.2	59.1	88	80
1R1_2_1206-1231	agatggggtttcaccatgttggccca	26	64.1	−34.8	44.6	53.8	80	80
1R2_2_1166-1187	acctcaggtgatccacctgcct	22	61.7	−29.5	44.2	59.1	82	80
1R3_2_1150-1174	cacctgcctcagcttcccaaagtgc	25	65.3	−34.2	40.6	60.0	84	84
1R4_2_1132-1157	caaagtgcttggattacaggcgtgag	26	61.0	−33.0	42.4	50.0	93	93
The separated output of a set of primer pair sequences with their theoretical PCR products								

(continued)

Table 3
(continued)

Primer ID	Sequence (5′–3′)	Length (nt)	T_m (°C)	dG, kcal/mol	T_m 3′end (°C)	CG %	*LC*	*PQ*	
Forward primer ID	Sequence (5′–3′)	T_m (°C)	PQ	Reverse primer ID	Sequence (5′–3′)	T_m (°C)	PQ	PCR Fragment Size (bp)	T^{opt} (°C)
1F1_1_8-27	gcggatcacttgaggtcagg	58.9	87	1R3_1_301-320	ctctggccatcccagttcaa	57.6	80	313	63
1F1_1_8-27	gcggatcacttgaggtcagg	58.9	87	1R4_1_291-310	cccagttcaagccatcccct	60.2	76	303	65
1F2_1_49-68	caacgtggagctaggtatgg	56.0	80	1R1_1_409-429	gaagagcctgagtgggcacaa	60.0	85	381	62
1F2_1_49-68	caacgtggagctaggtatgg	56.0	80	1R3_1_301-320	ctctggccatcccagttcaa	57.6	80	272	62
1F2_1_49-68	caacgtggagctaggtatgg	56.0	80	1R4_1_291-310	cccagttcaagccatcccct	60.2	76	262	62
1F3_1_109-130	caggagaatcacttcaacctgg	56.4	87	1R1_1_409-429	gaagagcctgagtgggcacaa	60.0	85	321	62
1F3_1_109-130	caggagaatcacttcaacctgg	56.4	87	1R3_1_301-320	ctctggccatcccagttcaa	57.6	80	212	62
1F3_1_109-130	caggagaatcacttcaacctgg	56.4	87	1R4_1_291-310	cccagttcaagccatcccct	60.2	76	202	62
1F4_1_134-154	gcggaggttgcagtgaaccga	62.6	90	1R1_1_409-429	gaagagcctgagtgggcacaa	60.0	85	296	66
1F4_1_134-154	gcggaggttgcagtgaaccga	62.6	90	1R2_1_311-333	tgtccttgctcagctctggccat	62.3	83	200	68
1F4_1_134-154	gcggaggttgcagtgaaccga	62.6	90	1R3_1_301-320	ctctggccatcccagttcaa	57.6	80	187	63
1F4_1_134-154	gcggaggttgcagtgaaccga	62.6	90	1R4_1_291-310	cccagttcaagccatcccct	60.2	76	177	65
1F5_1_153-172	gagaccgcgccactgcactc	64.3	73	1R2_1_311-333	tgtccttgctcagctctggccat	62.3	83	181	68
1F5_1_153-172	gagaccgcgccactgcactc	64.3	73	1R4_1_291-310	cccagttcaagccatcccct	60.2	76	158	65

5.9 Multiplex PCR Primer Design

Multiplex PCR is an approach commonly used to amplify several DNA target regions in a single reaction. The simultaneous amplification of many targets reduces the number of reactions that needs to be performed; multiplex PCR thus increases throughput efficiency. The design of multiplex PCR assays can be difficult because it involves extensive computational analyses of primer pairs for interactions. The multiplex PCR algorithm is based on the fast non-recursion method, with the software performing checks on product size and primers' thermodynamics parameters (enthalpy—**dH** and Gibb's free energy—**dG**) compatibility and cross-dimer formation for all primers. To achieve uniform amplification of the targets, the primers must be designed to bind with equal efficiencies to their targets. FastPCR can quickly design a set of multiplex PCR primers for all input sequences and/or multiplex targets within each sequence. PCR conditions may need to be adjusted, for example, by increasing or decreasing the annealing temperature so that all products are amplified equally efficiently.

To achieve uniform amplification, most existing multiplex primer design packages use primer melting temperature. In practical terms, the design of almost identical T_as and T_ms is very important. The melting temperatures of the PCR products are also important because these are related to annealing temperature values. The T_m of a PCR product directly depends on its CG content and its length; short products are more efficiently amplified at low PCR annealing temperatures (100 bp, 50–55 °C) than are long products (>3,000 bp, 65–72 °C). For most multiplex PCRs, there is usually a small variation (up to 5 °C) between the optimal T_as of all primer pairs. The annealing temperature must be optimal in order to maximize the likelihood of amplifying the target genomic sequences whilst minimizing the risk of nonspecific amplification. Further improvements can be achieved by selecting the optimal set of primers that maximize the range of common T_ms. Once prompted, FastPCR calculates multiplex PCR primer pairs for given target sequences. The speed of calculation depends on the numbers of target sequences and primer pairs involved.

An alternative way to design compatible multiplex PCR primer pairs is to use predesigned primers as references for the design of new primers. The user can select input options for the PCR products such as the minimum product size differences between the amplicons. Primer design conditions can be set individually for each given sequence or all primers can be designed using the same values; selected settings have higher priority for PCR primer or probe design than the general settings. The results include primers for individual sequences, compatible primers, product sizes, and annealing temperatures. Because clear differentiation of the products is dependent on using compatible primer pairs in the single reactions, the program recovers all potential variants of primer combinations for analyses of the chosen DNA regions and provides, in tabular form, their compatibility including information on primer-dimers, cross-hybridization, product size overlaps, and similar alternative primer pairs based on T_m. The user may choose those alternative compatible primer pair combinations that provide the desired product sizes. Using the program, researchers can select predesigned primer pairs from a target for their desired types of PCRs by changing the filtering conditions as mentioned above. For example, a conventional multiplex PCR requires differently sized (at least by 10 bp) amplicons for a set of target genes, so the value for the minimum size difference between PCR products can be selected.

In addition to avoid amplifying different amplicons of the same size, multiplex PCR must also minimize the generation of primer-dimers and secondary products, which becomes more difficult with increasing numbers of primers in a reaction. To avoid the problem of nonspecific amplification, FastPCR selects primer pairs that give the most likelihood of producing only the amplicons of the target sequences by choosing sequences which avoid repeats or other motifs. The program allows the user to design not only compatible pairs of primers but also compatible single primers for different targets or sequences. The input data can be either a single sequence with a minimum two internal tasks or many sequences with or without internal tasks. Most of the parameters on the interface are self-explanatory. Optionally, the user is asked to provide the sequence and select oligonucleotide designing parameters.

On the **PCR Primer Design** tab, the user clicks on the **Multiplex PCR** option. The user then selects the limit for the number of primer pairs (the default is 100), the minimal size difference between multiplex PCR products (the default 10 bp), and the maximal difference between the T_as of the PCR products (the default is ±5 °C). After specifying inputs and primer design options, the user can execute the primer design task. Once the design of the primer set is completed, the result will appear in two **Result** text editors: **PCR primer design result** and **Multiplex PCR compatible pair primers**. Figure 8 shows the access to the PCR primer design output. The result text editor **PCR primer design result** will display the individual PCR primer design data, including the primer list and the compatible primer pairs for all the sequences and their internal tasks. The second **Multiplex PCR compatible pair primers** text editor collects final search results that are presented as a list of the sets of the compatible primer pairs for multiplex PCR.

5.10 Group-Specific PCR Primers

Group-specific amplification, also called family-specific and sequence-specific amplification, is an important tool for comparative studies of related genes, sequences, and genomes that can be applied to studies of evolution, especially for gene families and for cloning new related sequences. Specific targets such as homologous genes or members of a transposable element family can be amplified to uncover DNA polymorphisms associated with these sequences or other genetic investigations. The overall strategy of designing group-specific PCR primers uses a hash index of 12-mers to identify common regions in the target sequences, followed by standard PCR primer design for the current sequence, and then testing the complementarity of these primers to the other sequences. FastPCR performs either multiple sequence alignment or accepts alignment sequence input, giving it the flexibility to use a different strategy for primer design. If required, it can design degenerate PCR primers to amplify the polymorphic region of all related sequences.

The FastPCR package designs large sets of universal primer pairs for each given sequence, identifies conserved regions, and generates suitable primers for all given targets. The steps of the algorithm are performed automatically and the user can influence the settings for the primer design options. The quality of primer design is dependent on sequence relationships, genetic similarity, and suitability of the consensus sequence for the design of good primers. The software is able to generate group-specific primers for each set of sequences independently, which are suitable for all sequences. Primer alignment parameters for group-specific PCR primers are similar to those used for in silico PCR. The user chooses **Group-specific PCR** on the **PCR Primer Design** tab. After specifying inputs and PCR primer design options, the user can execute the PCR primer design task. The program takes multiple separate DNA sequences in either FASTA or alignment formats.

Once the primer set design is complete, the result will appear in the **Result** text editor, as the **PCR primer design result**. Figure 6 shows the access to the PCR primer design output (Table 3). The result text editor **PCR primer design result** displays the individual group-specific PCR primer design data, including the primer list and compatible primer pairs for all the sequences and their internal tasks where suitable primers are found. In the case where an alignment has been input, the result text editor displays only one group-specific PCR primer design set, including degenerate primers, in the primer list as well compatible primer pairs for all the sequences.

5.11 Simple Sequence Repeat Locus Search and PCR Primer Design

SSRs or microsatellites are short tandem repeats of one or more bases. Microsatellites are ubiquitously distributed throughout eukaryotic genomes, often highly polymorphic in length, and thereby an important class of markers for population genetic studies. Our approach to locating SSRs is to analyze low-complexity regions in DNA by using linguistic sequence complexity. This method allows the detection of perfect and imperfect SSRs with a single, up to 10-base, repeat motif. Each entry sequence is processed for identification of SSRs and the SSR flanking regions are used to design compatible forward and reverse primers for their amplification by PCR.

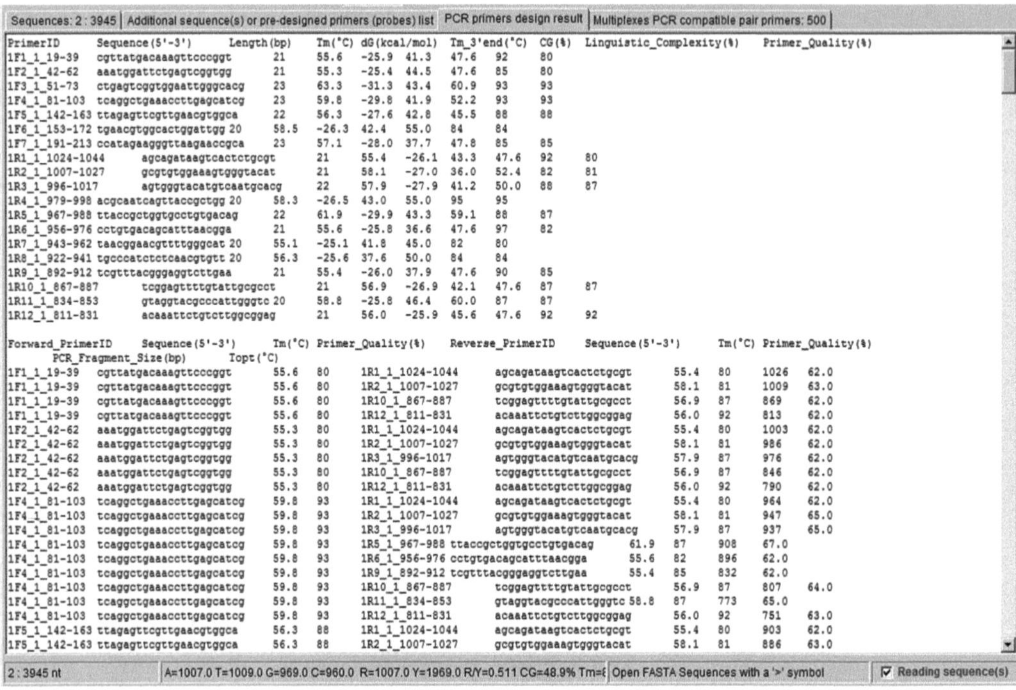

Fig. 6 An example of group-specific PCR results

FastPCR identifies all SSRs within each entry sequence and designs compatible PCR primer pairs for each SSR locus. The default PCR primer design parameters are that the primers must be within 100 bases from either side of the identified SSR. Often the sequences available around SSR loci are not suitable for designing good primers; the user can increase or decrease the distance from either side to find more efficient and compatible primer pairs. The capabilities of FastPCR make it a complete bioinformatics tool for the use of microsatellites as markers, from discovery through to primer design. For example, the user can specify PCR primer design to SSR loci within 200 bp around an SSR, with the command: "`-ssr/200.`" The software finds all SSR sites and then will design PCR primers and compatible primer pairs independently for each SSR locus.

5.12 Oligonucleotide Design for In Vitro Long Sequence Synthesis

The application to make long synthetic DNA molecules relies on the in vitro assembly of a set of short oligonucleotides, either for LCR [37] or for assembly PCR [38]. These oligonucleotides should be adjacent on the same strand and overlap the complementary oligonucleotides from the second strand. There are two major parameters for designing oligonucleotides for gene synthesis for LCR or assembly PCR. First, the oligonucleotides should share similar T_m values. Second, a given oligonucleotide sequence should be unique to avoid multiple nonspecific binding that may lead to incorrect assembly. The software must dynamically choose the length of the oligonucleotides to ensure both the specificity and the uniform T_m. The algorithm of FastPCR is able to design oligonucleotides for long sequences containing repeats and to minimize their potential nonspecific hybridization during 3′ end extension in PCR. For long sequence assembly, oligonucleotide design starts from the 5′ end of a given sequence; the oligonucleotide length is dynamically changed until a unique 3′ end has been found and the T_m of the oligonucleotide has reached the T_m threshold. All oligonucleotides are designed without gaps between them. The other strand is used for the design of the overlapping oligonucleotides using the same algorithm as above but with the T_m of the overlapping regions reaching the T_m −15 °C threshold. The composition of the sequence at the 3′ terminus is important because stability at the 3′ end of the double-stranded complexes will improve the specificity of extension by the polymerase. To reduce nonspecific polymerase extension (and ligation), the algorithm chooses only unique sequences for the 3′ terminus. Minimally, the last two nucleotides at the 3′ terminus must not be complementary to any nonspecific targets. Other complementary regions are less important for assembling multiple fragments by PCR and ligation.

The input data can comprise either a single or many sequences. Most of the parameters on the interface are self-explanatory. The user is asked to provide the sequence and select oligonucleotide design parameters. The user clicks on **Oligo options** on the **Oligos Assembly** tab, and chooses the minimal oligonucleotide length and T_m threshold, which by default are 40 nt and 60 °C, respectively. The interface allows changing T_m calculation parameters. The search process runs after pressing **F5** or from menu bar or toolbox. The research result is presented as a list of oligonucleotides for both strands. On each strand, all oligonucleotides are adjacent with no gap between neighboring primers. An oligonucleotide will overlap two oligonucleotides from the complementary strand. The algorithm pays attention to avoid nonspecific oligonucleotide hybridization to repeated regions. Where it is not possible to design primers outside of repeated sequences, it is likewise difficult to find short specific oligonucleotides. The solution to this problem is to divide the sequence into short segments, design a set of oligonucleotides for each segment independently, and then combine all these segments in the second PCR for final amplification.

5.13 Polymerase Extension PCR for Fragment Assembly

Sequence-independent cloning, including ligation-independent cloning, requires generation of complementary single-stranded overhangs in both the vector and insertion fragments. Similarly, multiple fragments can be joined or concatenated in an ordered manner using overlapping primers in PCR. Annealing of the complementary regions between different targets in the primer overlaps allows the polymerase to synthesize a contiguous fragment containing the target sequences during thermal cycling, a process called "overlap extension PCR" (OE-PCR) (Chapter 8) [39]. The efficiency depends on the T_m and on the length and uniqueness of the overlap. To achieve this, the program designs compatible forward and reverse primers at the ends of each fragment, and then extends the 5′ end of primers using sequences from the primers of the fragment that will be adjacent in the final product. The input sequence can be made of either a single or many sequences. The user needs to pay special attention to the preparation of the given sequences for assembly.

Users can specify the locations for both forward and reverse primers design using " [ ] " to bracket the region. The bracketed sequences will be used by the program for designing the overlapping primers. The program selects the overlapping area so that the primers from overlapping fragments are similar in size with optimal annealing temperatures. The program adds the required bases so that the T_m of the overlap is similar to, or higher than, the T_m of the initial primers. Primers are tested for dimer formation within the appropriate primer pairs. The user chooses **Polymerase extension cloning** (**OE-PCR**) on the **PCR Primer Design** tab and selects the limit for multiple-PCR-compatible combinations of primer pairs (default is 100). After specifying sequence inputs and PCR primer design options, the user can execute the search task. Once the design of the primer sets is complete, the result will appear in two text editors: **PCR primer design result** and **PCR fragments assembling compatible pair primers**. The text editor **PCR primer design result** window displays the individual PCR primer design data, including the primer list and the compatible primer pairs for all sequences whose primers are found. The **PCR fragments assembling compatible pair primers** text editor collects the final search result and presents it as a list of sets of compatible primer pairs for individual fragment amplification and assembly. Figure 7 shows a sample result visualization window.

5.14 In Silico PCR

Modelling the hybridization of primers to targeted annealing sites is the only way to predict PCR products [7, 24, 40–44]. The last 10–12 bases at the 3′ end of primers are important for binding stability; single mismatches can reduce PCR efficiency, the effect increasing with proximity to the 3′ terminus. FastPCR allows simultaneous testing of single primers or a set of primers designed for multiplex PCR. It performs a fast, gapless alignment to test the complementarity of the primers to the target sequences. For in silico PCR, a quick alignment to detect primer locations on the reference sequence is performed by analyses of both strands using

Fig. 7 An example of results for polymerase extension PCR for fragment assembly

a hash index of 7- to 12-mers (allowing up to one mismatch) and by calculating the local similarity for the whole primer. The parameters can be altered to allow different degrees of mismatches at the 3′ end of the primers. The parameters for quick alignment may be set: the minimum is 0–5 mismatches (default 2 mismatches) at 3′ end of primer. The program can also handle degenerate primers or probes, including those with 5′ or 3′ tail sequences. It includes the detection of non-Watson–Crick base-pairing in in silico PCR, e.g., the stable guanine mismatches G·G, G·T, and G·A. Probable PCR products can be found for both linear and circular templates in both standard and inverse PCR, as well as in multiplex PCR and using bisulphite-treated DNA. This in silico tool is useful for quickly analyzing primers or probes against target sequences, for determining primer location, orientation, and efficiency of binding, and for calculating primer T_m and annealing temperature in PCR.

The user must input a preexisting primer list into a second **Additional sequence(s) or pre-designed primers (probes) list** text editor. The number of preexisting primers is not limited; it can be as many as the user needs. The target sequences can be entered either as multiple separate DNA sequences or by opening files from the selected folders. For in silico PCR against whole genome(s) or a list of chromosomes, the user must specify the directory containing the input. The program will be consistent: it will look at each file to find the position of the primers. The user can execute the search task with **F5** on the **in silico PCR** tab or can specify search options including stringency and PCR product detection settings. For the stringency options, users can specify the number of mismatches that the primers are allowed at 3′ terminus. The default specificity settings allow a maximum two mismatches within the 3′ end region of the primers. These mismatches within the 3′ end of the primers should not be located close to each other. Once the primer set design is complete, the results will appear in the text editors **In silico PCR Result**.

In silico PCR Result text editor reports the specificity of the primers (locations, including target position, similarity, and T_m), a summary of primer pairs in relation to the PCR template, and detailed information on each primer pair, including its length and T_a. It will show the target-specific primers that have been found. The actual targets will be listed along with detailed alignments between primers and targets (Fig. 8).

6 Primer Analyses

Individual and sets of primers are evaluated using FastPCR or the online software. They calculate primer T_ms using default or other formulae for features of the primers including normal and degenerate nucleotide combinations, CG content, extinction coefficient, unit conversion (nmol per OD), mass (μg per OD), molecular weight, and linguistic complexity and consider primer PCR efficiency. Users can select either DNA or RNA primers (online: PrimerAnalyser, http://primerdigital.com/tools/PrimerAnalyser.html) with normal or degenerate oligonucleotides or modifications with various labels (for example, inosine, uridine, or fluorescent dyes). Tools allow the choice of other nearest neighbor thermodynamic parameters or non-thermodynamic T_m calculation formulae.

For LNA modifications the four symbols, dA = E, dC = F, dG = J, and dT = L, are used. Both programs perform analyses on-

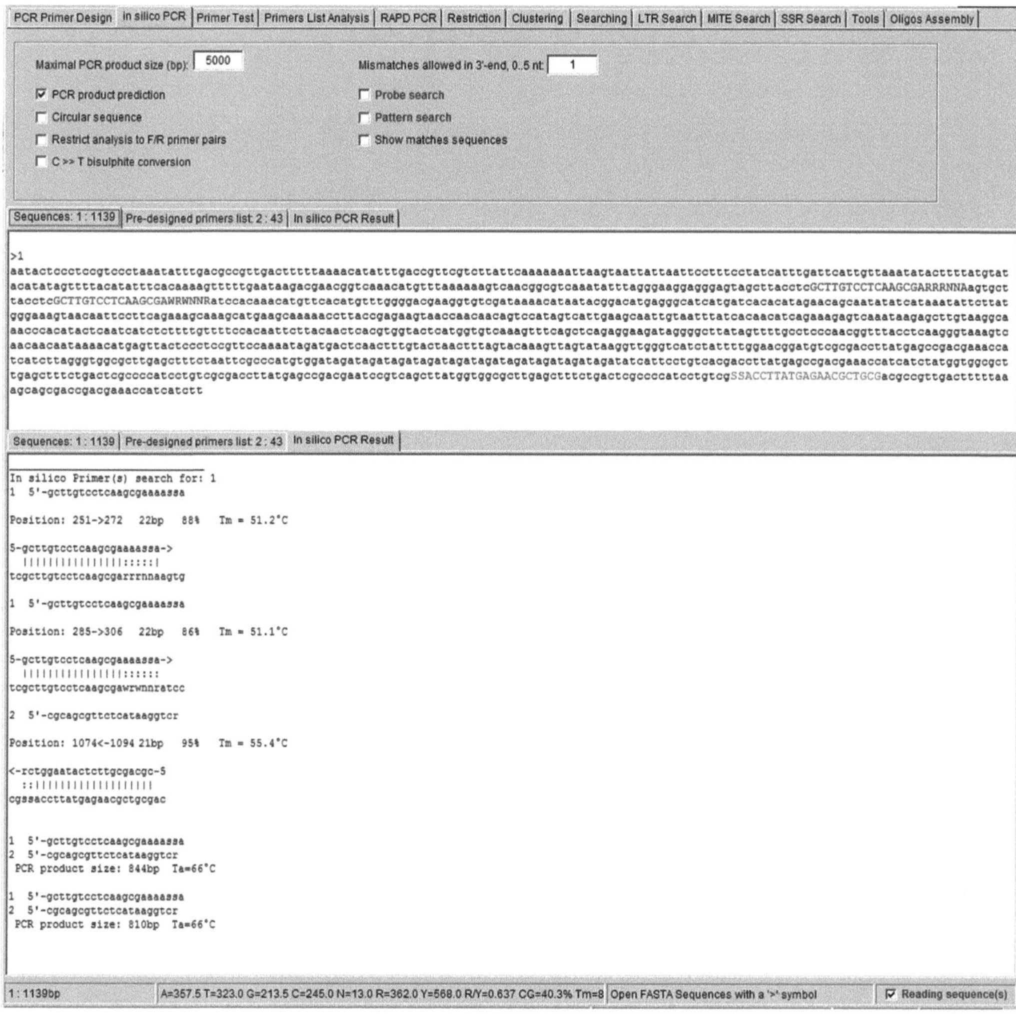

Fig. 8 An example of in silico PCR result

type, allowing users to see the results immediately on screen. They can also calculate the volume of solvent required to attain a specific concentration from the known mass (mg), OD, or moles of oligonucleotide.

All primers are analyzed for intra- and inter-primer interactions regarding formation of dimers. Primer(s) can efficiently hybridize using the 5′ end or the middle of the oligonucleotides. Even though such interactions are not efficiently extended by DNA polymerase, their formation reduces the effective primer concentration available for binding to the targets and their presence can strongly inhibit PCR because double-stranded DNA at high concentrations is a strong inhibitor of DNA polymerase (Fig. 9).

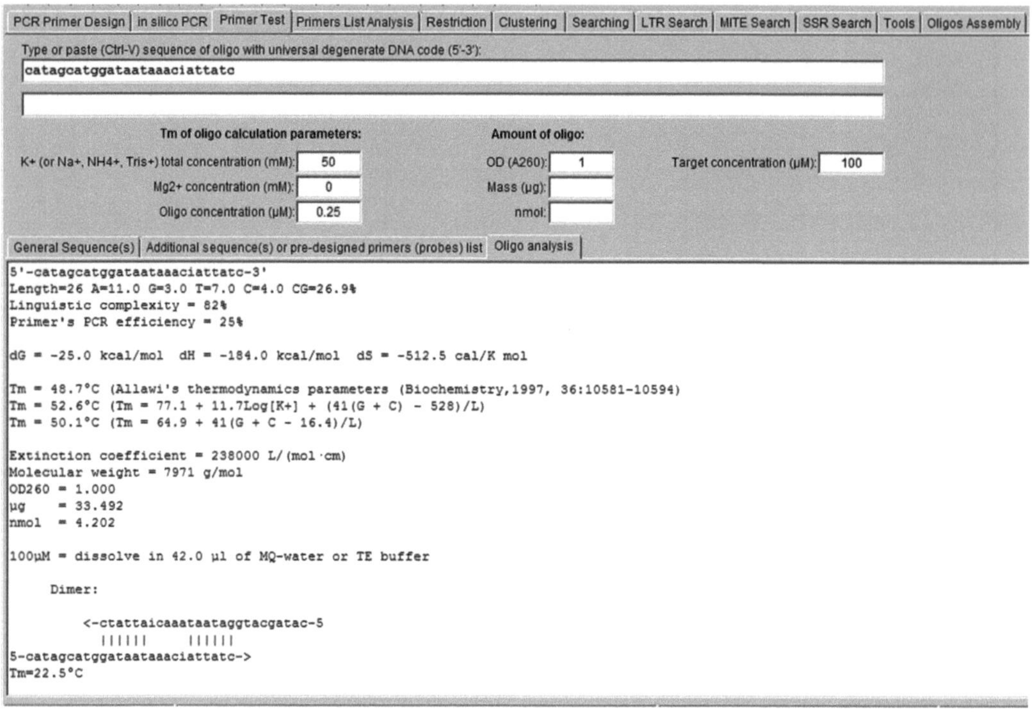

Fig. 9 Example result of the oligonucleotide analysis

7 Availability

The FastPCR software is available for download at http://primerdigital.com/fastpcr.html; the online version is available at http://primerdigital.com/tools/pcr.html. The program manual, licence agreement, and installation files can be found at http://primerdigital.com/fastpcr/. YouTube tutorial videos have been placed at http://www.youtube.com/user/primerdigital. Web tools are accessible at http://primerdigital.com/tools/.

Acknowledgments

Web tools are available free to academic institutions, provided that they are used for noncommercial research and education only. They may not be reproduced or distributed for commercial use. This work was partially supported by the companies PrimerDigital Ltd. and Oligomer Ltd. and by the Academy of Finland, Project 134079.

References

1. Untergasser A et al (2012) Primer3—new capabilities and interfaces. Nucleic Acids Res 40:e115. doi:10.1093/nar/gks596
2. Kalendar R, Lee D, Schulman AH (2009) FastPCR software for PCR primer and probe design and repeat search. Genes, Genomes and Genomics 3:1–14
3. Kalendar R, Lee D, Schulman AH (2011) Java web tools for PCR, *in silico* PCR, and oligonucleotide assembly and analysis. Genomics 98:137–144
4. Marshall OJ (2004) PerlPrimer: cross-platform, graphical primer design for standard, bisulphite and real-time PCR. Bioinformatics 20:2471–2472
5. Owczarzy R et al (2008) IDT SciTools: a suite for analysis and design of nucleic acid oligomers. Nucleic Acids Res 36:W163–9. doi:10.1093/nar/gkn198
6. Bekaert M, Teeling EC (2008) UniPrime: a workflow-based platform for improved universal primer design. Nucleic Acids Res 36:e56. doi:10.1093/nar/gkn191
7. Ye J et al (2012) Primer-BLAST: a tool to design target-specific primers for polymerase chain reaction. BMC Bioinformatics 13:134. doi:10.1186/1471-2105-13-134
8. Giegerich R, Meyer F, Schleiermacher C (1996) GeneFisher—software support for the detection of postulated genes. Proc Int Conf Intell Syst Mol Biol 4:68–77
9. Gadberry MD et al (2005) Primaclade—a flexible tool to find conserved PCR primers across multiple species. Bioinformatics 21:1263–1264. doi:10.1093/bioinformatics/bti134
10. National Center for Biotechnology Information, National Library of Medicine, Building 38A, Bethesda, MD, 20894. http://blast.ncbi.nlm.nih.gov/blastcgihelp.shtml
11. Nomenclature for incompletely specified bases in nucleic acid sequences (1984) http://www.chem.qmul.ac.uk/iubmb/misc/naseq.html
12. Allawi HT, SantaLucia J Jr (1997) Thermodynamics and NMR of internal G.T mismatches in DNA. Biochemistry 36:10581–10594. doi:10.1021/bi962590c
13. SantaLucia J (1998) A unified view of polymer, dumbbell, and oligonucleotide DNA nearest-neighbor thermodynamics. Proc Natl Acad Sci USA 95:1460–1465
14. Le Novere N (2001) MELTING, computing the melting temperature of nucleic acid duplex. Bioinformatics 17:1226–1227. doi:10.1093/bioinformatics/17.12.1226
15. Bolton ET, McCarthy BJ (1962) A general method for the isolation of RNA complementary to DNA. Proc Natl Acad Sci USA 48: 1390–1397
16. Guedin A et al (2010) How long is too long? Effects of loop size on G-quadruplex stability. Nucleic Acids Res 38:7858–68. doi:10.1093/nar/gkq639
17. Wallace RB et al (1979) Hybridization of synthetic oligodeoxyribonucleotides to ΦX 174 DNA: the effect of single base pair mismatch. Nucleic Acids Res 6:3543–57. doi:10.1093/nar/6.11.3543
18. von Ahsen N, Wittwer CT, Schutz E (2001) Oligonucleotide melting temperatures under PCR conditions: nearest-neighbor corrections for Mg^{2+}, deoxynucleotide triphosphate, and dimethyl sulfoxide concentrations with comparison to alternative empirical formulas. Clin Chem 47:1956–1961
19. Kypr J et al (2009) Circular dichroism and conformational polymorphism of DNA. Nucleic Acids Res 37:1713–25. doi:10.1093/nar/gkp026
20. Gabrielian A, Bolshoy A (1999) Sequence complexity and DNA curvature. Comput Chem 23:263–74. doi:10.1016/S0097-8485(99)00007-8
21. Orlov YL, Potapov VN (2004) Complexity: an internet resource for analysis of DNA sequence complexity. Nucleic Acids Res 32:W628–33. doi:10.1093/nar/gkh466
22. Gilson MK et al (1997) The statistical-thermodynamic basis for computation of binding affinities: a critical review. Biophys J 72:1047–69. doi:10.1016/S0006-3495(97)78756-3
23. Peyret N et al (1999) Nearest-neighbor thermodynamics and NMR of DNA sequences with internal A.A, C.C, G.G, and T.T mismatches. Biochemistry 38:3468–3477. doi:10.1021/bi9825091
24. Watkins NE Jr, SantaLucia J Jr (2005) Nearest-neighbor thermodynamics of deoxyinosine pairs in DNA duplexes. Nucleic Acids Res 33:6258–67. doi:10.1093/nar/gki918
25. Sen D, Gilbert W (1992) Guanine quartet structures. Methods Enzymol 211:191–199
26. Il'icheva IA, Florent'ev VL (1992) Four-stranded complexes of oligonucleotides-quadruplexes. Mol Biol (Mosk) 26:512–531
27. Shing HP (1994) The non-B-DNA structure of d(CA/TG)n does not differ from that of Z-DNA. Proc Natl Acad Sci USA 91:9549–9553

28. SantaLucia J Jr, Hicks D (2004) The thermodynamics of DNA structural motifs. Annu Rev Biophys Biomol Struct 33:415–40. doi:10.1146/annurev.biophys.32.110601.141800
29. Williams JGK et al (1990) DNA polymorphisms amplified by arbitrary primers are useful as genetic-markers. Nucleic Acids Res 18:6531–5. doi:10.1093/nar/18.22.6531
30. Welsh J, Mcclelland M (1990) Fingerprinting genomes using PCR with arbitrary primers. Nucleic Acids Res 18:7213–8. doi:10.1093/nar/18.24.7213
31. Kalendar R, Schulman A (2006) IRAP and REMAP for retrotransposon-based genotyping and fingerprinting. Nat Protoc 1:2478–84. doi:10.1038/nprot.2006.377
32. Chang RY, O'Donoughue LS, Bureau TE (2001) Inter-MITE polymorphisms (IMP): a high throughput transposon-based genome mapping and fingerprinting approach. Theor Appl Genet 102:773–781
33. Nelson DL et al (1989) Alu polymerase chain reaction: a method for rapid isolation of human-specific sequences from complex DNA sources. Proc Natl Acad Sci USA 86: 6686–6690
34. Sinnett D et al (1990) Alumorphs—human DNA polymorphisms detected by polymerase chain reaction using Alu-specific primers. Genomics 7:331–334
35. Jurka J et al (2005) Repbase update, a database of eukaryotic repetitive elements. Cytogenet Genom Res 110:462–7. doi:10.1159/000084979
36. TREP, the Triticeae Repeat Sequence Database (2008) http://wheat.pw.usda.gov/ITMI/Repeats/
37. Landegren U et al (1988) A ligase-mediated gene detection technique. Science 241:1077–1080
38. Higasa K, Hayashi K (2002) Ordered catenation of sequence-tagged sites and multiplexed SNP genotyping by sequencing. Nucleic Acids Res 30:E11
39. Quan J, Tian J (2009) Circular polymerase extension cloning of complex gene libraries and pathways. PLoS One 4:e6441. doi:10.1371/journal.pone.0006441
40. Cao YF et al (2005) Information theory-based algorithm for *in silico* prediction of PCR products with whole genomic sequences as templates. BMC Bioinformatics 6:190. doi:10.1186/1471-2105-6-190
41. Rubin E, Levy AA (1996) A mathematical model and a computerized simulation of PCR using complex templates. Nucleic Acids Res 24:3538–45. doi:10.1093/nar/24.18.3538
42. Lexa M, Valle G (2003) PRIMEX: rapid identification of oligonucleotide matches in whole genomes. Bioinformatics 19:2486–2488
43. Nishigaki K et al (2000) Whole genome sequence-enabled prediction of sequences performed for random PCR products of *Escherichia coli*. Nucleic Acids Res 28:1879–1884
44. Rotmistrovsky K, Jang W, Schuler GD (2004) A web server for performing electronic PCR. Nucleic Acids Res 32:W108–12. doi:10.1093/nar/gkh450

Index

Svein Valla and Rahmi Lale (eds.), *DNA Cloning and Assembly Methods*, Methods in Molecular Biology, vol. 1116, DOI 10.1007/978-1-62703-764-8, © Springer Science+Business Media New York 2014

MIX
Papier aus verantwortungsvollen Quellen
Paper from responsible sources
FSC® C105338

If you have any concerns about our products,
you can contact us on
ProductSafety@springernature.com

In case Publisher is established outside the EU,
the EU authorized representative is:
Springer Nature Customer Service Center GmbH
Europaplatz 3, 69115 Heidelberg, Germany

Printed by Libri Plureos GmbH
in Hamburg, Germany